Polysaccharide-Based Materials: Developments and Properties

Polysaccharide-Based Materials: Developments and Properties

Guest Editor

Andrés Gerardo Salvay

Basel • Beijing • Wuhan • Barcelona • Belgrade • Novi Sad • Cluj • Manchester

Guest Editor
Andrés Gerardo Salvay
Departamento de Ciencia y Tecnología
Universidad Nacional de Quilmes
Bernal
Argentina

Editorial Office
MDPI AG
Grosspeteranlage 5
4052 Basel, Switzerland

This is a reprint of the Special Issue, published open access by the journal *Polymers* (ISSN 2073-4360), freely accessible at: https://www.mdpi.com/journal/polymers/special_issues/X6050717F1.

For citation purposes, cite each article independently as indicated on the article page online and as indicated below:

Lastname, A.A.; Lastname, B.B. Article Title. *Journal Name* **Year**, *Volume Number*, Page Range.

ISBN 978-3-7258-6818-6 (Hbk)
ISBN 978-3-7258-6819-3 (PDF)
https://doi.org/10.3390/books978-3-7258-6819-3

Contents

About the Editor

Andrés Gerardo Salvay

Andrés Gerardo Salvay holds a master's degree in Physics and a PhD in Chemistry from the Universidad Nacional de La Plata, Argentina. He carried out postdoctoral research at the Universidade Estadual Paulista in São José do Rio Preto, Brazil, and at the Institut de Biologie Structurale in Grenoble, France. He also served as a Visiting Research Professor at Université Joseph Fourier in Grenoble, France. He is currently a Research Professor at the Universidad Nacional de Quilmes, Argentina. His research focuses on protein biophysics, biopolymer interactions, and the development of biopolymeric materials. His work is characterised by an interdisciplinary approach that integrates physical chemistry, molecular biophysics, and materials science, bridging fundamental research with technological and environmental applications. In addition, he is actively involved in the training of human resources, as well as in scientific evaluation and editorial activities.

Preface

The utilisation of bio-based materials has emerged as a promising and innovative alternative to materials derived from non-renewable and non-biodegradable resources. Against this background, increasing research efforts have been directed towards the study and development of polysaccharide-based materials. Polysaccharides are primarily obtained from biomass resources, including agricultural by-products, marine organisms such as algae and crustaceans, as well as fungi and microorganisms through metabolic pathways capable of producing exopolysaccharides. The overarching aim of this Reprint is to promote the design of novel polysaccharide-based materials and composites, the valorisation of bio-based resources, and the comprehensive investigation of their structure–property–function relationships for a wide range of applications. A major challenge that remains is the development of polysaccharide-based materials with functional performance comparable to that of conventional synthetic polymers. In this context, this Reprint brings together recent advances in the processing, modification, and functional characterisation of polysaccharide-based materials. The collected contributions highlight the exceptional versatility of polysaccharides as sustainable and functional building blocks for advanced material systems. Through innovative processing strategies, controlled chemical modifications, and synergistic combinations with other biopolymers, the studies demonstrate how representative polysaccharides, including cellulose, chitosan, alginate, hyaluronic acid, starch, and kefiran, can be engineered to obtain materials with tailorable mechanical, physicochemical, electrical and biological properties. The reported applications span a broad range of fields, including flexible sensors, biomedical hydrogels, sustainable packaging, wastewater treatment, and soil stabilisation. Overall, this Reprint offers valuable insights into the design of next-generation polymeric materials, supporting the advancement of sustainable and high-performance engineering solutions.

Andrés Gerardo Salvay
Guest Editor

polymers MDPI

Editorial

Polysaccharide-Based Materials: Developments and Properties

Andrés Gerardo Salvay

Laboratorio de Obtención, Modificación, Caracterización y Evaluación de Materiales, Departamento de Ciencia y Tecnología, Universidad Nacional de Quilmes, Roque Sáenz Peña 352, Bernal B1876BXD, Argentina; asalvay@unq.edu.ar

The growing concern for environmental sustainability and the urgent demand to reduce dependence on non-renewable resources have placed bio-based materials at the centre of materials science innovation. Within this context, polysaccharides have emerged as one of the most promising classes of biopolymers. Their natural abundance, structural versatility, chemical functionality, and intrinsic biodegradability make them ideal candidates for advanced, functional, and sustainable materials [1,2].

Polysaccharides are among the most abundant biopolymers on Earth, sourced from plants, marine organisms, and microorganisms [3]. Their extraction and utilisation represent an essential step towards a circular bioeconomy. The structural complexity of polysaccharides, arising from their monosaccharide composition, linkage types, branching patterns, and degrees of substitution, provides a rich foundation for tailoring their material properties [4]. This versatility enables their conversion into a wide variety of functional forms such as hydrogels [1], films [2], aerogels [5], and composites [6,7].

In recent years, increasing attention has been paid to the valorisation of agricultural and industrial by-products as sources of polysaccharides [8]. At the same time, microbial fermentation has gained relevance as a sustainable alternative route, capable of producing exopolysaccharides (EPS) with well-defined structures and properties [9]. Microbial EPS such as dextran, pullulan, gellan, and bacterial cellulose have been extensively investigated as renewable matrices for high-performance materials [10].

The conversion of polysaccharides into materials with targeted performance involves a complex interplay between molecular structure, processing strategy, and application requirements. Physical treatments, chemical modifications, crosslinking reactions, and blending with other biopolymers or plasticisers are often used to overcome intrinsic limitations such as brittleness, hydrophilicity, or poor water barrier properties [11,12]. Hybridization with nanomaterials has proven effective to reinforce polysaccharide matrices [13]. Crosslinking reactions, either covalent or non-covalent, have proven particularly effective in enhancing hydrogels' stability and functional performance [14]. From a biotechnological viewpoint, enzyme-mediated transformations and green synthesis routes are increasingly preferred over conventional chemical treatments, offering more environmentally benign processes [15].

The versatility of polysaccharide-based materials translates into an impressive range of applications. In biomedicine, polysaccharide hydrogels and films have been developed as carriers for controlled drug delivery, scaffolds for tissue regeneration, and wound-healing dressings [16]. Their intrinsic biocompatibility, non-toxicity, and tunable degradation profiles make them attractive for applications requiring close interaction with biological tissues. In the food and packaging industries, polysaccharide films and coatings serve as biodegradable alternatives to petroleum-based plastics, offering tunable permeability and mechanical properties [2]. Incorporation of natural antimicrobials or antioxidants

 https://doi.org/10.3390/polym17223028

into such films has led to the emergence of active packaging materials that can extend the shelf life of food products [7]. In environmental technologies, polysaccharide-based adsorbents and membranes are increasingly used for wastewater treatment and pollutant removal, demonstrating excellent performance due to their high surface area and functional group diversity [17]. Furthermore, the integration of polysaccharides into hybrid and composite materials opens new possibilities for advanced applications in sensors and flexible electronics [18].

Despite significant advances, key challenges remain before polysaccharide-based materials can achieve their full potential. Mechanical robustness, high hydrophilicity, and variability in properties arising from differences in natural sources or extraction methods continue to hinder large-scale adoption. Addressing these issues requires both improved formulation and processing techniques and the standardisation of feedstock characterisation. The design of hybrid materials, combining polysaccharides with other biopolymers, nanofillers, or synthetic polymers, represents one of the most promising strategies to overcome current limitations. Likewise, nanostructuration and self-assembly strategies offer innovative routes to control morphology and enhance performance at multiple scales. Computational modelling and molecular simulations are also expected to play an increasingly important role in predicting and optimising material behaviour. Moreover, true sustainability cannot be achieved without considering the entire lifecycle of materials. Future efforts should integrate environmental and economic assessments, ensuring that bio-based materials deliver genuine ecological advantages compared with conventional counterparts.

In this Special Issue, we have collected the most recent advances in developments and properties of polysaccharide-based materials, including ten original research articles. The research topics mainly cover the design of novel polysaccharide-based materials and composites, the valorisation of bio-based resources, and the exploration of functional properties for applications.

Polysaccharides are increasingly exploited as structural platforms for advanced functional materials. Kadsanit et al. [19] investigated the synthesis and modification of dialdehyde bacterial cellulose (DBC) through controlled periodate oxidation. Using response surface methodology, the authors established predictive correlations between the reaction conditions and the degree of oxidation, which strongly influenced morphology and aldehyde content. The resulting DBC acted as an efficient crosslinking and reinforcing agent for gelatine matrices, enhancing tensile strength and stability. Moreover, the ability to fine-tune the oxidation degree provides additional flexibility in designing materials with tailored sponge-like morphologies and mechanical properties suitable for biomedical applications. Li et al. [20] reported a sustainable strategy for developing high-performance ionic conductive flexible hydrogels from Camellia oleifera shells, an agricultural by-product. By extracting nanocellulose and introducing zwitterionic functionalities via atom transfer radical polymerisation, they fabricated nanocellulose–polyvinyl alcohol (NC-PSBMA/PVA) composite hydrogels with remarkable porosity, conductivity, and mechanical strength seven times greater than neat PVA. The resulting materials exhibit high sensitivity and rapid response, representing a promising route towards biodegradable and high-efficiency flexible electronic sensors.

In the field of energy storage, Legerstee et al. [21] examined magnesium transfer phenomena in aqueous alginate-based electrolytes using atomic force microscopy. Their findings demonstrated reversible magnesium deposition and stripping without dendrite formation, enabling safe and controlled operation. This work underscores the potential of polysaccharide-based electrolytes to improve the stability and environmental compatibility of emerging magnesium battery technologies.

Several contributions focused on circular-economy approaches and eco-friendly solutions derived from polysaccharide-rich wastes. Lamoudan et al. [22] proposed a pioneering method to transform cellulose-containing textile waste into multifunctional construction panels using a papermaking process. The integration of phosphorylated lignocellulosic fibres endowed the panels with enhanced mechanical resistance, thermal insulation, and fire-retardant properties. Meeting international construction standards, these panels exemplify how upcycling textile residues can simultaneously reduce waste and yield value-added materials for sustainable architecture. Complementing this perspective, Trivunac et al. [23] converted discarded cotton and mixed yarns into high-performance carbon adsorbents through hydrothermal carbonisation and KOH activation. The modified cellulose-based materials displayed large surface areas and abundant oxygen functionalities, leading to high adsorption capacity for heavy metals, especially lead. This approach not only valorises textile waste but also provides a scalable solution for wastewater treatment, reinforcing the environmental potential of polysaccharide-based sorbents.

Wang et al. [24] investigated chitosan as a green alternative to cement for soil stabilisation. Their systematic evaluation revealed that optimal acid concentration (0.5–1%) and curing temperature (45–65 °C) significantly enhance unconfined compression strength and durability. Microscopic analyses confirmed that chitosan forms hydrogen and electrostatic interactions with soil particles, filling voids and strengthening the matrix. The results establish chitosan as a sustainable and effective soil-binding agent for eco-friendly civil engineering.

Hydrogels, films, and coatings based on polysaccharides continue to attract attention for their biocompatibility and tunable properties. Gorroñogoitia et al. [25] designed semi-interpenetrating (semi-IPN) hydrogels combining alginate and sulphated hyaluronic acid for articular cartilage tissue engineering. By adjusting the alginate-to-hyaluronic acid ratio, the researchers achieved control over rheological and viscoelastic behaviour, optimising the balance between mechanical performance and biological functionality. These hydrogels effectively mimic extracellular matrix properties, demonstrating strong potential for regenerative medicine.

In line with food packaging and food preservation, Nitikornwarakul et al. [26] studied glutinous rice starch–chitosan (GRS/CS) blends as active coatings to extend mango shelf life. Increasing chitosan content enhanced mechanical strength, hydrophobicity, and resistance to fungal infection, maintaining fruit quality for up to ten days compared to two days for uncoated samples. The optimised coatings reduced dehydration and respiration, confirming their applicability as biodegradable, natural-based films for post-harvest preservation. Shlosman et al. [27] developed cellulose-encapsulated emulsions of thymol and eugenol essential oils through lyophilisation. The resulting powders retained antimicrobial and antioxidant properties, showed high encapsulation efficiency, and exhibited slow release and improved thermal stability. These characteristics make them suitable for integration into polymer matrices, particularly for agricultural and packaging applications requiring controlled release at elevated temperatures.

Further extending the exploration of microbial polysaccharides, Ramírez Tapias et al. [28] investigated films derived from integral milk kefir grain biomass and purified kefiran. Both materials formed homogeneous structures, yet kefiran films were more transparent and exhibited stronger polymeric interactions. The addition of glycerol acted as a plasticiser in integral milk kefir grain biomass-based films. On the other hand, an antiplasticisation effect at low glycerol levels in purified kefiran films points to complex polymer–plasticiser interactions. These findings emphasise the differences between materials derived from the integral milk kefir grains and those from purified kefiran, providing insights into their application potential.

The ten contributions collected in this Special Issue illustrate the remarkable versatility of polysaccharides as sustainable, functional building blocks for advanced materials. Through innovative processing, controlled chemical modification, and synergistic blending with other biopolymers, these studies reveal how cellulose, chitosan, alginate, hyaluronic acid, starch, and kefiran can yield materials with tunable mechanical, electrical, and biological properties. Applications range from flexible sensors and biomedical hydrogels to sustainable packaging, wastewater treatment, and soil stabilisation. Collectively, they demonstrate that polysaccharide-based systems can effectively bridge the gap between renewable resource utilisation and high-performance engineering solutions.

Funding: This work was funded by the Universidad Nacional de Quilmes (UNQ, Argentina) through the R&D programme PUNQ 745/25.

Conflicts of Interest: The author declares no conflicts of interest.

References

1. Berradi, A.; Aziz, F.; El Achaby, M.; Ouazzani, N.; Mandi, L.A. Comprehensive review of polysaccharide-based hydrogels as promising biomaterials. *Polymers* **2023**, *15*, 2908. [CrossRef]
2. Goyal, R.; Verma, S.; Vaish, M.; Shekhar, A. Polysaccharide: A biodegradable biopolymer for edible films and coating. *J. Current Res. Food Sci.* **2025**, *6*, 170–176. [CrossRef]
3. Benalaya, I.; Alves, G.; Lopes, J.; Silva, L.R. A review of natural polysaccharides: Sources, characteristics, properties, food, and pharmaceutical applications. *Int. J. Mol. Sci.* **2024**, *25*, 1322. [CrossRef]
4. Xu, B.-W.; Li, S.-S.; Ding, W.-L.; Zhang, C.; Rehman, M.U.; Tareen, M.F.; Wang, L.; Huang, S.-C. From structure to function: A comprehensive overview of polysaccharide roles and applications. *Food Front.* **2025**, *6*, 15–19. [CrossRef]
5. Muhammad, A.; Lee, D.; Shin, Y.; Park, J. Recent progress in polysaccharide aerogels: Their synthesis, application, and future outlook. *Polymers* **2021**, *13*, 1347. [CrossRef]
6. Lago, A.; Delgado, J.F.; Rezzani, G.D.; Cottet, C.; Ramírez Tapias, Y.A.; Peltzer, M.A.; Salvay, A.G. Multi-component biodegradable materials based on water kefir grains and yeast biomasses: Effect of the mixing ratio on the properties of the films. *Polymers* **2023**, *15*, 2594. [CrossRef] [PubMed]
7. Li, S.; Ren, Y.; Hou, Y.; Zhan, Q.; Jin, P.; Zheng, Y.; Wu, Z. Polysaccharide-based composite films: Promising biodegradable food packaging materials. *Foods* **2024**, *13*, 3674. [CrossRef] [PubMed]
8. Maqsood, S.; Khalid, W.; Kumar, P.; Benmebarek, I.E.; Rasool, I.F.U.; Trif, M.; Moreno, M.; Esatbeyoglu, T. Valorization of plant-based agro-industrial waste and by-products for the production of polysaccharides: Towards a more circular economy. *Appl. Food Res.* **2025**, *5*, 100954. [CrossRef]
9. Mouro, C.; Gomes, A.P.; Gouveia, I.C. Microbial exopolysaccharides: Structure, diversity, applications, and future frontiers in sustainable functional materials. *Polysaccharides* **2024**, *5*, 241–287. [CrossRef]
10. Coma, M.E.; Peltzer, M.A.; Delgado, J.F.; Salvay, A.G. Water kefir grains as an innovative source of materials: Study of plasticiser content on film properties. *Eur. Polym. J.* **2019**, *120*, 109234. [CrossRef]
11. Liu, T.; Ren, Q.; Wang, S.; Gao, J.; Shen, C.; Zhang, S.; Wang, Y.; Guan, F. Chemical modification of polysaccharides: A review of synthetic approaches, biological activity and the structure–activity relationship. *Molecules* **2023**, *28*, 6073. [CrossRef]
12. Janik, W.; Jakubski, L.; Kudła, S.; Dudek, G. Modified polysaccharides for food packaging applications: A review. *Int. J. Biol. Macromol.* **2024**, *258*, 128916. [CrossRef]
13. Trache, D.; Thakur, V.K.; Boukherroub, R. Cellulose nanocrystals/graphene hybrids—A promising new class of materials for advanced applications. *Nanomaterials* **2020**, *10*, 1523. [CrossRef] [PubMed]
14. Zhang, R.; Liu, X.; Zhang, W.; Cui, B.; Du, Y.; Huang, Y.; Li, W.; Liu, Q.; Ren, C.; Tang, Z. A review of polysaccharide-based hydrogels: From structural modification to biomedical applications. *Int. J. Biol. Macromol.* **2025**, *310*, 143519. [CrossRef] [PubMed]
15. Schmid, J.; Sieber, V. Enzymatic transformations involved in the biosynthesis of microbial exo-polysaccharides based on the assembly of repeat units. *ChemBioChem* **2015**, *16*, 1141–1147. [CrossRef] [PubMed]
16. Ma, C.; Du, L.; Guo, Y.; Yang, X. A review of polysaccharide hydrogels as materials for skin repair and wound dressing: Construction, functionalization and challenges. *Int. J. Biol. Macromol.* **2024**, *280*, 135838. [CrossRef]
17. Díaz Bukvic, G.; Rossi, E.; Errea, M.I. Polysaccharides as economic and sustainable raw materials for the preparation of adsorbents for water treatment. *Polysaccharides* **2023**, *4*, 219–255. [CrossRef]
18. Wang, Y.; Huang, Z.; Zhang, D.; Li, M.; Wen, X.; Lei, X.; Ye, Z.; Pang, J. High performance multifunctional bio-gel sensor prepared with konjac glucomannan and κ-carrageenan. *Carbohydr. Polym.* **2025**, *368*, 124215. [CrossRef]

19. Kadsanit, K.; Sriariyanun, M.; Phisalaphong, M.; Kirdponpattara, S. Tailoring dialdehyde bacterial cellulose synthesis for versatile applications. *Polymers* **2025**, *17*, 1836. [CrossRef]
20. Li, J.; Peng, W.; Lei, Z.; Jian, J.; Cong, J.; Zhao, C.; Wu, Y.; Su, J.; Han, S. Fabrication of zwitterionized nanocellulose/polyvinyl alcohol composite hydrogels derived from camellia oleifera shells for high-performance flexible sensing. *Polymers* **2025**, *17*, 1901. [CrossRef]
21. Legerstee, W.J.; Kiriinya, L.; Kwakernaak, M.; Kelder, E.M. Magnesium transfer between atomic force microscopy probes and metal electrodes in aqueous alginate electrolytes. *Polymers* **2024**, *16*, 1615. [CrossRef]
22. Lamoudan, H.; Abenghal, L.; Belosinschi, D.; Brouillette, F.; Dolez, P.; Panneton, R.; Fonrouge, C. Sustainable transformation of cellulose-containing textile waste into multifunctional panels with tailored fr-lignocellulosic fibres. *Polymers* **2024**, *16*, 3242. [CrossRef]
23. Trivunac, K.; Mihajlović, S.; Vukčević, M.; Maletić, M.; Pejić, B.; Kalijadis, A.; Perić Grujić, A. Modified cellulose-based waste for enhanced adsorption of selected heavy metals from wastewater. *Polymers* **2024**, *16*, 2610. [CrossRef]
24. Wang, R.; Ong, D.E.L.; Sadighi, H.; Goli, M.; Xia, P.; Fatehi, H.; Yao, T. Optimizing soil stabilization with chitosan: Investigating acid concentration, temperature, and long-term strength. *Polymers* **2025**, *17*, 151. [CrossRef] [PubMed]
25. Gorroñogoitia, I.; Olza, S.; Alonso-Varona, A.; Zaldua, A.M. The effect of alginate/hyaluronic acid proportion on semi-interpenetrating hydrogel properties for articular cartilage tissue engineering. *Polymers* **2025**, *17*, 528. [CrossRef] [PubMed]
26. Nitikornwarakul, C.; Wangpradid, R.; Rakkapao, N. Impact of molar composition on the functional properties of glutinous rice starch–chitosan blend: Natural-based active coating for extending mango shelf life. *Polymers* **2024**, *16*, 1375. [CrossRef] [PubMed]
27. Shlosman, K.; Rein, D.M.; Shemesh, R.; Cohen, Y. Lyophilized emulsions of thymol and eugenol essential oils encapsulated in cellulose. *Polymers* **2024**, *16*, 1422. [CrossRef]
28. Ramírez Tapias, Y.A.; Rezzani, G.D.; Delgado, J.F.; Peltzer, M.A.; Salvay, A.G. New Materials from the Integral Milk kefir grain biomass and the purified kefiran: The role of glycerol content on the film's properties. *Polymers* **2024**, *16*, 3106. [CrossRef]

polymers

MDPI

Article

Tailoring Dialdehyde Bacterial Cellulose Synthesis for Versatile Applications

Krittanan Kadsanit [1], Malinee Sriariyanun [2,3], Muenduen Phisalaphong [4] and Suchata Kirdponpattara [1,3,*]

1 Department of Chemical Engineering, Faculty of Engineering, King Mongkut's University of Technology North Bangkok, Bangkok 10800, Thailand; krittanankadsanit@gmail.com
2 The Sirindhorn International Thai-German Graduate School of Engineering, King Mongkut's University of Technology North Bangkok, Bangkok 10800, Thailand; malinee.s@tggs.kmutnb.ac.th
3 Biorefinery and Process Automation Engineering Center (BPAEC), King Mongkut's University of Technology North Bangkok, Bangkok 10800, Thailand
4 Department of Chemical Engineering, Faculty of Engineering, Chulalongkorn University, Bangkok 10330, Thailand; muenduen.p@chula.ac.th
* Correspondence: suchata.k@eng.kmutnb.ac.th; Tel.: +66-2-555-200

Abstract: Dialdehyde bacterial cellulose (DBC) has been implemented in versatile applications. DBC was prepared from bacterial cellulose (BC) through periodate oxidation with varying parameters, including the mole ratio of BC and $NaOI_4$, temperature, and reaction time. The relationship between the degree of oxidation (DO)/aldehyde content and these parameters was proposed as a quadratic equation to predict the oxidation conditions needed to achieve a specific DO using Response Surface Methodology (RSM). The chemical structure and morphology of DBC were influenced by DO. DBC with different DO levels was used as a crosslinker and a reinforcing agent for gelatin sponge fabrication. Results indicated that a high DO of DBC could enhance the tensile strength and structural stability of the gelatin matrix. Selecting the proper DO level could control the morphological structure of the gelatin sponge, which is crucial for biomedical applications.

Keywords: bacterial cellulose; degree of oxidation; dialdehyde bacterial cellulose; gelatin; optimization; RSM

1. Introduction

Bacterial cellulose (BC) or microbial nanocellulose is a polysaccharide with a linear chain of β-1,4 linked d-glucose units. Via metabolic pathway, pure cellulose nanofibrils are extruded through the cell wall of *Acetobacter xylinum* and then packed in the form of a jelly-like pellicle with high crystallinity, a high degree of polymerization, and high mechanical strength [1]. Consequently, BC has been widely used as a material-based composite and reinforcement for other polymers [2–4]. Moreover, with outstanding characteristics, including biocompatibility, non-toxicity, high porosity, and good liquid-holding capacity, BC has been applied in various biomedical applications [5], such as wound dressings [6], drug delivery [7], and scaffolds for tissue engineering [8].

To tailor BC for specific applications and broaden its utility, further modification of its properties is necessary. In the context of tissue engineering, the limited biodegradability of BC in the human body poses a significant restriction [9]. To address this challenge, both in situ and ex situ functionalization methods are commonly used. One of the most extensive functionalization techniques is periodate oxidation. Strong ions (IO_4^-) oxidize the C2–C3 bond of the glucopyranoside ring and form two aldehyde groups located at the C2 and C3 positions per glucose unit [10], named dialdehyde bacterial cellulose (DBC).

The dialdehyde functional group can crosslink polymers consisting of the amine functional group through the Schiff's base reaction. For example, keratin was crosslinked by DBC to enhance its mechanical strength for use as an adsorbent in the removal of dye and heavy metal ions [11]. To obtain an antibacterial and healing wound dressing without toxicity, chitosan was self-crosslinked with dialdehyde carboxymethyl BC instead of using glutaraldehyde or formaldehyde as a crosslinker [12].

In addition, the aldehyde functional groups modified on the cellulose polymer chain can develop BC characteristics, including water solubility, biodegradability, and antibacterial activity. Li et al. (2009) [9] prepared DBC with an aldehyde content of 60.3% and tested its biodegradation in vitro. DBC could degrade in DI water and phosphate-buffered saline, resulting in mass losses of approximately 50% and 80%, respectively, in 60 days. Furthermore, approximately 10% of the aldehyde content of DBC showed a 16% degradation in Tris-HCl-buffered synthetic body fluid over a 14-day period [13]. Moreover, cellophane was oxidized with sodium periodate by varying reaction times (2–6 h) to develop antimicrobial activity [14]. Aldehyde content or the degree of oxidation (DO) directly influences DBC properties. Most publications reported only a specific value of aldehyde content.

The DO via periodate oxidation is influenced by four independent factors: periodate concentration (the mole ratio of BC to periodate), temperature, reaction time, and pH. Achieving the desired characteristics of DBC regarding DO is quite challenging. Furthermore, minimizing reaction time and the costs for chemicals and utilities is essential. The effects of each parameter must be thoroughly examined, as the study of their impacts and interactions remains insufficient. In this recent work, BC was oxidized using $NaIO_4$ to produce DBC with varying DO levels. The Design of Experiment (DOE) with Central Composite Design (CCD) approach-based Response Surface Methodology (RSM) was employed to analyze the impact of three independent factors: mole ratio between BC and $NaIO_4$, temperature, and reaction time on DO and DBC properties. A quadratic equation analyzed through RSM was applied to predict the reaction conditions for generating DBC with specific DO levels of 20%, 40%, 60%, 80%, and 100%. The chemical and physical characteristics of DBCs were analyzed. To give an idea, DBCs with specific DO levels were used as reinforcement and crosslinkers for gelatin sponge fabrication. The chemical and physical characteristics of the gelatin–DBC (GDB) sponge, affected by DO levels, were examined to provide insights for further applications.

2. Materials and Methods

2.1. Bacterial Cellulose (BC)

BC cubes ($1 \times 1 \times 1$ cm^3) synthesized by *Acetobacter xylinum* were kindly supported by Fruitia Food Processing Co., Ltd., Nakorn Pathom, Thailand. The BC was initially treated with 1% NaOH and then rinsed with DI water to achieve a neutral pH. Next, to prepare the BC slurry, the BC cubes were homogenized using a blender (HR2096, Philips, Drachten, The Netherlands) at 1200 rpm for 3 min [15] and stored at 4 °C until use. The dry weight of the BC slurry was controlled at 1 g dry BC/100 g slurry.

2.2. Chemicals

Sodium periodate ($NaIO_4$) and hydroxylamine hydrochloride were purchased from Alfa-Aesar, Ward Hill, MA, USA. Gelatin Type A (300 bloom), 2,4,6-trinitrobenzenesulfonic acid (TNBS), and 7.5% sodium bicarbonate were supplied by Sigma-Aldrich, St. Louis, MO, USA.

2.3. Synthesis of Dialdehyde Bacterial Cellulose (DBC)

BC slurry was mixed with $NaIO_4$ in mole ratios of 1:1, 1:1.5, and 1:2. The mixture was stirred at 250 rpm in the dark at 40, 50, and 60 °C for 4, 8, and 12 h. Afterward, the DBC was obtained and rinsed with deionized (DI) water to remove excess $NaIO_4$ and $NaIO_3$ (byproduct). The DBC slurry was stored at 4 °C until use.

2.4. Experimental Design

Factorial design has been extensively applied to experimental design to study the impact of each independent variable on others. Many experiments must be conducted, wasting time and resources. Design-Expert software (Version 7.0, Stat-Ease Inc., Minneapolis, MN, USA) was used in this study to minimize the number of experimental runs. CCD was utilized to design a range of independent parameters, including the mole ratio of BC to $NaOI_4$ (X_1), temperature (X_2), and time (X_3). The total number of experimental runs designed was 19, comprising six axial points, eight factorial points, and five central points. The central point was repeated five times to ensure accuracy. The parameters for the central points included a mole ratio of 1:1.5, a temperature of 50 °C, and a reaction time of 8 h. Statistical analysis was performed using Analysis of Variance (ANOVA). RSM was applied to examine the interaction between parameters through three-dimensional (3D) surface plots and a quadratic equation (Equation (1)).

$$Y = \beta_0 + \sum_{i=1}^{n} \beta_i x_i + \sum_{i=1}^{n} \beta_{ii} x_i^2 + \sum_{i=1} \sum_{j=i+1}^{n} \beta_{ij} x_i x_j \quad (1)$$

where Y is a DO of DBC; β_0 is a constant; β_i, β_{ii}, and β_{ij} are linear, quadratic, and second-order interaction coefficients, respectively; x_i and x_j are independent variables; and n is the number of variables.

2.5. Fabrication of Gelatin–DBC (GDB) Sponge

A 15 wt.% gelatin solution was prepared at 60 °C and then mixed with DBC slurry in a 1:1 weight ratio while controlling the DBC concentration at 1% weight (dry basis). The mixture was stirred at 60 °C for 30 min. Afterwards, 50 g of the mixture was poured into a mold with a surface area of 6.5 × 15 cm^2. Finally, it was lyophilized at −40 °C until fully dried. The composite sponge was stored in a desiccator at room temperature until use.

2.6. Characterization of DBC and GDB Sponge

Five replicates were performed for all investigations, excluding FT-IR and SEM.

2.6.1. Fourier Transform-Infrared Spectrometer (FT-IR)

A freeze-dried sample was crushed into a fine powder and scanned using an FT-IR spectrometer (PerkinElmer, Spectrum One, Waltham, MA, USA) between 4000 and 400 cm^{-1} with a resolution of 4 cm^{-1}.

2.6.2. Scanning Electron Microscopy (SEM)

Morphologies of BC, DBCs, and GBD sponges were examined by SEM (JEOL, JSM-IT500HR, Tokyo, Japan). The BC and DBC slurries were freeze-dried before coating with a thin layer of gold. For the GBD sponges, they were directly coated with gold. Then, all the samples were observed at a voltage of 5 kV under high resolution.

2.6.3. Determination of Degree of Oxidation (DO)

Adapted from published work [16], 0.1 g dry weight (m) of a sample was mixed with 30 mL DI water and adjusted to pH 4 using 1 M NaOH or HCl. The mixture was added with 20 mL of 0.25 M hydroxylamine hydrochloride solution and then shaken at 150 rpm.

After 24 h, the supernatant from the mixture was titrated with 0.1 M NaOH (M_{NaOH}). A volume of 0.1 M NaOH used for the titration of the sample (V_s) was recorded and compared with that used for titration BC (V_b), which was used as a blank. The molecular weight of BC is 160 g/mol. The DO was calculated by Equation (2).

$$DO\ (\%) = 160 \times 0.1 \times M_{NaOH} \times (V_s - V_b)/m \quad (2)$$

2.6.4. Degree of Crosslinking

According to Thongsrikhem et al. (2022) [15], the quantitative percentage of occupied amino acid compared to the total amino acid of gelatin was examined using 2,4,6-trinitrobenzenesulfonic acid (TNBS) as a reagent. The aliquot obtained from the reaction was measured for optical density at 346 nm using a UV–visible spectrophotometer (PG Instruments T80+, Leicestershire, UK). Equation (3) was used to calculate the degree of crosslinking. A represents the optical density, and W denotes the weights of the sample (s) and the control (c), which is native gelatin sponge.

$$Degree\ of\ crosslinking\ (\%) = 1 - ((A_s/W_s)/(A_c/W_c)) \times 100 \quad (3)$$

2.6.5. Swelling and Weight Loss

The GDB sponge was cut into 1 × 1 cm^2 and weighed (W_i) before being immersed in distilled water. After 24 h, the immersed sponge had its excess water removed by paper wipe and reweighed (W_w) for calculating swelling ability using Equation (4). The same procedure was carried out to determine weight loss, but the immersion duration was extended to 7 days. Then, the sponge was removed from the water and dried at 60 °C for 24 h. The dried sponge was weighed again (W_f). The percentage weight loss was calculated using Equation (5).

$$Swelling\ (times) = (W_w - W_i)/W_i \quad (4)$$

$$Weight\ loss\ (\%) = 100 \times \left(W_i - W_f\right)/W_i \quad (5)$$

2.6.6. Porosity

The porosity of the sponge was determined using the liquid replacement technique [17]. A 1 × 1 cm^2 sponge was measured for dimensions (width × length × thickness = V_t), weighed (W_i), then immersed in 99.7% hexane (J.T. Baker, Phillipsburg, NJ, USA) under vacuum conditions, allowing for the complete penetration of hexane into the sponge's interior without causing any swelling or shrinkage. The sponge was reweighed (W_h) after reaching equilibrium. The porosity of the sponge was calculated using Equation (6).

$$Porosity\ (\%) = 100 \times (W_h - W_i)/(hexane\ density \times V_t) \quad (6)$$

2.6.7. Mechanical Test

Following ASTM D882, sponge (1 × 10 cm^2) strength was analyzed by a Universal Testing Machine (Instron model 1123, Norwood, MA, USA) with a tensile rate of 2 mm/min. Tensile strength and elongation at break were obtained.

2.7. Statistical Analysis

Regression analysis and Analysis of Variance (ANOVA) were used to analyze the experimental data by considering a significance level (α) of 0.05. The coefficient of determination (R^2), adjusted coefficient of determination (R^2_{adj}), and predicted coefficient of determination (R^2_{pred}) were employed to validate the obtained quadratic model. The optimization conditions were considered using the desirability function.

3. Results and Discussion

DBC was synthesized through periodate oxidation using $NaIO_4$ to oxidize the hydroxyl groups to create aldehyde groups within the BC structure. The experimental conditions were designed by DOE, and the simulated equation was employed to predict the conditions needed to achieve the specific DO of DBC. Model validation was conducted to ensure the accuracy of the obtained equation. After producing the DBC with varying levels of DO, it was incorporated into gelatin, serving as both a crosslinker and a reinforcement for fabricating the GDB sponges. The sponges were characterized based on their chemical and physical properties to reveal the influence of the DO level of DBC.

3.1. Experimental Design and Data Analysis

The experimental design consisted of 19 experiments, as summarized in Table 1, along with predicted and actual DO (Y) values and residuals (the difference between the actual and predicted DO). A second-order polynomial equation (Equation (7)) describes the relationship between DO and the process parameters. A high degree of agreement between the predicted and actual experimental DO is demonstrated in residual terms (Table 1). Most residuals were quite low, indicating the success of the correlation development between DO and the reaction parameters. According to the actual DO values of runs 8 to 12 (the central point), it was confirmed that the DBC synthesis was highly reproducible. Moreover, the experimental conditions, designed using a full factorial design (27 experimental runs), were also carried out. Results of DO and DBC yield were summarized in the Supplementary File (Table S1).

$$Y = -134.69 - 17.41X_1 + 5.05X_2 + 2.21X_3 + 0.82X_1X_2 + 2.04X_1X_3 - 0.007X_2X_3 - 3.52X_1^2 - 0.05X_2^2 - 0.08X_3^2 \quad (7)$$

Table 1. Degree of oxidation (DO) from prediction and experiment and their residuals.

Run	Conditions			DO		
	Mole Ratio of BC and $NaIO_4$ (X_1)	Temperature (X_2)	Time (X_3)	Predicted	Actual	Residual
1	1:1 (−1)	40 (−1)	4 (−1)	15.6	18.3	2.7
2	1:1 (−1)	40 (−1)	12 (+1)	37.7	36.3	−1.4
3	1:1 (−1)	50 (0)	8 (0)	42.1	40	−2.1
4	1:1 (−1)	60 (+1)	4 (−1)	34.7	36.4	1.7
5	1:1 (−1)	60 (+1)	12 (+1)	55.8	55	−0.8
6	1:1.5 (0)	40 (−1)	8 (0)	39.3	39.1	−0.2
7	1:1.5 (0)	50 (0)	4 (−1)	41.5	35.2	−6.3
8	1:1.5 (0)	50 (0)	8 (0)	57.6	57.7	0.1
9	1:1.5 (0)	50 (0)	8 (0)	57.6	57.6	0
10	1:1.5 (0)	50 (0)	8 (0)	57.6	56.7	−0.9
11	1:1.5 (0)	50 (0)	8 (0)	57.6	57.2	−0.4
12	1:1.5 (0)	50 (0)	8 (0)	57.6	57.8	0.2
13	1:1.5 (0)	50 (0)	12 (+1)	71.2	78	7.6
14	1:1.5 (0)	60 (+1)	8 (0)	66.1	66.8	−0.7
15	1:2 (+1)	40 (−1)	4 (−1)	28.5	29.2	0.7
16	1:2 (+1)	40 (−1)	12 (+1)	66.9	65.2	−1.7
17	1:2 (+1)	50 (0)	8 (0)	71.3	73.9	2.6
18	1:2 (+1)	60 (+1)	4 (−1)	64	65.3	1.3
19	1:2 (+1)	60 (+1)	12 (+1)	101.4	98.6	−2.8

ANOVA was used to assess the reliability of the quadratic model, the impact of each individual factor, and the combined effect of two factors that significantly influenced the experimental results. The data analysis from ANOVA is presented in Table 2 and in Table S2 in the Supplementary Materials (a full version). The regression model showed statistical significance with a high f-value (53.46) and a very low probability (<0.0001). Additionally, it was determined that X_1, X_2, and X_3 were highly significant terms in the model. However, X_1X_2 and X_1X_3 become significant terms when the probability was considered at <0.05. Nagpal et al. (2019) [18] suggested that the value of "Prob less than f″ < 0.05 indicated significant model terms.

Table 2. ANOVA data analysis for the model.

Source	Model	X_1	X_2	X_3	X_1X_2	X_1X_3	X_2X_3	X_1^2	X_2^2	X_3^2	Lack of Fit Data
Sum of Squares	6598.22	2137.44	1795.60	2211.17	133.66	133.66	0.5512	2.12	65.11	4.14	122.60
df	9	1	1	1	1	1	1	1	1	1	5
Mean Square	733.14	2137.44	1795.60	2211.17	133.66	133.66	0.5512	2.12	65.11	4.14	24.52
f-value	53.46	155.87	130.94	161.24	9.75	9.75	0.04	0.15	4.75	0.30	119.61
p-value	<0.0001	<0.0001	<0.0001	<0.0001	0.0110	0.0198	0.8120	0.7434	0.1045	0.4477	0.0002
R^2			0.9816			Adeq precision			31.9295		
R^2_{adj}			0.9633			C.V. (%)			6.87		
R^2_{pred}			0.8537								

Lack of fit was significant in this model, resulting in a 0.02% chance. The f-value of the lack of fit was this large due to noise. The R^2 of 0.9816 implied that 98.16% of the DO of DBC was affected by the variation in the independent variables. Based on general principles, the difference between R^2_{adj} and R^2_{pred} was less than 0.2, implying reasonable agreement between the two. Adeq precision represents the ratio between signal and noise.

From the 3D plot (Figure 1), the DO increased as the mole ratio, temperature, and reaction time rose. It had been reported that an increase in sodium periodate concentration enhances the number of periodate ions, leading to the rapid oxidation of secondary hydroxyl groups located on the carbon atoms in positions 2 and 3 in the structure of cellulose bacteria [16,19]. Considering Figure 1C, the shape of the 3D plot resembles a rhombus without curves. This suggests that the mole ratio and reaction time significantly influence DO, exhibiting a linear relationship. This aligns well with Figure 1D, which demonstrates a strong contribution from the mole ratio (X_1) and reaction time (X_3), at 32.2% and 32.1%, respectively. However, the impact of both mole ratio and temperature (X_1X_2) was slightly higher than that of both mole ratio and time (X_1X_3). There were no impacts from either temperature and time (X_2X_3) or squared mole ratio (X_1^2). This finding was also observed when the reaction temperature was controlled at 35°C; the reaction time needed to be prolonged to 26 h to obtain dialdehyde carboxymethyl cellulose with 78.6% DO [20].

Consequently, these three parameters all significantly influenced the DO level. Economically, operating costs can be dramatically reduced by minimizing $NaIO_4$ utilization, which can be accomplished through either increasing temperature or extending reaction duration. Conversely, high productivity is one of the most critical parameters for industry; therefore, the reaction time should be minimized through the utilization of a high molar ratio and elevated operational temperature. The simulated model could effectively assist in selecting operating conditions based on the investment cost.

Next, DBCs with specific DOs (20%, 40%, 60%, 80%, and 100%) were synthesized, designated as DBC20, DBC40, DBC60, DBC80, and DBC100, respectively. The second-order polynomial equation (Equation (7)) was used to predict the conditions, as shown in Table 3. The DBC was quantitatively analyzed for the DO level to validate the obtained model. The results revealed that the quadratic model had high precision, with an error value (a percentage of difference between DO_{actual} and $DO_{required}$) of $\leq$5.3%. Therefore, this model had great potential for predicting the reaction conditions to achieve the specific DO of DBC. The DBC was further analyzed for its characteristics.

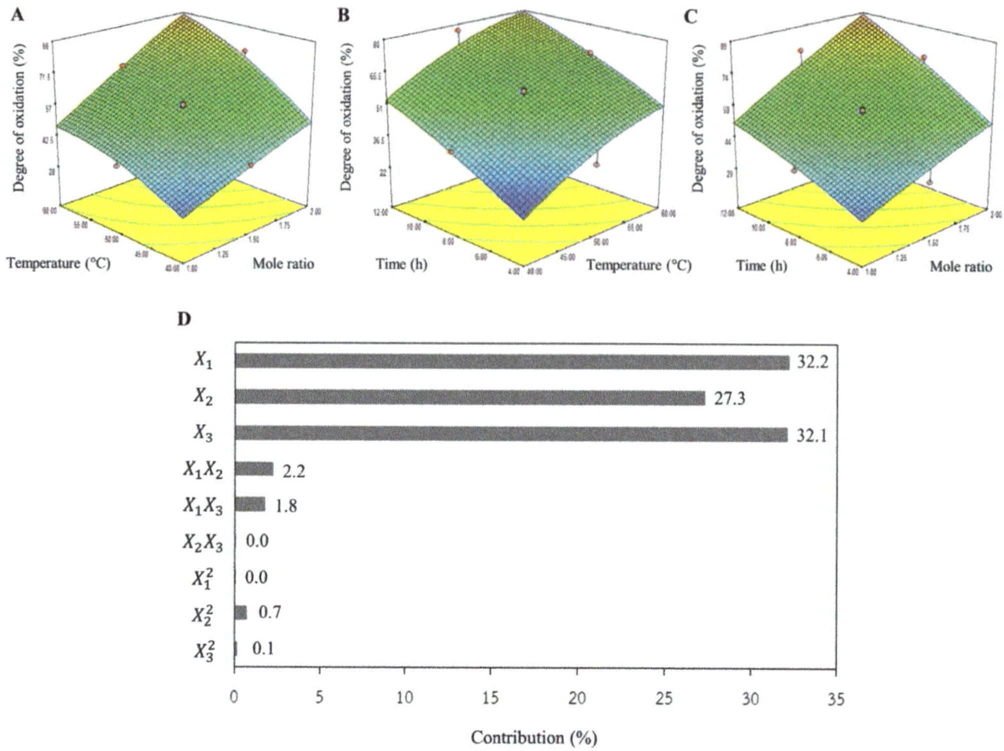

Figure 1. The 3D plots of aldehyde content for (**A**) temperature vs. mole ratio, (**B**) time vs. mole ratio, (**C**) time vs. temperature, and (**D**) percent contribution of each parameter.

Table 3. Predicted conditions for synthesizing the specified DO of DBC.

Symbol	Predicted Conditions			DO (%)		% Error
	Mole Ratio of BC and $NaIO_4$ (Time)	Temperature (°C)	Time (h)	Required	Actual	
DBC20	1:1.29	40	4	20	20.8 ± 0.1	4.0
DBC40	1:1.45	50	4	40	42.1 ± 0.2	5.3
DBC60	1:1.82	60	4	60	62.4 ± 0.5	4.0
DBC80	1:1.86	60	8	80	82.8 ± 0.9	3.5
DBC100	1:1.97	60	12	100	98.0 ± 0.6	2.0

% Error = $|(DO_{actual} - DO_{required})/DO_{required}) \times 100|$.

3.2. Characterization of DBC

3.2.1. FT-IR Analysis

FT-IR spectra of BC and DBC are displayed in Figure 2. The characteristic bands of BC were observed at 3341 cm^{-1} for OH stretching, 2896 cm^{-1} range for C-H, and 1107 cm^{-1} for C-O stretching [21]. All characteristic peaks of BC were also noticed in all DBCs.

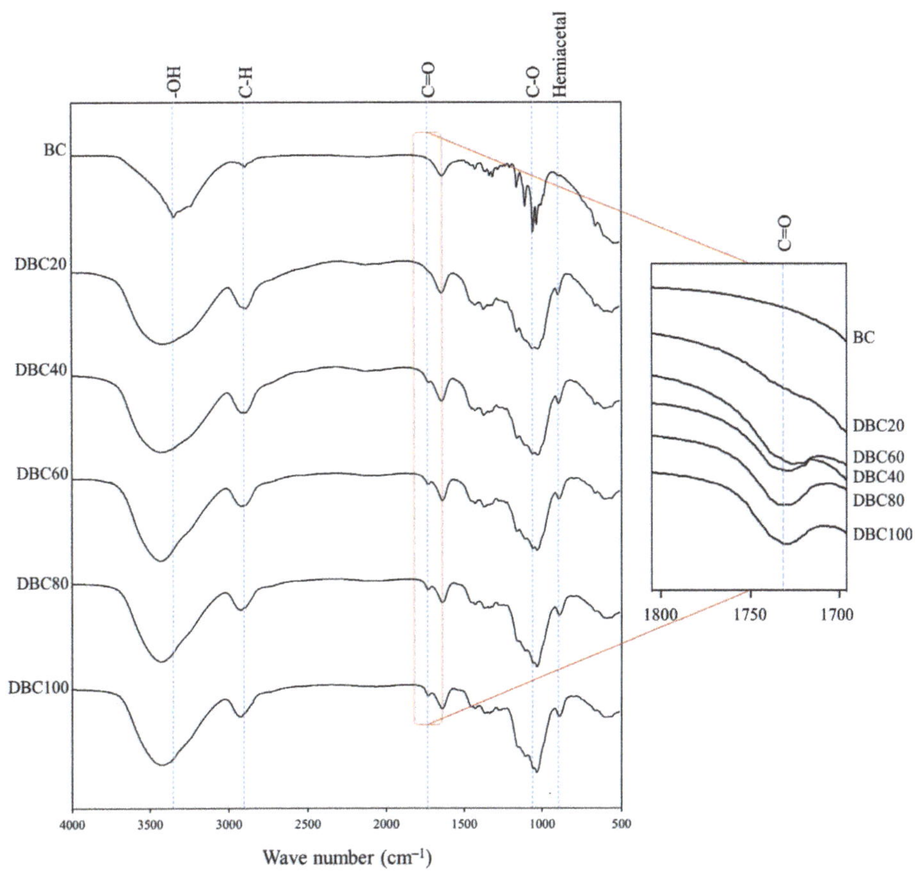

Figure 2. FT-IR spectrum of DBC with different DO levels.

New peaks at 1729 cm^{-1} and 886 cm^{-1} in DBC were observed, representing the aldehyde group (C=O) and the formation of hemiacetal bonds between aldehyde groups and neighboring hydroxyl groups, respectively. Wegrzynowska-Drzymalska et al. (2022) [22] also observed the aldehyde peak and hemiacetal bond peaks of dialdehyde cellulose at 1740 cm^{-1} and 880 cm^{-1}, respectively. In the zoomed-in figure at 1729 cm^{-1}, the intensities tended to be broadened with an increase in DO. It confirmed that the hydroxyl groups located at C2 and C3 of the D-glucose structure were oxidized by IO_4^- to form aldehyde groups, and this was in accordance with the DO level.

3.2.2. Morphologies of BC and DBCs

The structure of BC and DBCs is illustrated in Figure 3. A fibrous web-like network structure was observed in BC (Figure 3A). BC fibers were approximately 52.3 ± 6.6 nm in diameter (determined using ImageJ software, v1.54m). A similar morphology of BC was also reported in [23]. After BC underwent oxidation by periodate, ultrafine fibers, thinner than their original sizes, were observed in the following order: DBC20 (43.2 ± 6.0 nm), DBC40 (34.7 ± 6.5 nm), and DBC60 (27.3 ± 5.7 nm). It is also acknowledged in the synthesis of dialdehyde microcrystalline cellulose that the fiber length became shorter than its original length from 56.2 to 33.6 μm [24]. However, the fibrous structure could not be seen in 80DBC and 100DBC due to the harsh conditions: high mole ratio, high temperature, and prolonged reaction time, which strongly oxidized hydroxyl groups at C2 and C3 in glucose monomers, resulting in the weakening of the BC polymeric backbone and degradation of the fibrous structure. Tachai et al. (2024) [20] also found that periodate oxidation transformed a smooth surface of carboxymethyl cellulose into a pitted and non-smooth surface.

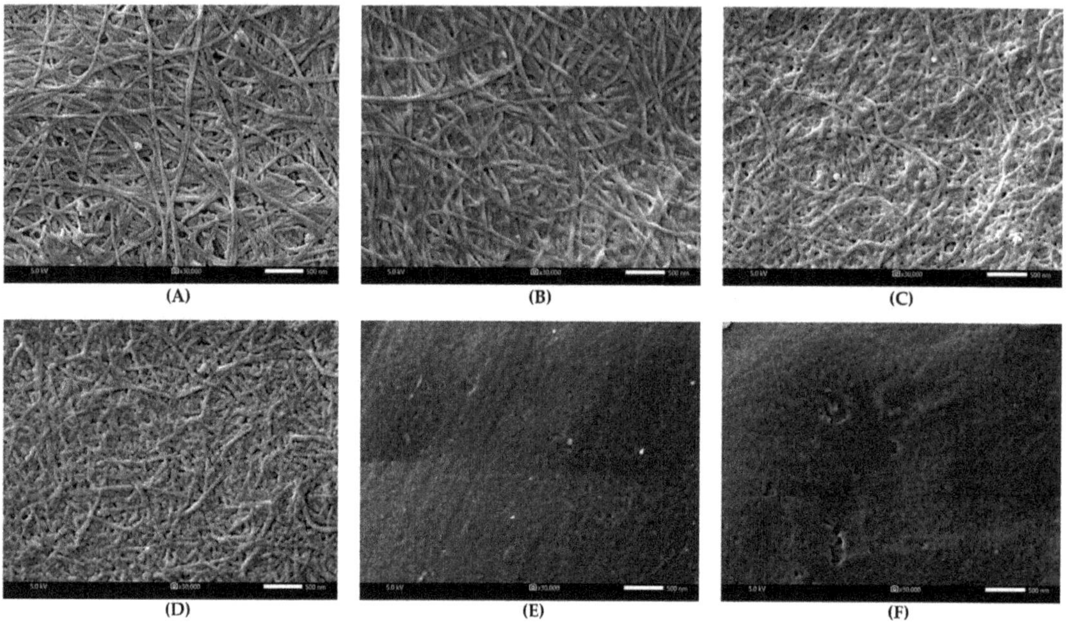

Figure 3. Morphologies of BC (**A**), DBC20 (**B**), DBC40 (**C**), DBC60 (**D**), DBC80 (**E**), and DBC100 (**F**).

The influence of various characteristics of DBC on the physical and mechanical properties of the gelatin crosslinked with DBC (GDB) sponge was further investigated.

3.3. Characterization of GDB Sponge

Gelatin, a protein biopolymer, is derived from the partial hydrolysis of collagen, exhibiting remarkable biocompatibility comparable to that of collagen, with more cost-effectiveness [25]. Due to the unique characteristics of gelatin, including its gel-forming ability, biodegradability, and promotion of cell attachment, proliferation, and differentiation [26], gelatin has been widely utilized in biomedical [27,28], tissue engineering [29,30], and pharmaceutical [31] applications. Gelatin is composed of various types of amino acids linked together by peptide bonds, with three primary amino acids: glycine (25–29%), proline (15–17%), and hydroxyproline (12–14%) [32]. Due to the high water solubility of gelatin, many applications require structural stability in the wet state; therefore, gelatin must be crosslinked through a Schiff's base reaction with substances containing aldehyde groups, such as glutaraldehyde and formaldehyde [33]. Nonetheless, these two chemicals are toxic to both animal and human cells, rendering them unsuitable for use in biomedical applications [34,35]. An alternative option is natural extracts such as cinnamaldehyde [15] and genipin [36], which are relatively expensive. Seeking a crosslinker with non-toxicity and biocompatibility is a challenging task.

Low mechanical strength is the weakest point of gelatin, which limits its applications in several areas [37]. Since neat gelatin is brittle and fragile, reinforcement with other polymers is necessary. For example, cellulose nanocrystals incorporated with electrospun gelatin nanofibers to improve their tensile strength for further utilization in biomedical fields [38]. Chitosan–gelatin hydrogel reinforced with cellulose nanofibrils showed enhancements in mechanical strength and thermal stability for diabetic wound healing applications [39]. Moreover, gelatin-starch was successfully integrated with polyvinyl alcohol to modify mechanical resilience [40]. The high potential of utilizing gelatin must include crosslinking and reinforcement steps. Consequently, the recent work proposed DBC simultaneously acting as a crosslinker and a reinforcing agent to improve the structural stability of gelatin sponge in both dry and wet states by controlling the DO level of DBC.

3.3.1. Degree of Crosslinking of GDB Sponge

DBC served simultaneously as a crosslinker and a reinforcement for the fabrication of gelatin sponges. The degree of crosslinking between the amino groups of gelatin and the aldehyde groups in DBC is summarized in Table 4. The results reveal that the degree of crosslinking increased with higher aldehyde content/DO. Due to the existing aldehyde content in DBC, it actively interacts with the amine functional groups of gelatin through Schiff's base reaction [41]. For the lowest aldehyde content (GDB20), the degree of crosslinking was only 12.6%. In contrast, the maximum degree of crosslinking achieved was nearly 50% using DBC100. However, the degree of crosslinking of the gelatin film crosslinked with dialdehyde cellulose was reported to be about 68–75% [41]. This might be attributed to the 9% lower gelatin concentration compared to recent work.

Table 4. Characteristics of gelatin (GL) and GDB sponges.

Sponge	Thickness (mm)	Porosity (%)	Degree of Crosslinking (%)	Weight Loss After 7 Days (%)	Swelling After 1 Day (Time)	Tensile Strength (MPa)	Elongation at Break (%)
GL	4.0 ± 0.4	23.9 ± 3.1	0.0 ± 0.0	52.3 ± 5.4	15.3 ± 2.9	5.9 ± 1.0	29.1 ± 5.6
GDB20	0.2 ± 0.0	3.5 ± 3.1	12.6 ± 2.8	33.0 ± 2.5	11.8 ± 0.6	2.8 ± 1.6	51.7 ± 8.8
GDB40	1.7 ± 0.3	32.3 ± 9.8	19.9 ± 2.7	15.0 ± 0.8	7.5 ± 0.8	4.6 ± 4.3	49.3 ± 5.0
GDB60	3.2 ± 0.6	41.3 ± 5.6	35.6 ± 2.3	11.4 ± 0.6	5.9 ± 1.3	4.9 ± 1.2	50.1 ± 6.8
GDB80	2.3 ± 0.4	28.3 ± 9.5	45.6 ± 1.6	10.6 ± 0.8	4.7 ± 0.2	6.8 ± 4.4	50.4 ± 8.6
GDB100	2.8 ± 0.2	47.7 ± 2.3	49.6 ± 1.6	9.6 ± 0.7	4.7 ± 0.2	9.0 ± 3.6	55.0 ± 2.6

FT-IR spectroscopy was used to characterize the sponges (Figure S1 in the Supplementary Materials). The characteristic bands of gelatin and DBC were present in all the GDB sponges. The vibration of the OH bond in gelatin and DBC peaks between 3100 and 3600 cm^{-1} was observed. After gelatin was crosslinked with DBC, the OH peaks shifted from 3284 cm^{-1} to 3279–3273 cm^{-1}. This shift could be attributed to the hydrogen bonding interaction between gelatin and DBC. Tohamy (2025) [42] also observed a shift in the OH peak from a higher (3336 cm^{-1}) to a lower (3330 cm^{-1}) wavenumber, indicating strong hydrogen bonding between dialdehyde cellulose and chitosan. This finding was also reported in the dialdehyde carboxylated cellulose crosslinked with soy protein [43]. For the chemical reaction (Schiff's base), a new peak of imine (C=N) around 1600–1700 cm^{-1} or a shift in the 1650–1600 cm^{-1} range [44] should be detected. However, this peak overlapped with the amide I of gelatin and the carbonyl peaks of DBC, leading to unclear identification. Moreover, the characteristic peaks (1729 and 886 cm^{-1}) of DBC disappeared in all GDB sponges, which was also reported in [41], indicating that these two peaks were utilized by the crosslinking process.

3.3.2. Weight Loss of GDB Sponge

The sponges were immersed in water for 7 days to determine the structural stability in terms of weight loss percentage, which directly relates to the degree of crosslinking. Generally, gelatin must be crosslinked to enhance its water resistance before being utilized in many applications. The percent weight loss of the GDB is presented in Table 4. Without crosslinking, the GL sponge had the highest weight loss compared to the GDB sponges. After crosslinking, the structure of the sponge was more stable in water, retaining more than 85% of its weight (GDB40, GDB60, GDB80, and GDB100) after 7 days. Due to the strong chemical reaction (Schiff's base reaction) between gelatin and DBC, DBC possessed the potential to be a crosslinker for gelatin. Consequently, DBC containing 40% DO could significantly improve the structural stability of gelatin.

3.3.3. Swelling of GDB Sponge

The swelling ability of the composite depends on several parameters, such as hydrophilicity/hydrophobicity, material structure, and interaction between polymer chains. Although gelatin and DBC both exhibit high hydrophilicity, after crosslinking, the swelling decreased as the DO of the DBC increased. Particularly with GDB80 and GDB100, the sponges could only swell to 4.7 times their original state. This may be due to the high aldehyde content in DBC effectively bonding with the gelatin polymer chains, resulting in a more rigid structure, which corresponds with the degree of crosslinking and percent weight loss. This finding aligns with gelatin gel crosslinked with dialdehyde cellulose nanocrystal [45] and dialdehyde starch [46].

3.3.4. Morphology of GDB Sponge

The fabrication technique employed in this research is lyophilization, which creates a three-dimensional porous structure suitable for various applications, including scaffolds for tissue engineering [47], wound dressings [48], and wastewater treatment [49]. After ice crystals are formed during the freezing step, they are sublimated during the drying phase, resulting in a porous structure, particularly in the interior of the sponge [17]. The thickness and porosity of the sponges are shown in Table 4, where the GL sponge exhibited the highest thickness. In contrast, the GDB20 structure, when gelatin crosslinked with DBC20, resulted in a densely non-porous film-like structure, as depicted in Figure 4 (second row). Due to the mild conditions, the structure of DBC20 (Figure 3B) retained fibers, similar to BC, without being destroyed by oxidation reactions. When gelatin was crosslinked with the DBC20 fibers, they tended to compact together, resulting in a film-like structure with a thickness of 0.2 mm. As the aldehyde content of DBC increased, the thickness and porosity of the GDB sponges reached their maximum at 60% DO. As seen in Figure 3, a fibrous network structure was observed in DBC20, DBC40, and DBC60, demonstrating a gradual trend of fiber degradation. With increasing DO, the fiber diameter and structure became thinner and looser, respectively. All these parameters substantially influenced the GDB sponge structure.

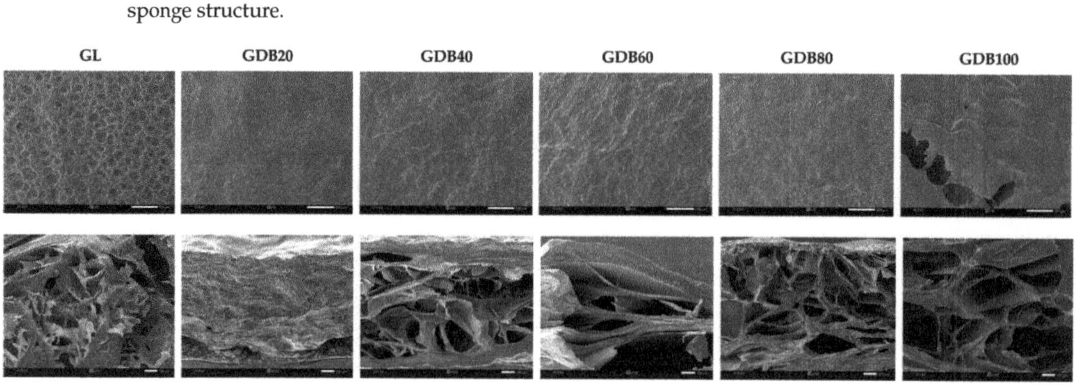

Figure 4. SEM images of the surface area with a magnification of ×40 (**above**) and cross-sectional area with a magnification of ×100 (**below**) of the GL and GDB sponges.

For the GDB80 sponge, the thickness and porosity decreased when compared to those of the GDB60. According to the oxidation conditions (Table 3), a doubled reaction time was operated for synthesizing DBC80 with nearly the same mole ratio and temperature, resulting in a high degree of degradation in the fibrous structure. The harsh conditions could damage the BC fibers. When comparing GDB80 and GDB100, greater thickness and porosity were observed in the gelatin crosslinked with DBC100. This suggests that DBC100 was well-incorporated into gelatin, resulting in a homogeneous composite sponge, which was obviously noticed on the smooth surface of GDB100 (Figure 4, first row).

The surface area of the sponges is shown in Figure 4 (1st row). A high roughness with numerous pits was noticed on the surface of the native gelatin (GL) sponge. This appearance was also observed in [17]. When gelatin was crosslinked with 20DBC (GDB20), all the pits disappeared and the surface became much smoother, implying a good distribution of 20DBC fibers within the gelatin. However, the GDB40 and GDB60 sponges exhibited greater surface roughness compared to the GDB20 sponge. This may be due to the smaller diameter of the fibers creating more roughness on the surface. For the non-fibrous structure of DBC80, the roughness on the surface of the GDB80 sponge was fine. In contrast, a homogeneous and smooth surface resembling an open-area sponge was only observed on the surface of the GDB100 sponge due to the finest and non-fibrous structure of DBC100.

3.3.5. Mechanical Strength of GDB Sponge

The tensile strength and elongation at break of the GL and GDB sponges are gathered in Table 4. Gelatin is recognized as a biopolymer with high strength but low elasticity. As shown in Table 4, the GL sponge exhibited quite high strength with the lowest elongation at break. Crosslinking with DBC, considering different DO and structure, enhanced the tensile strength of the GDB sponges as the DO of the DBC increased. The high degree of crosslinking contributed to the improved mechanical strength of the sponges. The structural integrity could be improved by the chemical reaction between amine and aldehyde groups, resulting in greater tensile strength [41]. Only the homogeneous sponge (GDB100) achieved the highest tensile strength of 9 MPa, representing a 52.5% improvement over that of the GL sponge.

After the gelatin sponges were crosslinked with DBC, the elongation at break showed huge improvement. This can be attributed to both the crosslinking and reinforcing processes between gelatin and DBC. This finding aligns with the results reported by Lin et al. (2019) [50]. The elasticity of all gelatin sponges crosslinked with DBCs improved by approximately 20–26% compared to the native GL sponge. An insignificant difference was observed in all GDB sponges. Generally, the enhancement of tensile strength conversely correlates with improved elasticity. Surprisingly, DBC could simultaneously enhance tensile strength and elongation at break.

3.4. Future Applications for DBC

DBC demonstrated significant potential as both a crosslinker and a reinforcing agent for protein biopolymers. DBC nanofiber was used to crosslink and reinforce the gelatin/carboxymethyl chitosan hydrogel for three-dimensional (3D) printing tissue engineering applications [51]. The 3D sponge-shaped collagen scaffolds were crosslinked with dialdehyde cellulose to enhance mechanical strength, and their ability to promote embryonic nerve cell culture was also demonstrated, confirming high biocompatibility [52]. To fabricate a rehydratable hydrogel for wound healing, quaternized chitosan was crosslinked with DBC to enhance its mechanical properties [53]. Not only are biomedical fields applied, but gelatin film was also crosslinked with dialdehyde cellulose to develop tensile strength, water resistance, and barrier properties for application in the packaging area [41].

For biological properties, the 6.59 mmol/g (100%) aldehyde content present in the dialdehyde microcrystalline cellulose (DAMC) structure exhibited a strong scavenging effect on 1,1-diphenyl-2-picrylhydrazyl (DPPH), (2,2′-Azino-bis-(3-ethyl-benzthiazoline-6-sylfonic acid (ABTS), and hydroxyl radicals, with half-inhibitory concentration (IC_{50}) values of 5.9, 5.6, and 8.1 mg/mL, respectively, indicating high antioxidant activity [24]. Along with its antibacterial activity, DAMC showed minimum inhibitory concentrations (MICs) of 15 mg/mL against *Staphylococcus aureus*, *Bacillus subtilis*, and *Escherichia coli*, respectively, and 30 mg/mL against *Salmonella typhimurium*. Gram-negative bacteria had greater sensitivity to DAMC compared to Gram-positive bacteria [54]. Moreover, DBC was biocompatible and promoted fibroblast attachment and proliferation on day 4 [55].

In conclusion, cellulose oxidized by periodate exhibited distinctive characteristics, including biocompatibility, biodegradability, and antioxidant and antibacterial activities, which may potentially be applied in various fields as a crosslinking and reinforcing agent to improve the mechanical properties and structural stability of protein-type biopolymers.

4. Conclusions

The preparation of DBC using a 1:2 molar ratio at 60 °C for 12 h yielded the highest aldehyde content of 98.3%. The molar ratio of BC and $NaIO_4$, as well as the reaction time, had a slightly greater effect on aldehyde content formation than temperature. The simulated quadratic model was highly accurate, with an error of only 5.3%. DO level significantly influenced the chemical and morphological structure of DBC. Gelatin was crosslinked and reinforced with DBCs containing DO/aldehyde content levels of 20%, 40%, 60%, 80%, and 100%. The GDB80 and GDB100 sponges exhibited excellent structural stability in water due to their high degree of crosslinking. The morphology of the sponge was influenced by both DO and the structure of the DBC. The strongest sponge was GDB100, featuring the highest tensile strength and elongation at break. The distinct characteristics of the GDB sponge will make it suitable for various applications, depending on the requirements, which can be adjusted by the DBC's DO.

Supplementary Materials: The following supporting information can be downloaded at: https://www.mdpi.com/article/10.3390/polym17131836/s1, Figure S1: FT-IR spectrum of GL and GDB sponges.; Table S1: ANOVA analysis for the response function Y.; Table S2: Degree of oxidation (DO) and yield of full factorial experimental design.

Author Contributions: Conceptualization, M.P. and S.K.; methodology, K.K. and S.K.; software, M.S.; validation, K.K. and S.K.; formal analysis, K.K.; investigation, S.K. and M.S.; resources, M.P.; data curation, S.K.; writing—original draft preparation, K.K., M.S., M.P. and S.K.; writing—review and editing, S.K. All authors have read and agreed to the published version of the manuscript.

Funding: This research received no external funding.

Institutional Review Board Statement: Not applicable.

Data Availability Statement: The original contribution presented in this study is included in the article. Further inquiries can be directed to the corresponding author.

Acknowledgments: The authors wish to express their sincere gratitude to the Department of Chemical Engineering, Faculty of Engineering, King Mongkut's University of Technology North Bangkok, for their generous support in providing all necessary facilities, which contributed significantly to the success of this project.

Conflicts of Interest: The authors declare no conflicts of interest.

References

1. Navya, P.V.; Gayathri, V.; Samanta, D.; Sampath, S. Bacterial cellulose: A promising biopolymer with interesting properties and applications. *Int. J. Biol. Macromol.* **2022**, *220*, 435–461. [CrossRef]
2. Cazon, P.; Velazquez, G.; Vazquez, M. Characterization of mechanical and barrier properties of bacterial cellulose, glycerol and polyvinyl alcohol (PVOH) composite films with eco-friendly UV-protective properties. *Food Hydrocoll.* **2020**, *99*, 105323. [CrossRef]
3. Park, D.; Kim, J.W.; Shin, K.; Kim, J.W. Bacterial cellulose nanofibrils-reinforced composite hydrogels for mechanical compression-responsive on-demand drug release. *Carbohydr. Polym.* **2021**, *272*, 118459. [CrossRef]
4. Styaningrum, P.; Indrianingsih, A.W.; Suryani, R.; Suratno; Bahmid, N.A.; Rahman, M.F.; Noviana, E. Synthesis, characterization, and evaluation of antibacterial properties of silver/carboxymethyl cellulose/bacterial cellulose/*Clitoria ternatea* extract aerogel composites. *Food Bioprod. Process.* **2024**, *148*, 527–537. [CrossRef]
5. Saleh, A.K.; Ray, J.B.; El-Sayed, M.H.; Alalawy, A.I.; Omer, N.; Abdelaziz, M.A.; Abouzeid, R. Functionalization of bacterial cellulose: Exploring diverse applications and biomedical innovations: A review. *Int. J. Biol. Macromol.* **2024**, *264*, 130454. [CrossRef] [PubMed]
6. Naomi, R.; Idrus, R.B.H.; Fauzi, M.B. Plant- vs. bacterial-derived cellulose for wound healing: A review. *Int. J. Environ. Res. Public Health* **2020**, *27*, 6830. [CrossRef]
7. Busuioc, C.; Isopencu, G.O.; Deleanu, I.-M. Bacterial cellulose-polyvinyl alcohol based complex composites for controlled drug release. *Appl. Sci.* **2023**, *13*, 1015. [CrossRef]
8. Unal, S.; Arslan, S.; Yilmaz, B.K.; Oktar, F.N.; Sengil, A.Z.; Gunduz, O. Production and characterization of bacterial cellulose scaffold and its modification with hyaluronic acid and gelatin for glioblastoma cell culture. *Cellulose* **2021**, *28*, 117–132. [CrossRef]
9. Li, J.; Wan, Y.; Li, L.; Liang, H.; Wang, J. Preparation and characterization of 2,3-dialdehyde bacterial cellulose for potential biodegradable tissue engineering scaffolds. *Mater. Sci. Eng. C* **2009**, *29*, 1635–1642. [CrossRef]
10. Fan, Q.G.; Lewis, D.M.; Tapley, K.N. Characterization of cellulose aldehyde using Fourier transform infrared spectroscopy. *J. Appl. Polym. Sci.* **2001**, *82*, 1195–1202. [CrossRef]
11. Peiravi-Rivash, O.; Mashreghi, M.; Baigenzhenov, O.; Hosseini-Bandegharaei, A. Producing bacterial nano-cellulose and keratin from wastes to synthesize keratin/cellulose nanobiocomposite for removal of dyes and heavy metal ions from waters and wastewaters. *Colloid. Surfaces A Physicochem. Eng. Asp.* **2023**, *656*, 130355. [CrossRef]
12. Xie, Y.; Qiao, K.; Yue, L.; Tang, T.; Zheng, Y.; Zhu, S.; Yang, H.; Fang, Z. A self-crosslinking, double-functional group modified bacterial cellulose gel used for antibacterial and healing of infected wound. *Bioact. Mater.* **2022**, *17*, 248–260. [CrossRef]
13. Favi, P.M.; Ospina, S.P.; Kachole, M.; Gao, M.; Atehortua, L.; Webster, T.J. Preparation and characterization of biodegradable nano hydroxyapatite–bacterial cellulose composites with well-defined honeycomb pore arrays for bone tissue engineering applications. *Cellulose* **2016**, *23*, 1263–1282. [CrossRef]
14. Gao, P.; Cha, R.; Luo, H.; Xu, Y.; Zhang, P.; Han, L.; Wang, X.; Zhang, Z.; Jiang, X. Development of antimicrobial oxidized cellulose film for active food packaging. *Carbohydr. Polym.* **2022**, *278*, 118922. [CrossRef] [PubMed]
15. Thongsrikhem, N.; Taokaew, S.; Sriariyanun, M.; Kirdponpattara, S. Antibacterial activity in gelatin-bacterial cellulose composite film by thermally crosslinking with cinnamaldehyde towards food packaging application. *Food Packag. Shelf Life* **2022**, *31*, 100766. [CrossRef]
16. Kim, U.J.; Wada, M.; Kuga, S. Solubilization of dialdehyde cellulose by hot water. *Carbohydr. Polym.* **2004**, *56*, 7–10. [CrossRef]
17. Kirdponpattara, S.; Philasaphong, M.; Kongruang, S. Gelatin-bacterial cellulose composite sponges thermally cross-linked with glucose for tissue engineering applications. *Carbohydr. Polym.* **2017**, *177*, 361–368. [CrossRef]
18. Nagpal, M.; Kaur, M.; Shama, D.; Baldi, A.; Chandra, R.; Madan, J. Optimization of sulfation of okra fruit gum for improved rheological and pharmacological properties. *Int. J. Biol. Macromol.* **2019**, *122*, 1–9. [CrossRef]
19. Zhang, Y.; Zhang, L.; Shao, S.; Chang, X.; Li, M. Periodate oxidation of carboxymethyl cellulose under controlled conditions. *ChemistrySelect* **2020**, *5*, 6765–6773. [CrossRef]
20. Tachai, K.; Deenu, A.; Kamthai, S. Short synthesis time and characterization of dialdehyde carboxymethyl cellulose (DCMC) from high bagasse carboxymethyl cellulose (CMCB) concentration. *J. Nat. Fibers* **2024**, *21*, 2371911. [CrossRef]
21. Keshka, S.M.; Ramadana, A.M.; Bondocka, S. Physicochemical characterization of novel Schiff bases derived from developed bacterial cellulose 2, 3-dialdehyde. *Carbohydr. Polym.* **2015**, *127*, 246–251. [CrossRef] [PubMed]
22. Wegrzynowska-Drzymalska, K.; Mlynarczyk, D.T.; Chelminiak-Dudkiewicz, D.; Kaczmarek, H.; Goslinski, T.; Ziegler-Borowska, M. Chitosan-gelatin films cross-linked with dialdehyde cellulose nanocrystals as potential materials for wound dressings. *Int. J. Mol. Sci.* **2022**, *23*, 9700. [CrossRef]
23. Potivara, K.; Phisalaphong, P. Development and characterization of bacterial cellulose reinforced with natural rubber. *Materials* **2019**, *12*, 2323. [CrossRef] [PubMed]
24. Zhang, L.; Ge, H.; Xu, M.; Cao, J.; Dai, Y. Physicochemical properties, antioxidant and antibacterial activities of dialdehyde microcrystalline cellulose. *Cellulose* **2017**, *24*, 2287–2298. [CrossRef]

25. Petros, S.; Tesfaye, T.; Ayele, M. A review on gelatin based hydrogels for medical textile applications. *J. Eng.* **2020**, *1*, 8866582. [CrossRef]
26. Kasi, P.B.; Serafin, A.; O'Brien, L.; Moghbel, N.; Novikov, L.N.; Kelk, P.; Collins, M.N. Electroconductive gelatin/hyaluronic acid/hydroxyapatite scaffolds for enhanced cell proliferation and osteogenic differentiation in bone tissue engineering. *Biomater. Adv.* **2025**, *173*, 214286. [CrossRef] [PubMed]
27. Abed, T.; Aly, S.H.; Salim, S.A.; Haikal, R.R.; Shams-Eldin, R.; EL-Moslamy, S.H.; Abdelazim, E.B.; Helmy, M.S.; Ali, A.A.; Eissa, N.G.; et al. Optimizing hydrogel performance composed of *Japanese pagoda* tree extract loaded-gelatin-sodium alginate-polyethylene oxide for biomedical applications: Influence incorporated calcium-based metal organic frameworks and zinc oxide NPs. *Int. J. Biol. Macromol.* **2025**, *310*, 143526. [CrossRef]
28. Chauhan, M.; Roopmani, P.; Rajendran, J.; Narayan, K.P.; Giri, J. Injectable, in-situ forming, tunable, biocompatible gelatin hydrogels for biomedical applications. *Int. J. Biol. Macromol.* **2025**, *285*, 138200. [CrossRef]
29. Miklosic, G.; Bektas, E.I.; Hangartner, A.; Pavan, M.; Garofolin, G.; Galesso, D.; Beninatto, R.; D'Este, M. Radical-free photopolymerizable composite of hyaluronic acid and gelatin for tissue engineering. *Acta Biomater.* **2025**, *197*, 121–134. [CrossRef]
30. Sonwane, S.; Bonde, S.; Bonde, C.; Chandarana, C. Advances in gelatin-based scaffolds for tissue engineering applications: A review. *J. Drug Deliv. Sci. Technol.* **2025**, *107*, 106789. [CrossRef]
31. Magar, H.S.; Fahim, A.M.; Hashem, M.S. Accurate, affordable, and easy electrochemical detection of ascorbic acid in fresh fruit juices and pharmaceutical samples using an electroactive gelatin sulfonamide. *RCS Adv.* **2024**, *14*, 39820–39832. [CrossRef]
32. Hassan, M.; Kanwal, T.; Siddiqui, A.J.; Ali, A.; Hussain, D.; Musharraf, S.G. Authentication of porcine, bovine, and fish gelatins based on quantitative profile of amino acid and chemometric analysis. *Food Control* **2025**, *168*, 110909. [CrossRef]
33. Prata, A.S.; Zanin, M.H.A.; Re, M.I.; Grosso, C.R.F. Release properties of chemical and enzymatic crosslinked gelatin-gum Arabic microparticles containing a fluorescent probe plus vetiver essential oil. *Colloids Surf. B Biointerfaces* **2008**, *67*, 171–178. [CrossRef] [PubMed]
34. Wang, Y.; Wu, Q.; Ren, B.; Muskhelishvili, L.; Davis, K.; Wynne, R.; Rua, D.; Cao, X. Subacute pulmonary toxicity of glutaraldehyde aerosols in a human in vitro airway tissue model. *Int. J. Mol. Sci.* **2022**, *23*, 12118. [CrossRef]
35. Bernardini, L.; Barbosa, E.; Charao, M.F.; Brucker, N. Formaldehyde toxicity reports from *in vitro* and *in vivo* studies: A review and updated data. *Drug Chem. Toxicol.* **2022**, *45*, 972–984. [CrossRef] [PubMed]
36. Zi, Y.; Shi, C.; Kan, G.; Peng, J.; Gong, H.; Wang, X.; Zhong, J. Two types of pH-responsive genipin-crosslinked gelatin conjugates with high surface hydrophobicity for emulsion stabilization. *Food Hydrocoll.* **2024**, *150*, 109718. [CrossRef]
37. Liu, Q.; Liu, J.; Qin, S.; Pei, Y.; Zheng, X.; Tang, K. High mechanical strength gelatin composite hydrogels reinforced by cellulose nanofibrils with unique beads-on-a-string morphology. *Int. J. Biol. Macromol.* **2020**, *164*, 1776–1784. [CrossRef]
38. Hivechi, A.; Bahrami, S.H.; Siegel, R.A. Investigation of morphological, mechanical and biological properties of cellulose nanocrystal reinforced electrospun gelatin nanofibers. *Int. J. Biol. Macromol.* **2019**, *124*, 411–417. [CrossRef] [PubMed]
39. Razack, S.A.; Lee, Y.; Shin, H.; Duraiarasan, S.; Chun, B.-S.; Kang, H.W. Cellulose nanofibrils reinforced chitosan-gelatin based hydrogel loaded with nanoemulsion of oregano essential oil for diabetic wound healing assisted by low level laser therapy. *Int. J. Biol. Macromol.* **2023**, *226*, 220–239. [CrossRef]
40. Qiao, D.; Huang, Y.; Hou, X.; Ye, F.; Wu, K.; Jiang, F.; Zhao, G.; Zhang, B.; Xie, F. Enhancing thermal stability and mechanical resilience in gelatin/starch composites through polyvinyl alcohol integration. *Carbohydr. Polym.* **2024**, *344*, 122528. [CrossRef]
41. Khin, M.N.; Easdani, M.; Aziz, T.; Shami, A.; Alharbi, N.K.; Al-Asmari, F.; Lin, L. Schiff's base crosslinked gelatin-dialdehyde cellulose film with gallic acid for improved water resistance and antimicrobial properties. *Food Hydrocoll.* **2025**, *166*, 111331. [CrossRef]
42. Tohamy, H.A.S. Microwaved schiff base dialdehyde cellulose-chitosan hydrogels for sustained drug release with DFT calculations. *BMC Chem.* **2025**, *19*, 114. [CrossRef]
43. Jiang, Y.; Miao, J.; Jiang, Q.; Zhang, Y.; Mo, L.; Gao, W.; Qin, Z. Dialdehyde carboxylated cellulose green crosslinked soybean protein isolate based on Schiff base reaction to improve the emulsification performance and stability of emulsion. *React. Funct. Polym.* **2025**, *212*, 106244. [CrossRef]
44. Kwak, H.W.; Lee, H.; Park, S.; Lee, M.E.; Jin, H.-J. Chemical and physical reinforcement of hydrophilic gelatin film with di-aldehyde nanocellulose. *Int. J. Biol. Macromol.* **2020**, *146*, 332–342. [CrossRef] [PubMed]
45. Jiang, Y.; Zhou, J.; Yang, Z.; Liu, D.; Xv, X.; Zhao, G.; Shi, H.; Zhang, Q. Dialdehyde cellulose nanocrystal/gelatin hydrogel optimized for 3D printing applications. *J. Mater. Sci.* **2018**, *53*, 11883–11900. [CrossRef]
46. Cui, T.; Sun, Y.; Wu, Y.; Wang, J.; Ding, Y.; Cheng, J.; Guo, M. Mechanical, microstructural, and rheological characterization of gelatin-dialdehyde starch hydrogels constructed by dual dynamic crosslinking. *LWT-Food Sci. Technol.* **2022**, *161*, 113374. [CrossRef]
47. Yilmaz, P.; Demirhan, E.; Ozbek, B. Eco-friendly fabrication of bioactive chitosan-hydroxyapatite sponge-like scaffolds functionalized with Ficus carica Linn leaf extract for enhanced antioxidant and antimicrobial properties. *Food Biosci.* **2025**, *69*, 106901. [CrossRef]

48. Huang, F.; Yang, D.; Bai, G.; Su, D.; Zhu, Z.; Long, Y. Rapid, organic-solvent-free freeze-thaw fabrication of pure chitosan sponges for scalable potential wound dressing application. *Carbohydr. Polym. Technol. Appl.* **2025**, *10*, 100809. [CrossRef]
49. Gouda, M.; Taleb, M.F.A. Sustainable cocoa biowaste-based carboxymethyl cellulose/chitosan sponges for wastewater treatment applications. *Int. J. Biol. Macromol.* **2025**, *309*, 142927. [CrossRef]
50. Lin, J.; Pan, D.; Sun, Y.; Ou, C.; Wang, Y.; Cao, J. The modification of gelatin films: Based on various cross-linking mechanism of glutaraldehyde at acidic and alkaline conditions. *Food Sci. Nutr.* **2019**, *7*, 4140–4146. [CrossRef]
51. Heidarian, P.; Kouzani, A.Z. A self-healing nanocomposite double network bacterial nanocellulose/gelatin hydrogel for three dimensional printing. *Carbohydr. Polym.* **2023**, *313*, 120879. [CrossRef] [PubMed]
52. Drobnik, J.; Pietrucha, K.; Kudzin, M.; Mader, K.; Szymanski, J.; Szczepanowska, A. Comparison of various types of collagenous scaffolds applied for embryonic nerve cell culture. *Biologicals* **2017**, *46*, 74–80. [CrossRef] [PubMed]
53. Deng, L.; Li, F.; Han, Z.; Qu, X.; Li, J.; Zhou, Z.; Chen, S.; Wang, H.; Lv, X. Bacterial cellulose-based hydrogel with regulated rehydration and enhanced antibacterial activity for wound healing. *Int. J. Biol. Macromol.* **2024**, *267*, 131291. [CrossRef] [PubMed]
54. He, X.; He, Z.; Li, Y.; Yu, H.; Zhang, L.; Ge, H.; Man, S.; Dai, Y. Modeling of the bacterial inactivation kinetics of dialdehyde cellulose in aqueous suspension. *Int. J. Biol. Macromol.* **2018**, *116*, 920–926. [CrossRef]
55. Lacin, N.T. Development of biodegradable antibacterial cellulose based hydrogel membranes for wound healing. *Int. J. Biol. Macromol.* **2014**, *67*, 22–27. [CrossRef]

polymers

MDPI

Article

Fabrication of Zwitterionized Nanocellulose/Polyvinyl Alcohol Composite Hydrogels Derived from Camellia Oleifera Shells for High-Performance Flexible Sensing

Jingnan Li [1,2,†], Weikang Peng [1,2,†], Zhendong Lei [1,2,†], Jialin Jian [1,2], Jie Cong [1,2], Chenyang Zhao [1,2], Yuming Wu [1,2], Jiaqi Su [1,2,*] and Shuaiyuan Han [1,2,3,4,*]

1 State Key Laboratory of Woody Oil Resources Utilization, Northeast Forestry University, Harbin 150040, China; lijingnan@nefu.edu.cn (J.L.); weikangp@nefu.edu.cn (W.P.); leizhendong@nefu.edu.cn (Z.L.); jjl1677292503@163.com (J.J.); congjie@nefu.edu.cn (J.C.); zcy61580909@163.com (C.Z.); wuyuming@nefu.edu.cn (Y.W.)
2 College of Material Science and Engineering, Northeast Forestry University, Harbin 150040, China
3 Engineering Research Center of Advanced Wooden Materials, Ministry of Education, Northeast Forestry University, Harbin 150040, China
4 Key Laboratory of Biobased Material Science and Technology, Ministry of Education, Northeast Forestry University, Harbin 150040, China
* Correspondence: sujiaqi@nefu.edu.cn (J.S.); hanshuaiyuan@nefu.edu.cn (S.H.)
† These authors contributed equally to this work.

Abstract: To address the growing demand for environmentally friendly flexible sensors, here, a composite hydrogel of nanocellulose (NC) and polyvinyl alcohol (PVA) was designed and fabricated using *Camellia oleifera* shells as a sustainable alternative to petroleum-based raw materials. Firstly, NC was extracted from *Camellia oleifera* shells and modified with 2-chloropropyl chloride to obtain a nanocellulose-based initiator (Init-NC) for atomic transfer radical polymerization (ATRP). Subsequently, sulfonyl betaine methacrylate (SBMA) was polymerized by Init-NC initiating to yield zwitterion-functionalized nanocellulose (NC-PSBMA). Finally, the NC-PSBMA/PVA hydrogel was fabricated by blending NC-PSBMA with PVA. A Fourier transform infrared spectrometer (FT-IR), proton nuclear magnetic resonance spectrometer (^1H-NMR), X-ray diffraction (XRD), scanning electron microscope (SEM), transmission electron microscope (TEM), universal mechanical testing machine, and digital source-meter were used to characterize the chemical structure, surface microstructure, and sensing performance. The results indicated that: (1) FT-IR and ^1H NMR confirmed the successful synthesis of NC-PSBMA; (2) SEM, TEM, and alternating current (AC) impedance spectroscopy verified that the NC-PSBMA/PVA hydrogel exhibits a uniform porous structure (pore diameter was 1.1737 μm), resulting in significantly better porosity (15.75%) and ionic conductivity (2.652 $S \cdot m^{-1}$) compared to the pure PVA hydrogel; and (3) mechanical testing combined with source meter testing showed that the tensile strength of the composite hydrogel increased by 6.4 times compared to the pure PVA hydrogel; meanwhile, it showed a high sensitivity (GF = 1.40, strain range 0–5%; GF = 1.67, strain range 5–20%) and rapid response time (<0.05 s). This study presents a novel approach to developing bio-based, flexible sensing materials.

Keywords: cellulose-based atrp initiator; nanocellulose extraction; zwitterionized nanocellulose; ATRP; cellulose-based hydrogels; flexible sensor

1. Introduction

With the rapid advancement of intelligent medical and wearable devices [1,2], there is an increasing demand for flexible and high-sensitivity sensing materials in biosensing applications, such as motion monitoring and cardiac detection [3,4]. Hydrogels, as soft materials with a three-dimensional network structure [5], exhibit excellent ionic conductivity [6] and tunable mechanical properties [7], making them ideal candidates for constructing flexible biosensors [8–10]. Among the numerous hydrogel systems, poly (vinyl alcohol) (PVA) hydrogels exhibit significant potential for biosensing applications due to their readily available raw materials, simple preparation process, and outstanding mechanical properties [11,12]. Nevertheless, PVA hydrogels possess inherent limitations: their single physical cross-linking network makes it challenging to balance mechanical strength with toughness [13,14]. Furthermore, the absence of ion-conducting groups limits their performance in high-sensitivity sensing applications [15,16].

To overcome the performance limitations of PVA hydrogels, researchers have modified them through physical blending [17,18] and chemical cross-linking [16]. Although physical blending is convenient and straightforward, the weak interactions between additives and the PVA matrix often lead to unstable material properties. Nanoparticles can be used as reinforced fillers in physical blending. For example, Tsou C. et al. [19] prepared PVA/CNT-NZnO composite hydrogels by ultrasonic solution mixing and freeze–thaw cycles. However, the use of nanoparticles often involves the phenomenon of agglomeration [20], which can cause a significant decline in the hydrogel's conductive properties over time. While chemical cross-linking enhances the network structure of hydrogels, the introduction of cross-linking agents can compromise their biodegradability. Zhao J. et al. [21] employed Titanium bis(triethanolamine) diisopropoxide (TE) as a cross-linking agent for the preparation of PVA hydrogels [22]. The presence of chemical cross-linkers renders hydrogels cytotoxic, produces unpleasant odors, and makes PVA non-biodegradable, adversely affecting some of its properties [23]. This presents challenges for natural degradation and makes PVA a less ideal green material. And notably, atomic transfer radical polymerization (ATRP) was used in this study for the surface functionalization of nanocellulose (NC), which has advantages compared to other processes' commercial materials [24]. Compared with conventional chemical cross-linking methods, such as the use of toxic cross-linking agents such as glutaraldehyde, the ATRP process can significantly reduce by-product generation and lower purification costs by virtue of modular reaction control and high atom economy. In terms of feedstock cost, nanocellulose extracted from oil tea husk is free of charge, which is much lower than commercial nanocellulose (5–10 USD/g) and synthetic conductive polymers (e.g., PEDOT:PSS, 20–50 USD/g). In terms of process scalability, ATRP technology is adapted to continuous flow reactors with batch yields of 50–100 g/h per unit [25], which is comparable to the industrial productivity of polyvinyl alcohol (PVA) hydrogels. In contrast, commercially available flexible sensors based on metal nanowires (e.g., silver nanowires, 300–500 USD/g) or carbon nanotubes (100–200 USD/g) are not only expensive in terms of material cost but also face agglomeration challenges in scale-up processing. Therefore, developing a green and efficient strategy for modifying PVA hydrogels with excellent overall performance remains a critical challenge.

Nanocellulose (NC), a nanomaterial derived from biomass [26,27] with high strength and a high specific surface area, is easily modified [28–30], exhibits excellent mechanical properties [31], and is biodegradable [32,33], making it particularly advantageous for hydrogel reinforcement. However, the absence of conducting groups on the NC surface limits its direct application in ion sensing [34]. Thanks to the robust interaction between PVA and NC, PVA hydrogels enhanced with pure NC exhibit outstanding mechanical strength. However, the absence of effective ion transport pathways limits its ionic conductivity, making it unsuitable for applications such as motion monitoring and electrocardiogram detection. Therefore, imparting the conductivity of NC and achieving synergistic enhancement with PVA has become a key focus in overcoming the performance limitations of traditional PVA hydrogels.

In this study, a green preparation strategy for utilizing biomass waste resources to produce ionized gel enhancers was proposed. Nanocellulose was first extracted from *Oleifera Shells*, and then SBMA amphiphilic groups were grafted onto the surface of nanocellulose using atom transfer radical polymerization (ATRP) to construct an ion-conducting network. Finally, the modified nanocellulose was compounded with polyvinyl alcohol (PVA) to form a hydrogel. The resulting hydrogel demonstrated a high signal-to-noise ratio (SNR) in both sports and electrocardiogram (ECG) monitoring, providing a sustainable alternative material for flexible electronic devices and promoting the high-value utilization of agricultural and forestry waste. This strategy aims to integrate the advantages of different components to endow the hydrogel with excellent performance, providing a sustainable solution for flexible electronic devices and promoting the high-value utilization of agricultural and forestry wastes.

2. Materials and Methods

2.1. Chemicals and Materials

Camellia oleifera shells were offered from Hunan Qiyi Biotechnology Co., Ltd. (Jiaxing, China) Anhydrous tetrahydrofuran (THF, 99.5%), anhydrous triethylamine (TEA, 99.5%), 2-chloropropionyl chloride (98%), anhydrous *N,N*-dimethylacetamide (DMAc, 99.8%), dichloromethane (DCM, analytical grade), polyvinyl alcohol (PVA, Mw~88,000), and *N,N,N,N,N*-pentamethyl diethylenetriamine (PMDETA, 98%) were sourced from Shanghai Titan Technology Co., Ltd. (Shanghai, China) Sulfobetaine methacrylate (SBMA, 98%) was obtained from Shanghai Macklin Biochemical Co., Ltd. (Shanghai, China) Copper(I) bromide (CuBr, 99.5%) was acquired from Shanghai Bide Pharmatech Co., Ltd. (Shanghai, China) Acetone (analytical grade) was procured from Sinopharm Chemical Reagent Co., Ltd. (Shanghai, China) Ethanol (analytical grade), methanol (analytical grade), and acetic acid (analytical grade) were purchased from Tianjin Tianli Chemical Reagent Co., Ltd. (Tianjin, China) Ammonia solution (analytical grade) was sourced from Tianjin Fuyu Fine Chemical Co., Ltd. (Tianjin, China) Sodium chlorite (analytical grade) was obtained from Tianjin Aopu Chemical Co., Ltd. (Tianjin, China) Toluene (analytical grade) was purchased from Xilong Scientific Co., Ltd. (Guangzhou, China)

2.2. Extraction of Nanocellulose

A total of 15.0 g of *Camellia oleifera* shell was accurately weighed and pulverized into 80-mesh powder using a reciprocating ball mill (30 Hz, MM400, Retsch, Haan, Germany). The powder was added to a Soxhlet extractor and extracted at 120 °C for 8 h using a 2:1 solution of toluene and ethanol by volume to remove the majority of the extractables, and the solid sample was collected after filtration and separation. The lignin was removed from the samples by washing with a sodium chlorite solution for 1 h at 75 °C, adjusting with glacial acetic acid to maintain a pH 3–5 environment, and repeating the washing five times to obtain hemicellulose. The removal of the remaining hemicellulose was then accomplished by washing for 2 h at 90 °C with a freshly configured 2 wt% sodium hydroxide solution. The crude cellulose was treated in the same acidic sodium chlorite system for 1 h. After separation, it was treated with a 5 wt% sodium hydroxide solution at 90 °C for 2 h to prepare alkalized cellulose, which was then made into a 2 wt% cellulose aqueous suspension. Finally, *Camellia oleifera* shell nanocellulose was produced by ultrasonic crushing for 30 min using an ultrasonic plant cell crusher (1200 W, JY98-IIIDN, Ningbo Scientz Biotechnology Co., Ltd., Ningbo, China).

2.3. Chlorination Modification of Nanocellulose

Solvent Centrifugal Replacement Method: An aqueous suspension of nanocellulose (100 g, 5.3 wt%, and 32.7 mmol) was centrifuged using acetone as the solvent at a speed of 5000 r/min. The supernatant was discarded to obtain a white transparent colloid. The centrifugation process was repeated, followed by sequential washing with tetrahydrofuran and ultra-dry tetrahydrofuran, each time centrifuging at a speed of 5000 r/min to obtain a suspension of nanocellulose in the anhydrous tetrahydrofuran system. The suspension was transferred to a reaction flask and anhydrous tetrahydrofuran was added under a nitrogen atmosphere. A total of (9.11 mL, 65.4 mmol) of anhydrous triethylamine was added while stirring, and 2-chloropropionyl chloride (8.31 mL, 65.4 mmol) was added dropwise under an ice water bath. After stirring at 40 °C for 16 h, a small amount of reaction mixture was taken and precipitated in water to obtain small brown solid particles. The precipitated liquid was centrifuged at a speed of 5000 r/min and washed sequentially with tetrahydrofuran, acetone, and distilled water to obtain a light yellow flocculent solid, which was dried as a cellulose initiator (Init-NC, 3.42 g, 15.2 mmol). The equation for Zwitterionized Nanocellulose is shown in Figure 1.

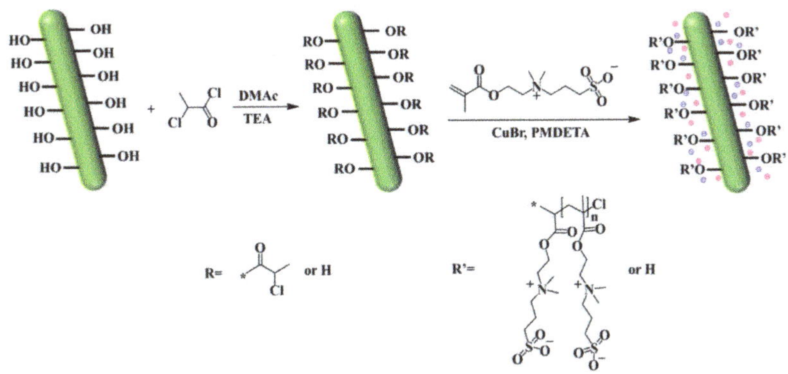

Figure 1. Equation for Zwitterionized Nanocellulose.

2.4. ATRP Polymerization

ATRP polymerization of SBMA: The Init-NC (0.377 g, 1.67 mmol) was dispersed in water to initiate the polymerization of SBMA. SBMA (4.6902 g, 21 mmol) and PMDETA (0.175 g, 1.01 mmol) were added in a reaction flask under a nitrogen atmosphere, and purified CuBr (0.24 g, 1.67 mmol) was added after 15 min, which was operated quickly, and nitrogen continued to be blown into the flask for 2–3 min to prevent the oxidation of Cu^+. The reaction was conducted at 60 °C for 20 h 50 min to obtain a blue turbid solution. The mixture was then centrifuged at 8000 r/min, and the light blue solid precipitate was collected. This precipitate was centrifuged multiple times with distilled water, causing the blue color to gradually fade. The precipitate was taken again and washed four times using a 12.5 wt% ammonia solution as the solvent, followed by centrifugation. The precipitate turned gray-white, and vacuum filtration was performed to remove the ammonia solution, yielding the solid product. Finally, the washed solid was homogeneously dispersed in water and dialyzed using a dialysis bag with a molecular weight cut-off of 8000–14,000. The retained supernatant was also poured for dialysis. After dialysis and freeze-drying, two types of white solids were obtained, with the supernatant and precipitate labeled as PSBMA and NC-PSBMA, respectively.

2.5. Preparation of PVA Gels

PVA hydrogels, and all hydrogels below, were obtained by the freeze–thaw method. A total of 0.4 g of PVA was mixed with 4 mL distilled water and stirred at 95 °C for 2 h to gradually dissolve the white solid into a colorless, clear solution. The prepared aqueous PVA solution was cryogenically frozen at −25 °C and thawed at room temperature for cycling. The PVA solution was transformed into a hydrogel through physical temperature variation.

2.6. Preparation of PSBMA/PVA Gels

A total of 0.4 g of PVA was mixed with 7 mL distilled water and stirred at 95 °C for 2 h. As the white solid gradually dissolved, the system became colorless and transparent. Then 0.3 g of PSBMA was added to the system and stirred at 95 °C for 1 h. The prepared solution was subjected to repeated cycles of cryogenic freezing at −25 °C and thawing at room temperature. The hydrogel of PSBMA/PVA solution was formed through the physical variable of temperature.

2.7. Preparation of NC-PSBMA/PVA Gels

A total of 0.4 g of PVA was mixed with 7 mL of distilled water and stirred at 95 °C for 2 h. The white solid gradually dissolved and the system became colorless and transparent. Then 0.3 g of NC-PSBMA was added to the system and stirred at 95 °C for 1 h, and the solution gradually turned into a light yellow transparent state. The prepared solution was subjected to repeated cycles of cryogenic freezing at −25 °C and thawing at room temperature. The hydrogel of NC-PSBMA/PVA solution was formed through the physical variable of temperature.

2.8. Instrumentation and Characterization

The functional group structure of the sample was analyzed using a Fourier Transform Infrared Spectrometer (TENSOR II, Bruker, Ettlingen, Germany) with a scanning range of 4000~500 cm^{-1}, a resolution of 4 cm^{-1}, and 32 scans. The sample's structure was examined using a Nuclear Magnetic Resonance Spectrometer (AVANCE III HD, Bruker, Zurich, Switzerland) at a frequency of 600 MHz. X-ray diffraction (40 mA, 10 kev) was used to analyze the crystalline properties of the samples using an X-ray diffractometer (XRD, ADVANCE A25, Bruker, Ettlingen, Germany). A Scanning Electron Microscope (Apreo S HiVac, Thermo Fisher, Prague, Czech Republic) was employed to observe the microscopic surface structure of the sample. Before testing, the sample was gold-sputtered, and the pores were binarized using Avizo software (Avizo2020.1) to facilitate the subsequent measurement and analysis of pore parameters. A PC-80D ECG meter (Shenzhen Carewell Electronics Co., Ltd., Shenzhen, China) was used for four-lead ECG acquisition.

An electrochemical workstation (CHI660F, Shanghai Chenhua Instrument Co., Ltd., Shanghai, China) was utilized to test the conductivity of the sample. The hydrogel sample, cut into dimensions of 10 mm × 10 mm × 1.27 mm, was clamped between two copper sheets with a thickness of 0.01 mm. The conductivity of the hydrogel sample was tested using Electrochemical Impedance Spectroscopy (EIS) in the frequency range of 10^1~10^5 Hz. The ionic conductivity of the hydrogel was calculated with the following Equation (1):

$$\sigma = \frac{L}{R_0 \times S} \tag{1}$$

where L was the height of the hydrogel (m); R_0 was the hydrogel resistance (Ω); and S was the total contact area between the hydrogel and the electrode sheet (m^2).

The stress–strain curve of the hydrogel was measured using a universal mechanical testing machine (DR-507AS, Dongguan Dongri Instrument Co., Ltd., Dongguan, China) at a speed of 50 mm/min. The sensing performance of the samples was evaluated using a digital source meter (2611B, Tektronix, Inc., Beaverton, OR, USA) with a current of 1 mA. The hydrogel sensor sensitivity was characterized by the strain sensitivity factor G_F, which was calculated with the following Equation (2):

$$GF = \frac{R - R_0}{R_0} \cdot \frac{1}{\varepsilon} \tag{2}$$

where R was the resistance of the hydrogel sensor at stretching (Ω); R_0 was the initial resistance of the hydrogel sensor (Ω); and ε was the tensile strain of the hydrogel (%).

The delay time was determined by measuring the change in resistance of the compressed hydrogel at room temperature during the deformation process. The stretching distance was 2.94 mm, and the stretching speed was 100 mm/min.

3. Results and Discussion

3.1. Micro-Morphological Analysis of Camellia Oleifera Shell Nanocellulose

The nanocellulose extracted from the *Camellia oleifera* shells exhibited a high aspect ratio (greater than 80), with a single fiber diameter of about 24.84 nm and a length of more than 2 μm, which was in line with the size of a typical nanocellulose, which was conducive to enhancing the mechanics and network construction in hydrogels. These fibers self-assembled into cluster aggregates with widths ranging from 20 to 50 nm and lengths reaching several micrometers (Figure 2). Notably, some individual fibers protruded from the surface of the aggregate, forming a distinctive "dendritic" structure. This particular morphology offered the following advantages: on the one hand, the abundance of surface hydroxyl groups and the dendritic structure provided sufficient active sites for chemical modification, allowing functional groups to be incorporated through chlorine modification. On the other hand, the multi-level ordered structure can enhance the interfacial interaction with the PVA matrix. The improvement in the mechanical strength of the hydrogel was achieved through the physical entanglement of the fiber network and the hydrogen bonding between ionic groups and PVA hydroxyl groups.

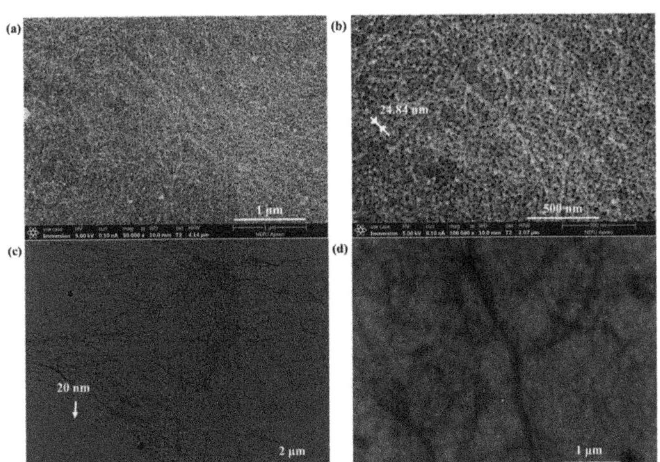

Figure 2. (**a**) 50,000× SEM image of NC. (**b**) 100,000× SEM image of NC. (**c**) 6000× TEM image of NC. (**d**) 10,000× TEM image of NC.

3.2. Structural Characterization

The FT-IR spectrum of freeze-dried NC and Init-NC showed (Figure 3a) a stretching vibrational peak of -CH_2- on NC at 2890 cm^{-1}. After the chlorination reaction, the Init-NC curve methylene was shifted, and the telescopic vibration peak of methyl-CH_3 appeared in the region of 2970–2800 cm^{-1}. Compared to NC, Init-NC showed a stretching vibration peak of the carbon–oxygen double bond at 1750 cm^{-1}, demonstrating the successful introduction of the carbon group from 2-chloropropionyl chloride, and thereby achieving the chlorination modification of NC. Based on the FT-IR analysis above, the two-step preparation of chlorinated cellulose using the solvent centrifugal displacement method was feasible and successfully produced a chlorine-modified nanocellulose initiator.

The O-H stretching vibration peak significantly weakened in NC-PSBMA compared with NC, while the C=O stretching vibration peak at 1725 cm^{-1} was stronger (Figure 3a). This indicated that most of the hydroxyl groups on its surface were involved in forming new chemical bonds, and NC-int successfully initiated the polymerization of monomers. In the ^1H-NMR spectrum of NC-PSBMA (Figure 3b), the peaks at 0.8–1.2 ppm attributed to methyl-CH_3 (25, 27, 46, 82, 63, and 67) indicated that the number of methyl groups increased in NC-PSBMA. The peaks at 3.2–3.5 ppm attributed to the methyl group attached to N^+ (36, 37, 54, 55, 75, and 76), and the presence of peaks of hydrogen on the cellulose carbon backbone (3.5–4.5 ppm) indicated that the NC-PSBMA was the product of the grafting polymerization that occurred. It was demonstrated that the NC-SBMA was synthesized successfully.

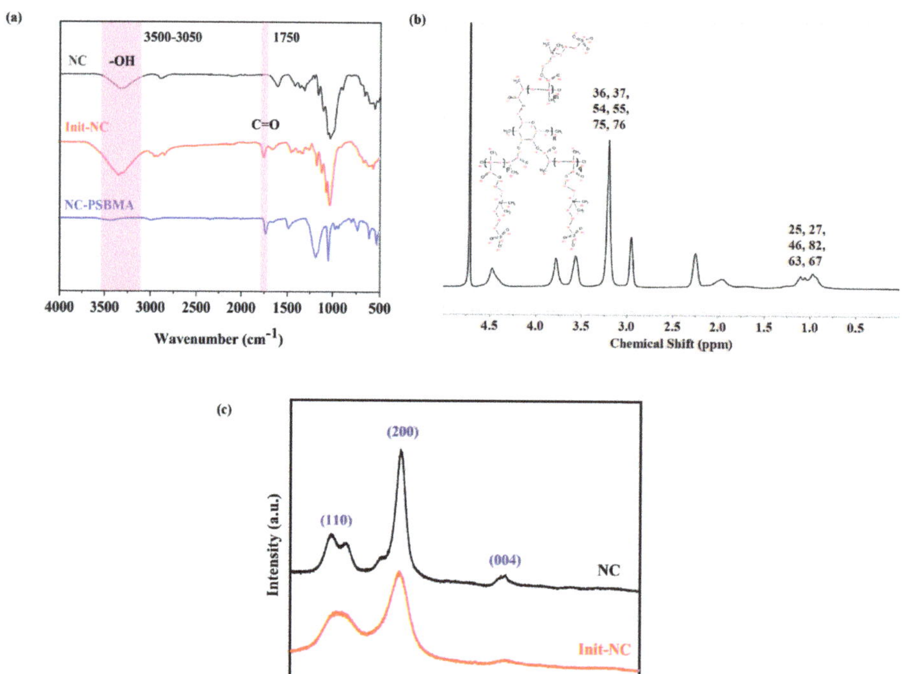

Figure 3. FT-IR spectrum of: (**a**) NC, Init-NC, and NC-PSBMA. (**b**) ^1H-NMR spectrum of NC-PSBMA. (**c**) XRD spectrum of NC and Init-NC.

Both the Init-NC and the NC exhibited characteristic cellulose diffraction peaks at $2\theta = 16.5°, 22.5°$, and $34.5°$, corresponding to the (001), (200), and (004) planes, respectively (Figure 3c). A slight decrease in the crystallinity of Init-NC compared to NC was observed, likely due to chlorination modification and swelling effects. The crystalline regions of cellulose remained largely intact, with no significant changes in crystalline morphology or adverse effects on mechanical properties.

3.3. Micro-Morphological Analysis of NC-PSBMA/PVA Hydrogels Surface

SEM images of the NC-PSBMA/PVA hydrogel surfaces at different magnifications revealed a relatively uniformly distributed porous network at low magnification. At high magnification, the internal pore structure appeared to be non-uniformly distributed, dense, and highly interconnected (Figure 4a,b). A highly smooth hole wall structure suggested that the block fiber network of NC-PSBMA and the PVA molecular chains were fully interpenetrated through hydrogen bonding and physical entanglement. Additionally, the presence of hierarchical porous structures in certain regions further enhanced the mechanical strength of the hydrogel via the "stress dispersion effect". This complex architecture not only promoted efficient ion transport but also simultaneously improved the tensile properties of the composite hydrogel.

Statistical analysis of the SEM images of NC-PSBMA/PVA hydrogels (Figure 4d,e) indicated an average pore size of 1.1737 ± (1.03731) μm with the size distribution from 0.253689 μm to 9.60339 μm. The specific surface area was 1.926 ± (4.377) μm^2, and the porosity was 15.75%. This size range indicated that the pore structure was at the micrometer level, which was advantageous for the diffusion and penetration of ions or water, as well as provided sufficient spatial selectivity for ion adsorption and preventing the desorption of conductive ions.

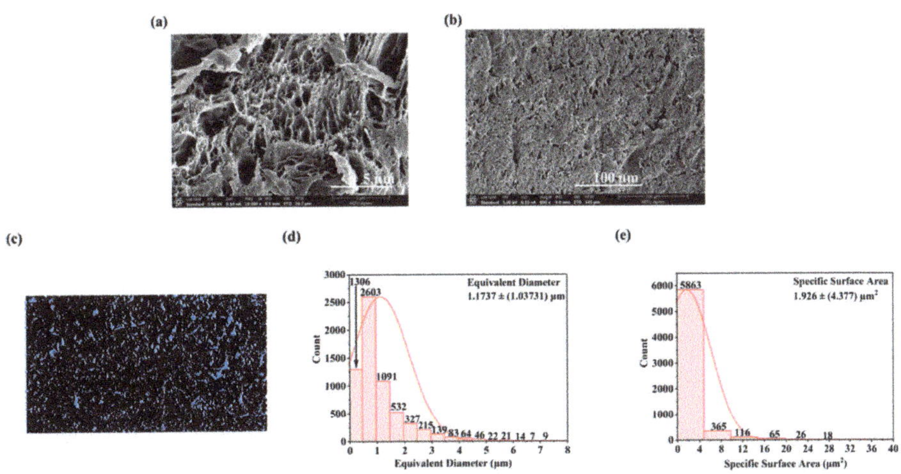

Figure 4. SEM images of NC-PSBMA/PVA hydrogels (**a**) 10,000× and (**b**) 600×. (**c**) Presentation effects of NC-PSBMA/PVA hydrogel pore binarization images after threshold segmentation. (**d**) Equivalent diameter distribution image of NC-PSBMA/PVA hydrogels' pores. (**e**) Distribution image of pore-specific surface area of NC-PSBMA/PVA hydrogels.

3.4. Mechanical Strength Characterization of Hydrogels

Figure 5 presents the stress–strain curves of NC-PSBMA/PVA, PSBMA/PVA, and pure PVA hydrogels. All three exhibited similar profiles characterized by elastic deformation regions followed by strain hardening without a distinct yield point, indicating that deformation is primarily elastic throughout the stretching process. Among them, the NC-PSBMA/PVA hydrogel exhibited a narrower elastic deformation range, primarily due to the rigid, brush-like architecture of NC-PSBMA and its strong interactions with the PVA matrix, which resulted in an earlier onset of strain hardening. Nevertheless, this hydrogel achieved a remarkable elongation at a break of approximately 300%, demonstrating excellent stretchability. In comparison, pure PVA hydrogel remained in the elastic region up to failure but exhibited a lower elongation at break, with only 250% elongation, attributed to weaker intermolecular interactions. Incorporating ionic polymers (PSBMA and NC-PSBMA) significantly enhanced the mechanical strength of PVA hydrogels. The addition of PSBMA increased the tensile strength by approximately 300% to 0.2 MPa. In contrast, the inclusion of NC-PSBMA—owing to its intrinsic strength and brush-like morphology—further amplified interactions with the PVA network, boosting the tensile strength by nearly sevenfold to 0.32 MPa.

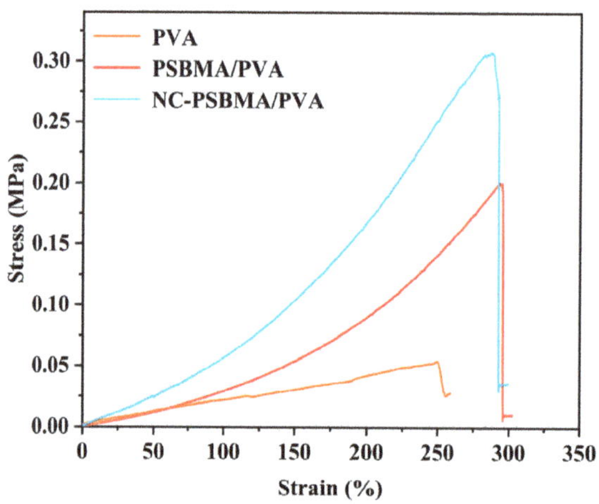

Figure 5. Stress–strain curve of NC-PSBMA/PVA, PSBMA/PVA, and pure PVA hydrogels.

3.5. Characterization of Conductivity for Hydrogels

The conductivity of all samples was measured using two-electrode AC impedance spectroscopy. So, the fitted equivalent circuit model for the impedance spectra is illustrated in Figure 6a. In this model, R_0 represented the intrinsic resistance of the hydrogel film. R_1 and R_2 corresponded to the contact resistance between the film and the two electrodes. CPE_1 and CPE_2 were constant phase elements, and L_0 denoted the inductance of the overall circuit.

The impedance spectra of the NC-PSBMA/PVA, PSBMA/PVA, and pure PVA hydrogels are shown in Figure 6b. Based on the equivalent circuit fitting results and the conductivity equation in the Section 2, the conductivities of the three hydrogels were determined to be 2.652 $S \cdot m^{-1}$, 0.363 $S \cdot m^{-1}$, and 0.135 $S \cdot m^{-1}$, respectively. These results indicated that the incorporation of ionic polymers substantially enhanced the conductivity of pure PVA hydrogels. Notably, the conductivity of the NC-PSBMA/PVA hydrogel was 19.64 times higher than that of the pure PVA hydrogel, primarily due to the additional mobile charge carriers introduced by the ionic polymers.

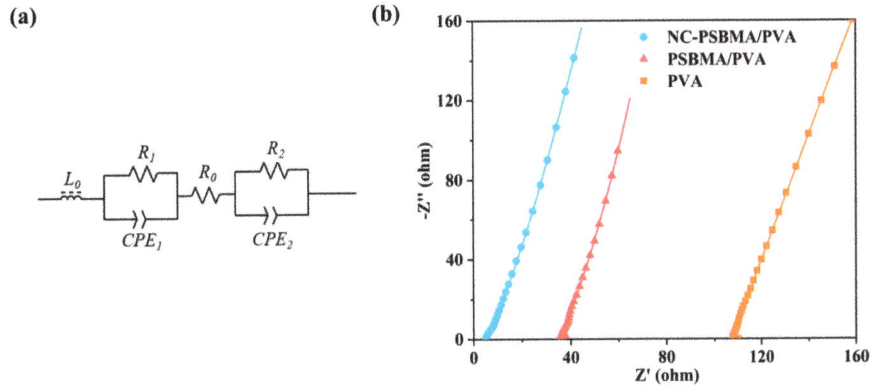

Figure 6. (**a**) Equivalent circuit. (**b**) Impedance spectra of NC-PSBMA/PVA, PSBMA/PVA, and pure PVA (the curves are the best fits to the equivalent circuit).

Interestingly, when introduced at the same mass fraction, the conductivity of the NC-PSBMA/PVA hydrogel was approximately seven times greater than that of the PSBMA/PVA hydrogel. This enhancement cannot be explained solely by the difference in ionic group content (since PSBMA contained more ionic groups per unit mass). Instead, it was attributed to differences in molecular architecture: compared with linear PSBMA, brush-like NC-PSBMA had densely packed side chains that formed a larger steric repulsion layer within the PVA matrix, significantly improving polymer dispersibility. Moreover, its brush-like structure minimizes chain entanglement and reduces the tendency for crystallization. As a result, NC-PSBMA was more uniformly dispersed in the PVA matrix, creating more efficient ion conduction pathways and thereby dramatically enhancing the overall conductivity.

3.6. Sensing Performance Characterization of NC-PSBMA/PVA Hydrogel

Employing copper electrodes, the ionogel (NC-PSBMA/PVA hydrogel) was systematically evaluated as a flexible sensing detector (Figure 7a). Under continuous cyclic strain within the low-strain regime (1–5%), the material demonstrated excellent linearity, reproducibility, and high symmetry in its electrical response, as characterized by the $(R - R_0)/R_0$ curves (Figure 7b), highlighting its superior electrical sensitivity. As the strain increased to 50%, the hydrogel sensor continued to maintain a stable and reliable response curve (Figure 7c), demonstrating its effectiveness across a wide range of strains. The hydrogels maintained consistent signal outputs under fixed strains of 10%, 20%, 30%, 40%, and 50% (Figure 7d). The time-delay measurements (Figure 7e) showed that the resistance change lags behind the strain change by approximately 0.05 s, indicating a response time of 0.05 s for the hydrogel. This suggested that the ion-conducting network within the hydrogel can respond rapidly to external deformation. The gauge factors (GF) were characterized under both small-strain ($\leq$5%) and large-strain (5–20%) conditions (Figure 7f). The GF measured 1.40 at strains below 5% and exhibited a moderate increase to 1.67 within the 5–20% strain range. This performance was comparable to that of state-of-the-art ionic conductive hydrogels, meeting the operational requirements for conventional strain sensing. During the cyclic stability experiments, the resistance of the samples showed a minor increase after 300 stretching cycles, while the resistance change sensitivity remained stable, indicating that the NC-PSBMA/PVA hydrogel sensors were suitable for repeated use and were stable for long-term use. This excellent stability may be attributed to the reversible hydrogen bonding and entangled network, which prevented structural damage during repeated deformation and ensured a consistent sensing performance. In practical applications, the hydrogel was affixed to the surface of a finger to monitor its bending motion (Figure 7h). The corresponding relative resistance change response curve (Figure 7i) showed that the peak signals exhibit distinct changes, accurately capturing the trajectory of the finger's bending motion. These results demonstrated that the NC-PSBMA/PVA hydrogel strain sensor possessed high sensitivity, making it suitable for wearable, flexible sensors with the strong potential for real-time monitoring of human motion.

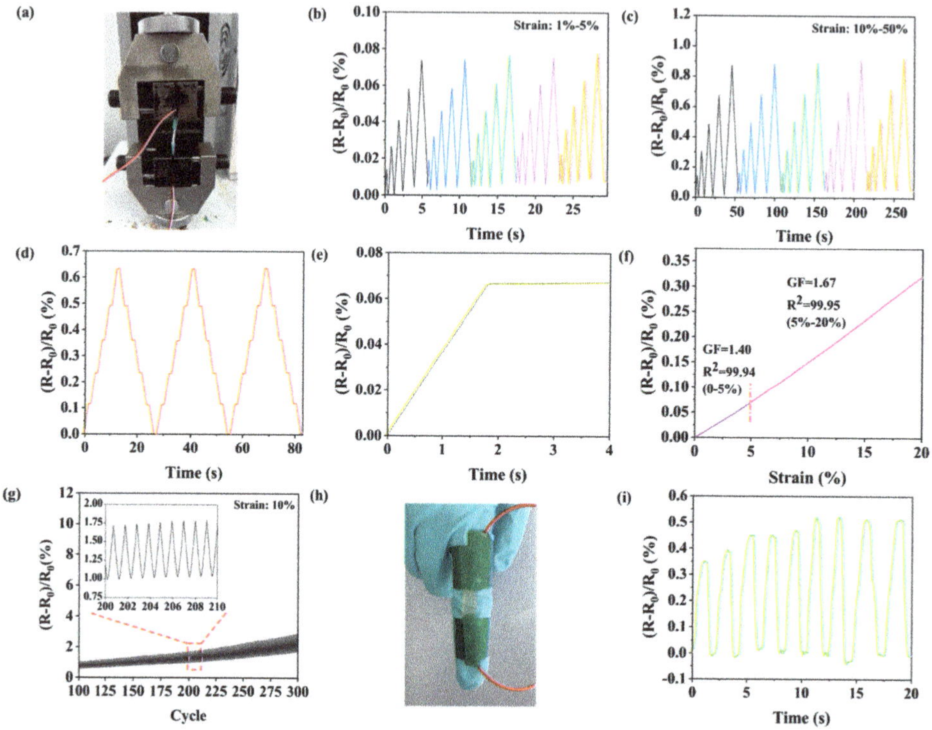

Figure 7. (**a**) Photograph of the testing device. Sensing resistance change curves (**b**) small strain, (**c**) large strain, and (**d**) at different strains (10%, 20%, 30%, 40%, and 50%). (**e**) Curve of resistivity with time. (**f**) Linear fitting curve of relative resistivity to strain. (**g**) Curve for 300 cycles. Sensing for finger bending, (**h**) photo of finger sensing and (**i**) curve of resistance with time.

3.7. ECG Signal Acquisition Testing for NC-PSBMA/PVA Hydrogels

During the ECG collection experiment, NC-PSBMA/PVA hydrogel electrodes were placed on the subject's right arm, left arm, right leg, and left leg, and a four-lead ECG was recorded using the PC-80D ECG meter. According to the electrocardiogram, the P-wave represented atrial depolarization, while the QRS complex reflected ventricular depolarization. At rest, the waveforms exhibited regular and orderly morphology and spacing (Figure 8a), meeting the characteristics of a normal sinus rhythm, thereby confirming the signal acquisition stability of the hydrogel electrode in static conditions. During exercise, the heart rate increased, leading to a noticeable rise in the frequency of the P-wave and QRS complexes, which reflected the influence of exercise-induced stress on cardiac electrophysiological activity [35]. As shown in Figure 8b, the ECG waveform remained clearly discernible even during physical activity. This was primarily attributed to the superior skin adhesion and elasticity of the NC-PSBMA/PVA hydrogel electrode. This was attributed to the NC-PSBMA/PVA hydrogel electrodes possessing excellent skin-fitting properties. The hydrogel electrode minimized motion artifacts at the electrode–skin interface through synchronous deformation with cutaneous tissue. This conformal contact suppressed interference from relative movement, enhancing ECG signal fidelity during ambulatory monitoring. No residue adhered to the electrodes after they were peeled from the skin surface, indicating that the interfacial forces in contact with the skin were reversible, thus avoiding adhesive residue while accurately and stably monitoring the ECG. Continuous wear tests showed that the electrodes did not trigger adverse skin reactions, ensuring reliable ECG monitoring, accuracy, and safety.

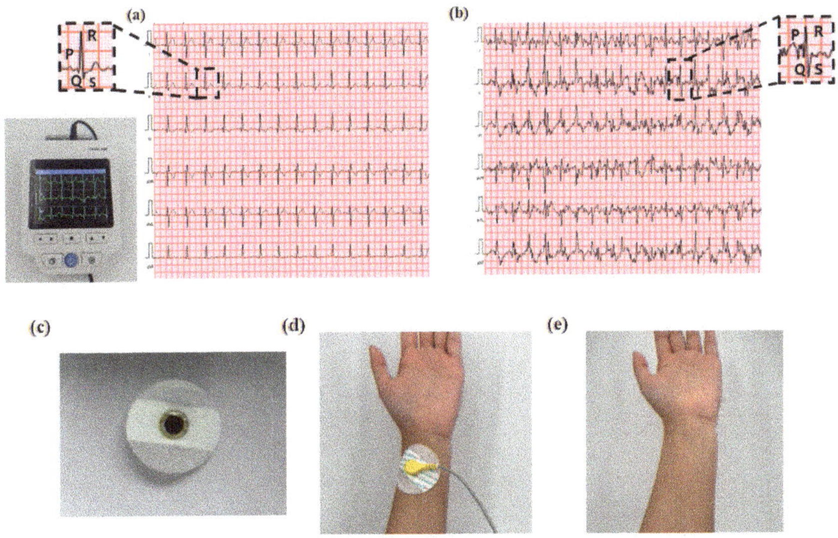

Figure 8. ECG acquisition signal during (**a**) rest and (**b**) motion. (**c**) NC-PSBMA/PVA hydrogel as an electrode for ECG monitoring. (**d**) Schematic diagram of the fit state of the hydrogel electrode to the skin surface. (**e**) Skin condition after 3 h of continuous wear testing.

4. Conclusions

This study successfully developed a high-performance ionic conductive hydrogel by complexing *Camellia oleifera* shell-derived and ionized nanocellulose (NC- PSBMA) with PVA. The iconic morphology of NC (diameter: ~24.84 nm, length: greater than 2 μm) enabled effective chlorination and graft polymerization to synthesize NC-PSBMA, as confirmed by FT-IR, NMR, and XRD analyses. Due to physical entanglement and hydrogen bonding, the NC-PSBMA/PVA hydrogel exhibited increased mechanical strength, with a tensile strength of 3.2 MPa (seven times higher than pure PVA) and an elongation of 300%. Optimized ion pathways enhanced ionic conductivity (2.652 S m^{-1}, 19.64 times higher than pure PVA). As a flexible sensor, the hydrogel demonstrated a high sensitivity (GF: 1.4–1.67), rapid response (0.05 s), and cyclic stability (50 cycles) in strain sensing.

The excellent performance of this hydrogel all stems from the triple synergistic effect of the NC-g-SBMA and PVA composite hydrogel: (1) Amphiphilic ionic bonding enhances the cross-linking density of the polymer network through multiple hydrogen bonding and electrostatic interactions, which significantly improves the mechanical properties; (2) the formation of $-SO_3^-$/$-N^+(CH_3)_2$ ion pairs to stable ion migration channels, which ensures the high sensitivity of the sensing performance; (3) surface charge modulation effectively reduces the contact impedance at the skin–sensor interface, which significantly improves the fit between the sensor and the skin and the signal transmission stability. As an artifact-free interference electrode, the hydrogel is capable of accurately monitoring electrocardiograms at rest and during exercise. This work not only transforms agricultural waste into advanced functional materials but also presents a sustainable design strategy for next-generation wearable bioelectronics and biomedical interfaces. In the future, we will further explore the in vivo application of hydrogels and promote the further development of hydrogels in wearable bioelectronic devices.

Author Contributions: Writing—original draft, formal analysis, investigation, J.L.; investigation, validation, software, W.P.; investigation, software, Z.L.; conceptualization, methodology, J.J.; investigation, J.C.; conceptualization, C.Z.; investigation, software, Y.W.; writing—review and editing, supervision, formal analysis, J.S.; writing—review and editing, conceptualization, supervision, project administration, funding acquisition, S.H. All authors have read and agreed to the published version of the manuscript.

Funding: This work was funded by the Fundamental Research Funds for the Central Universities (2572023CT05-03), China Postdoctoral Science Foundation (2024M760383), and Key R&D Program of Heilongjiang Province (GZ20220134).

Institutional Review Board Statement: Not applicable.

Data Availability Statement: The original contributions presented in this study are included in the article. Further inquiries can be directed to the corresponding authors.

Conflicts of Interest: The authors declare no conflicts of interest.

References

1. Mo, F.; Lin, Y.; Liu, Y.; Zhou, P.; Yang, J.; Ji, Z.; Wang, Y. Advances in ionic conductive hydrogels for skin sensor applications. *Mater. Sci. Eng. R Rep.* **2025**, *165*, 100989.
2. Wang, L.; Xu, T.; Zhang, X. Multifunctional conductive hydrogel-based flexible wearable sensors. *TrAC Trends Anal. Chem.* **2021**, *134*, 116130.
3. Wu, H.; Qi, H.; Wang, X.; Qiu, Y.; Shi, K.; Zhang, H.; Zhang, Z.; Zhang, W.; Tian, Y. Stretchable, sensitive, flexible strain sensor incorporated with patterned liquid metal on hydrogel for human motion monitoring and human–machine interaction. *J. Mater. Chem. C* **2022**, *10*, 8206–8217.
4. Chen, H.; Zhou, J.; Cao, H.; Liang, D.; Chen, L.; Yang, Y.; Wang, L.; Xie, J.; Duan, H.; Fu, Y. Thermo-Responsive and Phase-Separated Hydrogels for Cardiac Arrhythmia Diagnosis with Deep Learning Algorithms. *Biosens. Bioelectron.* **2025**, *276*, 117262.
5. Agrawal, A.; Hussain, C.M. 3D-printed hydrogel for diverse applications: A review. *Gels* **2023**, *9*, 960. [CrossRef]
6. Li, Y.; Liu, J.; Zhang, Q.; Hu, N.; Jiang, Z.; Kan, Q.; Kang, G. Growth of double-network tough hydrogel coatings by surface-initiated polymerization. *ACS Appl. Mater. Interfaces* **2024**, *16*, 10822–10831.
7. Zhang, M.; Xu, S.; Wang, R.; Che, Y.; Han, C.; Feng, W.; Wang, C.; Zhao, W. Electrospun nanofiber/hydrogel composite materials and their tissue engineering applications. *J. Mater. Sci. Technol.* **2023**, *162*, 157–178.
8. Lyu, J.; Zhou, Q.; Wang, H.; Xiao, Q.; Qiang, Z.; Li, X.; Wen, J.; Ye, C.; Zhu, M. Mechanically strong, freeze-resistant, and ionically conductive organohydrogels for flexible strain sensors and batteries. *Adv. Sci.* **2023**, *10*, 2206591.
9. Wang, T.; Xu, B.; Yu, T.; Yu, Y.; Fu, J.; Wang, Y.; Gao, X.; Xue, Z.; Li, R.; Chang, G. PVA/chitosan-based multifunctional hydrogels constructed through multi-bonding synergies and their application in flexible sensors. *Carbohydr. Polym.* **2025**, *350*, 123034.
10. Li, M.; Pu, J.; Cao, Q.; Zhao, W.; Gao, Y.; Meng, T.; Chen, J.; Guan, C. Recent advances in hydrogel-based flexible strain sensors for harsh environment applications. *Chem. Sci.* **2024**, *15*, 17799–17822.
11. Zhong, Y.; Lin, Q.; Yu, H.; Shao, L.; Cui, X.; Pang, Q.; Zhu, Y.; Hou, R. Construction methods and biomedical applications of PVA-based hydrogels. *Front. Chem.* **2024**, *12*, 1376799.
12. Kumar, A.; Han, S.S. PVA-based hydrogels for tissue engineering: A review. *Int. J. Polym. Mater. Polym. Biomater.* **2017**, *66*, 159–182.
13. Leif, K. Effective visco-elastic models of tough, doubly cross-linked, single-network polyvinyl alcohol (PVA) hydrogels. *Contin. Mech. Thermodyn.* **2021**, *33*, 2315–2329.
14. Hu, Y.; Luo, J.; Luo, S.; Fei, T.; Song, M.; Qin, H. High-strength polyvinyl alcohol-based hydrogel by vermiculite and lignocellulosic nanofibrils for electronic sensing. *e-Polymers* **2023**, *23*, 20230081.

15. Di, X.; Ma, Q.; Xu, Y.; Yang, M.; Wu, G.; Sun, P. High-performance ionic conductive poly (vinyl alcohol) hydrogels for flexible strain sensors based on a universal soaking strategy. *Mater. Chem. Front.* **2021**, *5*, 315–323.
16. Bai, H.; Chen, D.; Zhu, H.; Zhang, S.; Wang, W.; Ma, P.; Dong, W. Photo-crosslinking ionic conductive PVA-SbQ/$FeCl_3$ hydrogel sensors. *Colloids Surf. A Physicochem. Eng. Asp.* **2022**, *648*, 129205.
17. Hu, D.; Liu, D.; Hu, Y.; Wang, Y.; Lu, Y.; Bai, C.; Hossain, K.R.; Jiang, P.; Wang, X. Dual-physical network PVA hydrogel commensurate with articular cartilage bearing lubrication enabled by harnessing nanoscale crystalline domains. *Nano Res.* **2024**, *17*, 9784–9795.
18. Zhang, X.; Zhou, S.; Wang, Z.; Wei, X.; Zhang, S.; Jin, J. Facile preparation of hydrogel-coated surfaces with antifouling and salt resistance for efficient solar-driven water evaporation. *ACS Appl. Mater. Interfaces* **2023**, *15*, 50196–50205.
19. Tsou, C.H.; Chen, S.; Li, X.; Chen, J.C.; De Guzman, M.R.; Sun, Y.L.; Zhang, Y. Highly resilient antibacterial composite polyvinyl alcohol hydrogels reinforced with CNT-NZnO by forming a network of hydrogen and coordination bonding. *J. Polym. Res.* **2022**, *29*, 412.
20. Karimzadeh, Z.; Mahmoudpour, M.; Rahimpour, E.; Jouyban, A. Nanomaterial based PVA nanocomposite hydrogels for biomedical sensing: Advances toward designing the ideal flexible/wearable nanoprobes. *Adv. Colloid Interface Sci.* **2022**, *305*, 102705.
21. Zhou, Y.; Chu, R.; Fan, L.; Meng, X.; Zhao, J.; Wu, G.; Sun, F. Study on the mechanism and performance of polymer gels by TE and PVA chemical cross-linking. *J. Appl. Polym. Sci.* **2022**, *139*, 52043.
22. Gautam, L.; Warkar, S.G.; Ahmad, S.I.; Kant, R.; Jain, M. A review on carboxylic acid cross-linked polyvinyl alcohol: Properties and applications. *Polym. Eng. Sci.* **2022**, *62*, 225–246.
23. Martínez-Ramírez, A.P.; Rincón-Ortiz, S.A.; Baldovino-Medrano, V.G.; Blanco-Tirado, C.; Combariza, M.Y. Influence of reaction variables on the surface chemistry of cellulose nanofibers derived from palm oil empty fruit bunches. *RSC Adv.* **2023**, *13*, 36117–36129. [PubMed]
24. Wang, Q.; Feng, X.; Liu, X. Functionalization of nanocellulose using atom transfer radical polymerization and applications: A review. *Cellulose* **2023**, *30*, 8495–8537.
25. Chan, N.; Cunningham, M.F.; Hutchinson, R.A. Reducing ATRP catalyst concentration in batch, semibatch and continuous reactors. *Macromol. React. Eng.* **2010**, *4*, 369–380.
26. Abdelhaleem, A.F.; Abdel Goad, M.A.-H. Nanocellulose preparation, production modeling and characterization. *J. Adv. Eng. Trends* **2024**, *43*, 79–85.
27. Wang, H.; Wu, J.; Huang, B.; Lu, Q.-L. One-pot synthesis of UPy-functionalized nanocellulose under mechanochemical synergy for high-performance epoxy nanocomposites. *Polymers* **2022**, *14*, 2428.
28. Kumar, A.; Durand, H.; Zeno, E.; Balsollier, C.; Watbled, B.; Sillard, C.; Fort, S.; Baussanne, I.; Belgacem, N.; Lee, D. The surface chemistry of a nanocellulose drug carrier unravelled by MAS-DNP. *Chem. Sci.* **2020**, *11*, 3868–3877.
29. James, A.; Rahman, M.R.; Mohamad Said, K.A.; Kanakaraju, D.; Sueraya, A.Z.; Kuok, K.K.; Bin Bakri, M.K.; Rahman, M.M. A review of nanocellulose modification and compatibility barrier for various applications. *J. Thermoplast. Compos. Mater.* **2024**, *37*, 2149–2199.
30. Wang, H.; Wu, J.; Lian, Y.; Li, Y.; Huang, B.; Lu, Q. Zirconium phosphate assisted phosphoric acid Co-catalyzed hydrolysis of lignocellulose for enhanced extraction of nanocellulose. *Polymers* **2023**, *15*, 447. [CrossRef]
31. Garrido-Miranda, K.A.; Pesenti, H.; Contreras, A.; Vergara-Figueroa, J.; Recio-Sánchez, G.; Chumpitaz, D.; Ponce, S.; Hernandez-Montelongo, J. Nanocellulose/Nanoporous Silicon Composite Films as a Drug Delivery System. *Polymers* **2024**, *16*, 2055. [CrossRef] [PubMed]
32. Srivastava, S. Advancements in Nanocellulose Derived from Plant Waste: Strategies for Sustainable Innovations and Applications. *J. Solid Waste Technol. Manag.* **2024**, *50*, 543–563.
33. Gou, J.; Liu, W.; Tang, A. A renewable and biodegradable nanocellulose-based gel polymer electrolyte for lithium-ion battery. *J. Mater. Sci.* **2020**, *55*, 10699–10711.
34. Abushammala, H.; Mao, J. Novel electrically conductive cellulose nanocrystals with a core-shell nanostructure towards biodegradable electronics. *Nanomaterials* **2023**, *13*, 782. [CrossRef]
35. Nielsen, J.B.; Kühl, J.T.; Pietersen, A.; Graff, C.; Lind, B.; Struijk, J.J.; Holst, A. G, P-wave duration and the risk of atrial fibrillation: Results from the Copenhagen ECG Study. *Heart Rhythm* **2015**, *12*, 1887–1895.

polymers

MDPI

Article

Magnesium Transfer between Atomic Force Microscopy Probes and Metal Electrodes in Aqueous Alginate Electrolytes

Walter J. Legerstee [1,*], Lindah Kiriinya [1,2], Mark Kwakernaak [1] and Erik M. Kelder [1]

1 Department Storage of Electrochemical Energy, Reactor Institute Delft, Delft University of Technology, Mekelweg 15, 2629 JB Delft, The Netherlands
2 Department of Electrical and Information Engineering, Faculty of Engineering, University of Nairobi, Harry Thuku Street, Nairobi 00100, Kenya
* Correspondence: w.j.legerstee@tudelft.nl

Abstract: The upcoming energy transition requires not only renewable energy sources but also novel electricity storage systems such as batteries. Despite Li-ion batteries being the main storage systems, other batteries have been proposed to fulfil the requirements on safety, costs, and resource availability. Moving away from lithium, materials such as sodium, magnesium, zinc, and calcium are being considered. Water-based electrolytes are known for their improved safety, environmentally friendliness, and affordability. The key, however, is how to utilize the negative metal electrode, as using water-based electrolytes with these metals becomes an issue with respect to oxidation and/or dendrite formation. This work studied magnesium, where we aimed to determine if it can be electrochemically deposited in aqueous solutions with alginate-based additives to protect the magnesium. In order to do so, atomic force microscopy was used to research the morphological structure of magnesium deposition at the local scale by using a probe—the tip of a cantilever—as the active electrode, during charging and discharging. The second goal of using the AFM probe technology for magnesium deposition and stripping was an extension of our previous study in which we investigated, for lithium, whether it is possible to measure ion current and perform nonfaradaic impedance measurements at the local scale. The work presented here shows that this is possible in a relatively simple way because, with magnesium, no dendrite formation occurs, which hinders the stripping process.

Keywords: magnesium deposition; magnesium alginate; atomic force microscopy; aqueous electrolyte

1. Introduction

Due to the transition to green energy sources, the storage of electrical energy is becoming increasingly important. As a result, the development of lithium-ion batteries has accelerated in the last decade, which have found applications in various devices from mobile phones and laptops to electric bicycles and vehicles. State-of-the-art commercial Li-ion batteries are approaching their theoretical capacities, and, in order to bring these batteries to their practical maximum, solutions are being sought to implement the negative electrode of pure lithium, allowing an ultra-high specific anode capacity (3862 $mA \cdot g^{-1}$) [1]. Their development is especially complicated by the dendrite-forming property of lithium when it is electrochemically deposited, which, after years of research, still poses a major safety challenge [2]. Although this development will further increase energy density, the increasing demand for energy storage solutions is a direct result of the energy transition and will lead to a shortage of the limited raw materials needed to assemble lithium-based batteries in the near future. In addition, a number of disadvantages of lithium-based batteries, such as high cost and safety, give significant impetus to the development of new types of batteries. As a result, scientists are looking for alternatives to lithium as an energy carrier, with special attention to the possibilities of metals such as sodium, aluminum, and magnesium [3]. The latter two in particular are interesting candidates for use as pure metal electrodes because experiments have shown that neither of them has a strong tendency to develop dendrites during the deposition process [4,5]. Of these two lithium alternatives, magnesium has the lowest atomic mass (Mg 24.305, Al 26,982) and the lowest reduction potential after lithium (Li—3 V, Mg—2.37 V, Al—1.68 V). In addition, the volumetric capacity of magnesium is high compared to that of lithium (Mg 3833 $mAhcm^{-3}$, Li 2046 $mAhcm^{-3}$; in contrast, aluminum is superior, at 8046 $mAhcm^{-3}$). These properties, together with the high abundance (magnesium is the eighth most abundant element) and the low cost, make secondary magnesium batteries attractive alternatives to lithium-ion batteries. However, in addition to the development of high-performance electrode materials, the challenges for Mg-based batteries mainly include finding a suitable electrolyte and overcoming the formation of passivation layers on magnesium electrodes. In contrast to the solid electrolyte interphase (SEI) that forms on the anode surfaces in lithium-ion batteries, a layer forms on metallic magnesium electrodes that completely blocks ions [6]. Preventing the formation of a passivation layer is therefore a crucial factor in the composition of electrolytes for magnesium batteries. The development of nonaqueous electrolytes is hindered by their poor cathode kinetics and complex chemistry [7]. The presence of a very small amount of H_2O in these electrolytes can decompose the electrolyte; on the other hand, recent research showed that the presence of water can improve the kinetics at the cathode [8]. These conflicting findings make it interesting to look at aqueous electrolytes and to consider the presence of water in the electrochemical system of a magnesium battery. The alternative we studied is an aqueous electrolyte based on alginate, a natural hydrophilic polymer extracted from seaweed. We used this to electrochemically deposit Mg on different substrates. Atomic force microscopy (AFM) was then employed to research the reversible deposition of magnesium and investigate the morphological structure of the deposited magnesium at the local scale. As we worked with lithium in a previous study [9], we used the probe, the tip of the cantilever, as an active electrode. In the study presented here, the tip of the probe functioned as a point source, providing insight into the distribution of the deposited magnesium over an (infinitely) large surface area. We then used the results obtained under these highly controlled conditions at the macro scale and performed stripping and deposition experiments on different substrates.

1.1. Alginate Structure

Alginic acid is a polysaccharide that is abundantly available, environmentally friendly, cost-effective, and nontoxic. It is found in the cell walls of brown algae and is composed of two anionic monomers: (1,4) linked α–L guluronic acid (G) and (1,4) linked β–D–mannuronic acid (M) (Figure 1). The carboxyl group in the G monomer has the same orientation as the hydroxyl group, whereas in the M monomer the carboxyl group is oriented perpendicular to the hydroxyl group. Alginic acid is well known for its ability to bind multivalent cations, very efficiently forming alginate hydrogels in aqueous environments [10]. Upon deprotonation, the negatively charged carboxylate (COO^-) can chelate with cations. Multivalent cations can ultimately crosslink the alginate polymer chains, increasing the viscosity of the solution and, in most cases, resulting in the formation of a hydrogel. Gel formation was shown to be greatly influenced by the interactions of the G blocks [11]. The linkage of two G monomers creates a cavity, making it an ideal place (cage) for a multivalent cation to reside. The crosslinking of the G blocks by multivalent cations creates a tightly held junction, popularly referred to as an 'egg-box model' [11].

The binding of the cations to alginate is influenced by the cation's properties: its charge, affinity to water, ionic radius, and chemical affinity with the alginate. Bivalent alkaline earth cations (Mg^{2+}, Ca^{2+}, and Sr^{2+}) typically form ionic bonds. In contrast, bivalent transition metal ions (Mn^{2+}, Co^{2+}, Cu^{2+}, Fe^{2+}, and Zn^{2+}) and trivalent metal cations (Fe^{3+}, Cr^{3+}, Al^{3+}, Ga^{3+}, Sc^{3+}, and La^{3+}) usually form complex uronates via strong coordination–covalent bonds [12]. Alginate's affinity for these bivalent cations was shown to increase as follows: Pb > Cu > Cd > Ba > Sr > Ca > Co, Ni, Zn, Mn > Mg [11]. Mg-alginate has long been considered as a nongelling alginate; however, studies found that Mg induces gelation in high-G-content polymers with longer gelation times [13]. Mg-alginates gels are, however, not stable in water and dissolve quickly. The present work investigated the transfer of magnesium between the AFM probes and magnesium metal electrodes in an aqueous Mg-alginate electrolyte.

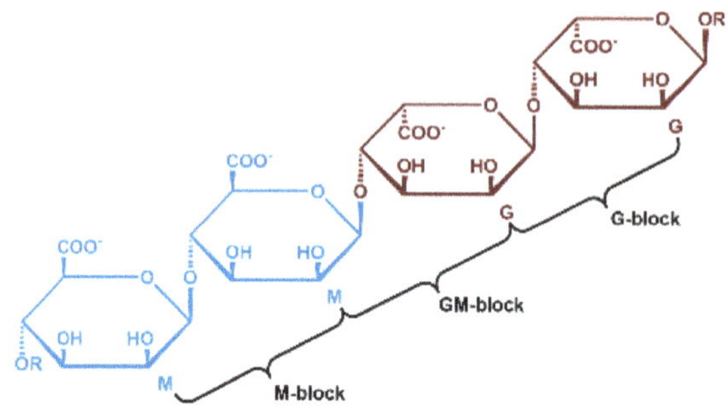

Figure 1. Structure of alginate polymers with monomer mannuronic (M) and guluronic (G) units.

1.2. Magnesium Surface Reactions

In general, magnesium dissolves in aqueous environments via the following anodic reaction,

$$Mg_{(s)} \rightarrow Mg^{2+}_{(aq)} + 2e^{-} \tag{1}$$

which is followed by

$$2H_2O_{(l)} + 2e^{-} \rightarrow H_{2\,(g)} + 2(OH)^{-}_{(aq)} \tag{2}$$

Due to the rise in the pH on magnesium's surface, the following reaction forms magnesium hydroxide, which can precipitate on magnesium's surface [14]:

$$Mg^{2+}_{(aq)} + 2(OH)^{-}_{(aq)} \rightarrow Mg(OH)_{2\,(s)} \tag{3}$$

The result is perceived as a dark layer covering the surface of the magnesium [14,15].

2. Experimental Details

2.1. Mg-Alginate Electrolyte Synthesis

Magnesium alginate was synthesized using alginic acid (Alg-acid; VWR International, Amsterdam, The Netherlands) and magnesium hydroxide ($Mg(OH)_2$; (Merck Sigma, Amsterdam, The Nethelands). Alg acid powder was first weighed, and 20 mL of methanol was added to one equivalent quantity of the Alg acid. The addition of methanol served to dissolve the Alg acid. To deprotonate the Alg acid, ammonia was added, producing a thick slurry. In a separate beaker, $Mg(OH)_2$ was weighed, and 5 mL of methanol was added to ean quivalent quantity of $Mg(OH)_2$. The mixture was then put in a sonication bath to accelerate the dissolution. The powder was then added to the Alg acid slurry and stirred for one day at a constant temperature of 40 °C in order to minimize alginate decomposition. After decanting, the residue was air-dried for 4 days and then vacuum-dried for another 2 days. The vacuum-drying process yielded a solid Mg-alginate. The solid Mg-alginate was ball-milled to a powder in a planetary ball mill (Fritsch, Pulverisette 7, Ede, The Netherlands). The Mg-alginate was crushed in the ball mill for 15 h at 240 RPM. The resulting Mg-alginate powder (Mg-Alg-P) was used to prepare the electrolyte. In order to do so, 2 w/o Mg-Alg-P was dissolved in water, together with 5 w/o NaCl and 5 w/o $MgCl_2 \cdot 6H_2O$, where the Cl^- ions acted as the conducting electrolyte ions for charge compensation.

2.2. Sample Preparation for Deposition and Stripping Experiments

Deposition experiments were performed on flat substrates to avoid the preferred growth of magnesium on the seed-like features on the surface of the sample. As a non-magnesium substrate, we used single-sided highly polished aluminum (thickness 0.8 mm, Ra < 0.05 μm). An accurate and flat sample of pure magnesium was obtained by means of sputter deposition. A silicon wafer (Sigma-Aldrich, St. Louis, MO, USA; <100>, 50.8 mm diameter, 0.5 mm thickness, essentially without dopants) was cleaned with acetone and isopropanol and placed into a sputter deposition system (AJA Int. ATC 1800, Hingham, MA, USA) with a pressure of $<10^{-8}$ mBar and a high-purity Mg target (Kurt J. Lesker, East Sussex, UK), 99.95% pure, 2″ diameter × 0.25″ thick). After plasma cleaning (5 min in argon at 20 mBar and an RF power of 24 W),a layer of magnesium of 500 nm was applied (26 min in argon at 5 mBar and an RF power of 100 W). The silicon wafer sputtered with magnesium was then transferred to a glove box in an airtight container and stored there until needed for the experiments.

2.3. Electro-Chemical Experiments Using AFM

In our previous work [9], we described how we used AFM to perform local electrochemical measurements on the electrodes of lithium-ion batteries at the submicron to nano scales. We developed several methods to provide the tip of the AFM probe with a small amount of lithium so that nonfaradaic measurements can be performed. In the work presented here, we used one of these techniques, the thin-film method, to investigate reversible magnesium deposition between the probe tip and substrate with the use of the prepared alginate-based electrolyte.

We used a similar AFM system (NT-NDT NTegra P9, Moskow, Russia) as in our previous study, with the main difference being that this setup was not housed in a protected glovebox environment due to conducting experiments with aqueous solutions. The used AFM had an accessible design for sample handling and was coupled with a galvanostat/potentiostat (Metrohm Autolab PG-STAT302F, Utrecht, The Netherlands) to perform the electrochemical experiments between the tip and substrate. All deposition experiments with the AFM probe as an electrode described here were performed with a diamond-coated conductive probe (Nanosensors GmbH CDT-NCHR, Neuchatel, Switzerland) with a nominal resonance frequency of around 400 kHz and a nominal force constant of around 80 $N{\cdot}m^{-1}$.

For detecting the presence of magnesium on the probe tip, we used the method described in our previous work [9]. Similarly, in the work presented here, the addition or removal of magnesium to or from the tip was monitored by measuring the resonance frequency of the cantilever after the experiment. For this purpose, the probe was lifted from the Mg-alginate film, and the AFM instrument was set in tapping mode. The resonance frequency could be determined automatically by a frequency sweep executed by the oscillating system. Since our goal was a qualitative analysis of the deposition of magnesium, we did not determine the amount of Mg (as we performed in our previous work) but used the resonance shift as an indication to determine whether deposition or tripping had occurred.

The magnesium transfer experiments were performed as shown schematically in Figure 2. For deposition to the probe tip, this probe tip was connected as the working electrode and the magnesium substrate as the counter electrode (Figure 2a) and vice versa for stripping magnesium from the probe tip to the substrate (Figure 2b). Mg-alginate was applied to the substrate by moving a piece of tissue material soaked with Mg-alginate over the surface, leaving a thin film on the substrate. The remaining film layer had a thickness of between 5 and 8 μm and completely covered the substrate surface. With the AFM setup operating in contact mode, the probe was moved slowly to the surface of the Mg-alginate film while accurately monitoring the moment of bending of the probe with the detector of the laser reflection signal. At the moment a small deviation of the probe was detected, the approach was immediately stopped, and the distance between sample and probe (z axis) was manually adjusted to improve the contact between the probe tip and electrolyte. When contact between the probe tip and electrolyte stabilized, a bias voltage was applied between tip and substrate to start the deposition process. The deposition was stopped by removing the bias voltage or lifting the tip of the probe from the electrolyte surface.

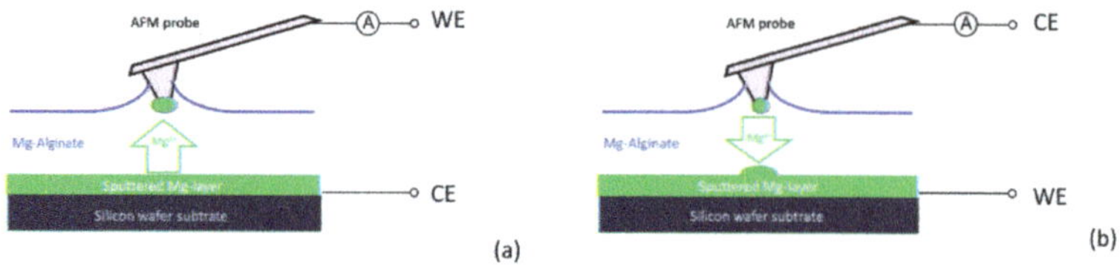

Figure 2. Active probe AFM experiments: (**a**) stripping the substrate and deposit on probe tip and (**b**) stripping the probe tip to deposit on substrate.

For the reversable deposition experiment, we started with a fresh AFM probe and a flat and unused magnesium-sputtered wafer substrate carefully covered with Mg-alginate. After a short deposition time, the presence of a small amount of magnesium on the probe tip was confirmed by the shift in the resonance frequency of the cantilever. The substrate was then rinsed with demi water to remove the Mg-alginate solution, and the surface was investigated using AFM in imaging mode. A new substrate was mounted and carefully covered with Mg-alginate for the stripping experiment. Stripping of the tip was achieved by simply reversing the polarity between the tip and substrate (Figure 2b). Again, a shift in the resonance frequency of the cantilever was used to determine whether magnesium had been removed from the tip. After detecting a significant shift in the resonance frequency, the substrate was carefully rinsed clean with deionized water and then examined via scanning electron microscopy (SEM).

The deposition results on the substrates were examined using a different probe than used for electrochemical experiments: a more accurate nonconductive probe (Nano sensors PPP-NCHR, Neuchatel, Switzerland, resonance frequency 330 kHz, force constant 42 N/m, tip radius < 10 nm). Scanning electron microscopy (SEM; JEOL JSM-IT100, Zaventem, Belgium) and energy-dispersive spectroscopy (EDS) were used for investigating the results of deposition on the probe tip.

2.4. Bulk Stripping and Deposition Measurements

Figure 3 provides a schematic representation of the bulk experiments performed in a dismountable laboratory cell (Figure 3c) with high-purity magnesium foil (Sigma Aldrich; purity 99.99%, thickness 0.015 mm) as the counter and reference electrode and the deposition substrate as the working electrode. The electrodes were separated by a porous glass fiber membrane (thickness 0.25 mm) in the first experiments. Because the glass fiber appeared to have an influence on the deposition (see results below), the separator material was removed in follow-up experiments, and the spacing was obtained by applying a Teflon ring (thickness 0.25 mm, inner diameter 6 mm) between the layers, as shown in Figure 3a.

To investigate the shape and distribution of the deposition more closely, a symmetrical cell (Mg plate on one side, sputtered Mg substrate on the other side) was provided with an extra mask, as shown in Figure 3b. On the working electrode side, an insulating mask (Kapton foil, thickness 50 μm) was provided with a pinhole (d = 350 μm) in the middle and stacked together with a separation ring (Kapton spacer, thickness 300 μm) and the deposition substrate (counter electrode).

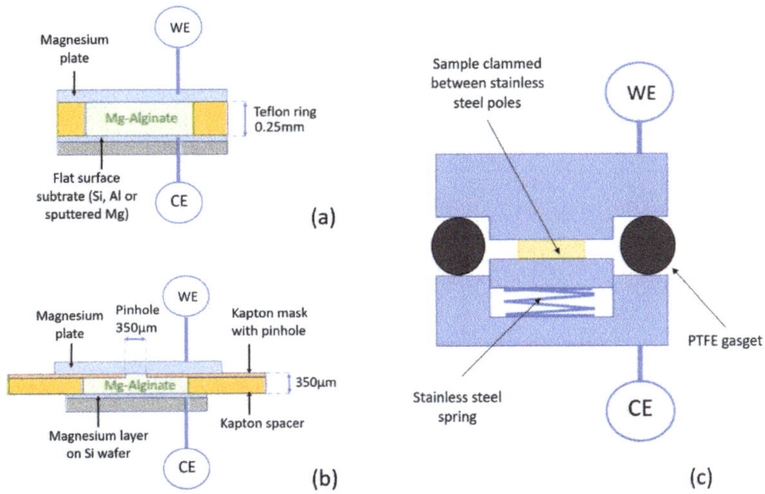

Figure 3. Bulk deposition experiments using Mg-alginate as an electrolyte. (**a**) Mg-alginate is situated between a magnesium plate as working electrode and a flat substrate as the counter electrode where deposition takes place. (**b**) Symmetrical cell with a mask on the stripping side. (**c**) Airtight laboratory cell consisting of airtight-mounted stainless-steel poles separated by a PTFE gasket where the sample is clamped between poles with a stainless-steel spring.

For stripping/deposition experiments, the laboratory cell was connected to a Maccor^{TM} battery tester (Maccor 4000), which was used under constant-current conditions (galvanostatic). Deposition on the substrate side was carried out with a current density of 100 μA.cm^{-2} for 15 min, with a fixed potential limit of 0.2 V vs. Mg for all samples. After deposition, the laboratory cells were carefully dismounted and samples were rinsed with demi- water to remove residual alginate on the surface, which was followed by drying under a vacuum for 30 min. After this cleaning procedure, the result of deposition on the sample surface was investigated via SEM. For the non-Mg substrates, the presence of magnesium was confirmed with EDS.

3. Results and Discussion

3.1. Characterization of Mg-Alginate Solutions

The various Mg alginates and Mg-alginate aqueous solutions were characterized by X-ray diffraction (XRD: X'Pert Pro PANalytical, Malvern, UK) and Fourier transform infrared (FTIR: Thermo Scientific Nicolet iS50 FTIR, Bleiswijk, The Netherlands) spectroscopy, which showed that the solid samples did not contain any residual $Mg(OH)_2$ or alginic acid (Figure 4c,d).

3.2. Electrochemical Characterization

The ionic conductivity of the alginate samples was measured at different temperatures, as shown in Figure 4a. The conductivity for all concentrations of electrolyte generally increased with temperature. The Mg-alginate concentration influenced the conductivity strongly, with lower concentrations (2 wt.% and 25 wt.%) exhibiting higher conductivities, which can be understood by the strong dependency of Mg-ion mobility on concentration. Hence, the electrolyte conductivity seemed to rely more on the ionic mobility than the concentration of the Mg ions, e.g., addition of more charge carriers (Mg-alginate) did not improve the conductivity of the electrolyte.

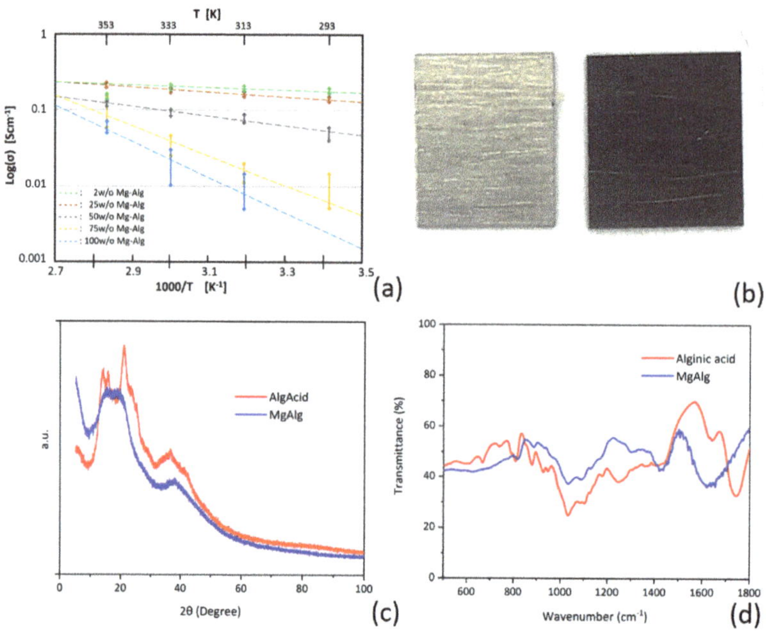

Figure 4. (**a**) Conductivity of electrolytes at different concentrations; (**b**) bare magnesium surface compared with surface covered by black layer after contact with alginate solution; (**c**) X-ray diffraction analysis of alginate acid (red) and magnesium alginate (blue); (**d**) Fourier transform infrared spectroscopy analysis of alginate acid (red) and magnesium alginate (blue).

The Mg-alginate concentrations with the most promising conductivity (2 wt.%, 25 wt.%, and 50 wt.%) were then chosen to perform cyclic voltammetry (CV) and electrochemical impedance spectroscopy (EIS) measurements (Metrohm Autolab PGSTAT12). For characterization, EIS analyses were used to measure the internal resistance of the electrolyte and the interfacial resistance between the electrolyte and the magnesium electrodes. While running EIS and CV, a dark layer was observed on the pristine magnesium electrodes (Figure 4b). This layer became more apparent with each experiment, so the effect of the layer on the system was investigated. The EIS profiles using 2 w/o (Figure 5a), 25 w/o (Figure 5b), and 50 w/o (Figure 5c) show a significant change in the overall impedance. This seemed to reduce the charge transfer resistance, causing a subsequent reduction in the impedance but introduced a new process with its own time constant, which was clearly seen from the additional semicircle formed in the high-frequency range (Figure 5a–c).

In Figure 5c–e, the CV curves for 2 w/o, 25 w/o, and 50 w/o are presented, respectively. Before formation of the black layer, the CV showed a low cycling efficiency due to the high impedance. After the formation of the black layer, the impedance reduced, causing the cycling efficiency to improve. However, it looked like it was limited by the number of charge carriers. The same behavior can be seen in Figure 5d, where cycling efficiency improves after the formation of the black layer. The CV curve for the 25 w/o electrolyte, however, did not seem to be affected by electronic conduction since the electrolyte contained more charge carriers. Figure 5f (black curve, before formation of black layer) presents a very low cycling efficiency due to the high impedance (see Figure 5c). The formation of the black layer did not seem to significantly improve the efficiency due to the increased interfacial and charge transfer resistances (see Figure 5c).

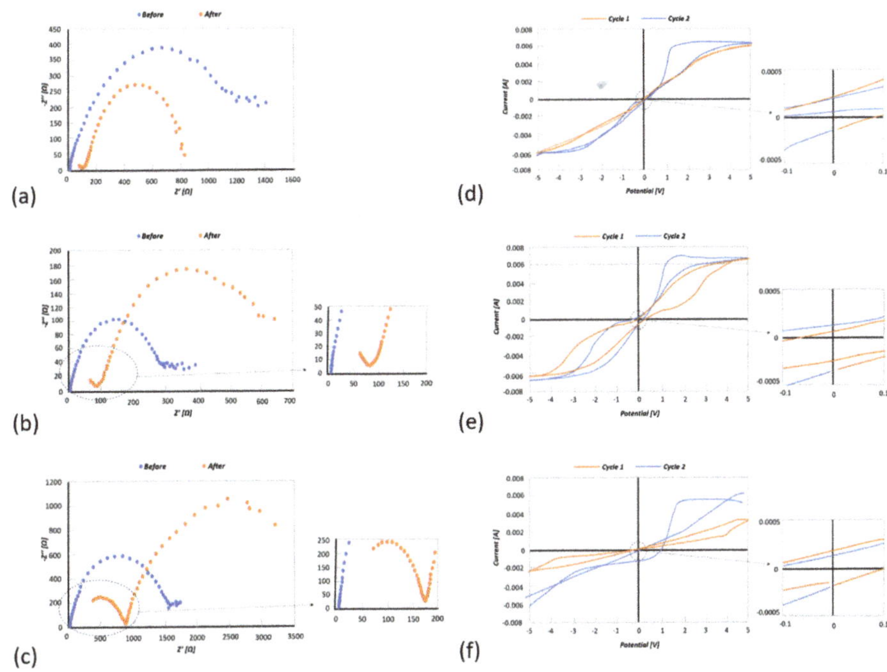

Figure 5. (**a**) EIS results for the 2 wt.% Mg-alginate electrolyte before (blue) and after (red) forming of the black layer; (**b**) EIS results for the 25 wt.% Mg-alginate electrolyte before (blue) and after (red) forming of the black layer; (**c**) EIS results for the 50 wt.% Mg-alginate electrolyte before (blue) and after (red) forming of the black layer; (**d**) CV curve for 2 wt.% Mg-alginate electrolyte before (blue) and after (red) forming of the black layer; (**e**) CV curve for 25 wt.% Mg-alginate electrolyte before (blue) and after (red) forming of the black layer and; (**f**) CV curve for 50 wt.% Mg-alginate electrolyte before (blue) and after (red) forming of the black layer.

3.3. Electrochemical Deposition of Magnesium Using AFM Probe as an Electrode

Magnesium deposition on the probe tip was performed as shown in Figure 2a. It should be noted that the performance of this process strongly depends on both the ion current and the contact area with the electrolyte. The latter is practically indeterminable in advance because the interaction between the probe tip and the electrolyte surface is influenced by a number of interdependent or variable parameters, such as the capillary forces between tip and liquid, the viscosity of the electrolyte, the deflection of the cantilever, and the setpoint of the control loop of the AFM. Therefore, the correct settings were found via trial and error until the right conditions occurred and magnesium was deposited in the desired manner on the probe tip. Figure 6a,b indicate two extremes of deposition on the probe tip. Figure 6a shows the situation when the probe landed on the electrolyte with a very sensitive adjustment so that the tip just touched the electrolyte and was not adjusted further. Only a small part of the liquid was in contact with the tip, resulting in a deposition of magnesium as a cone-shaped structure. This formation can be understood by the wetting of the tip, the deposition rate of magnesium, and the exhausting of Mg^{2+} ions in the solution in the formed cone at the tip, as depicted in the schematic in Figure 6a. It further proves that the deposition must be metallic magnesium; otherwise, electrons would not find their way to the rim of the cone. Figure 6b shows the situation when the probe had sunk just too deep into the electrolyte, with part of the cantilever in contact with the liquid. The deposition in the SEM photo shows complete coverage of the cantilever and the tip. Figure 6c shows the desired situation, where the tip just breaks through the liquid surface and deposition only takes place at the end of the probe tip. Once this situation is achieved, subsequent depositions can be reproduced with similar settings.

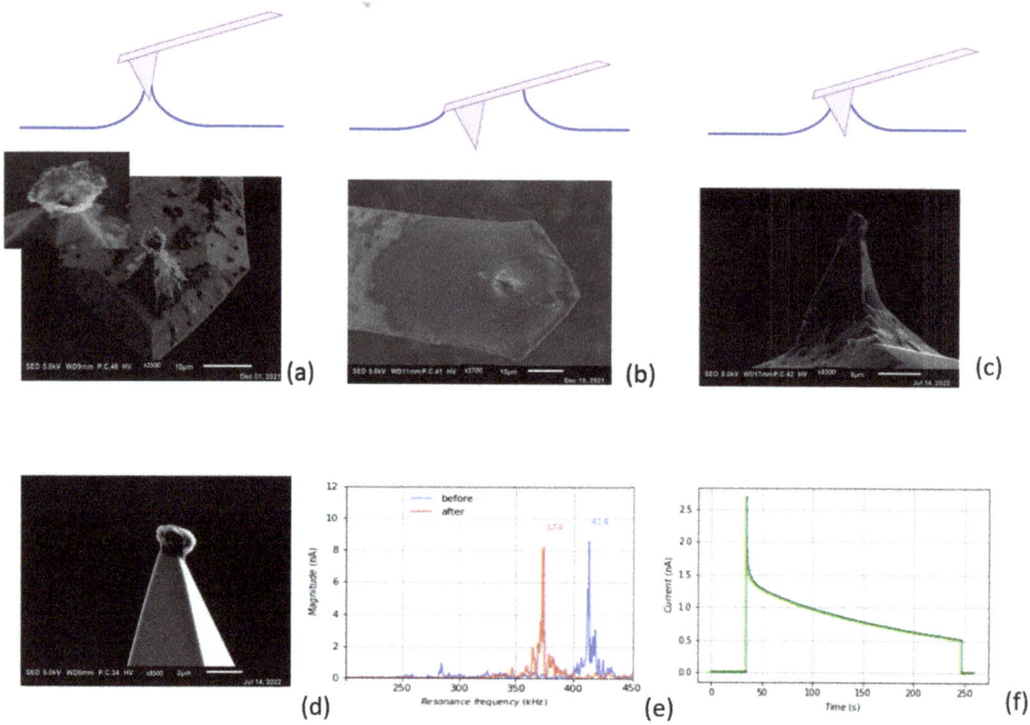

Figure 6. Fine tuning of the AFM settings for optimal deposition of Mg on the very end of the probe tip. (**a**) Very small electrolyte contact area results in cone-formed Mg deposition. Inset: Cone in detail (up) and a schematical impression on how the Mg cone is growing due to exhausting of Mg^{2+} ions; (**b**) electrolyte contact area covers part of the cantilever, which results in overall coverage of Mg on the cantilever; (**c**) correct settings resulting in a deposition on the tip-end; (**d**) subsequent deposition with the same settings as in (**c**); (**e**) shift in the resonance frequency of the cantilever caused by deposition of the tip; (**f**) registration of the current during deposition.

The AFM settings used for the deposition in Figure 6c appeared to give the best results and were also reproducible, as can be seen in one of the subsequent depositions with these settings (Figure 6d). Figure 6e shows a graph of the resonance frequency of the cantilever before and after deposition, and Figure 6f shows the measurement of the current between the tip and substrate. Both indicate the successful deposition during the experiment.

AFM offers good opportunities to take a closer look at the morphological structure of a deposit, because the tip of the probe can be regarded as a point source. Deposition on a tip results in a freely positioned spherical shape that can be viewed all around with SEM. To obtain an impression of the morphological structure of the deposited magnesium when using the Mg-alginate electrolyte, a long-term magnesium deposition was performed, as can be seen in Figure 7a. For comparison, we show a similar deposition with lithium in Figure 7b (taken from our previous study, [9]).

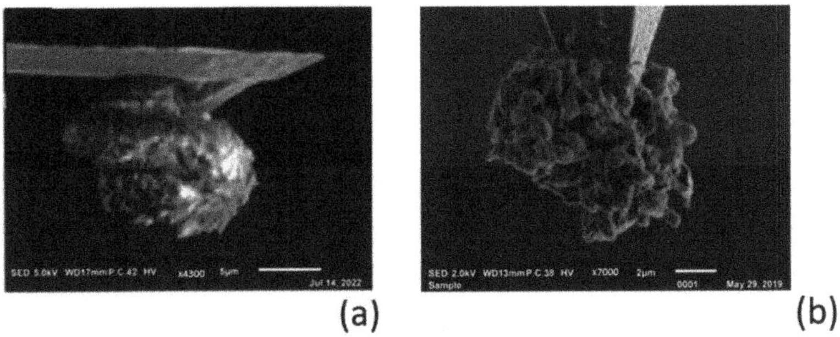

Figure 7. (**a**) Large amount of magnesium deposited on the tip; (**b**) lithium deposition on the tip [9].

The difference in deposition morphologies between lithium and magnesium can be clearly seen in the side-view SEM pictures of the probe tip: Figure 7a shows a massive spherical magnesium deposit with a compact dendrite-free structure, while Figure 7b shows a lithium deposit with a porous and dendrite-shaped spherical structure. These experimental findings are theoretically supported by the modeling and DFT calculations in the works of Jäckle and co-authors [16,17], which describe how magnesium tends to grow with a smooth surface because it exhibits lower diffusion barriers than lithium. The dendrite formation that occurs with the electrodeposition of lithium is also influenced by the electrode/electrolyte interface upon long-term cycling. However, since the origin of dendrite formation is based on the characteristic differences between Li and Mg with regard to the elemental properties relevant for growth, it can be concluded that the properties of the electrolyte in the case of magnesium do not significantly influence dendrite growth.

3.4. Reversable Magnesium Deposition between Probe Tip and Substrate

In our previous study [9], we observed that it is not possible to electrochemically remove (strip) lithium from a shape in a dendrite structure (as shown in Figure 7b). Inspection with SEM showed that the magnesium deposited on the tip had a dense structure without any dendrite formation, which allowed for reversible magnesium deposition. First, magnesium was taken from the substrate and deposited on the probe tip, as schematically shown in Figure 2a. After this experiment, the surface of the substate was scanned with AFM using the accurate probe that was selected for surface investigation. The image in Figure 8a shows the spot where the magnesium was taken during the experiment. Obviously, as can be seen in this picture, magnesium was taken from a very small area, the size of which was comparable to the size of the probe tip. The probe with magnesium deposited on the tip was used for further experiments. Stripping of magnesium from the probe tip was performed as schematically shown in Figure 2b. Figure 8b shows an image of the magnesium feature that was electrochemically removed from the probe tip and deposited on the substrate. Again, the size and shape of the deposited material were the same as those of the probe tip, meaning that the ion current followed a very straight path through the electrolyte. Figure 8c gives the shift in the resonance frequency of the probe cantilever measured in tapping mode after each experiment. Although the distortion of the peak made it difficult to determine the precise shift, the clearly discernible change in resonance frequency gave a good indication of the addition and removal of mass at the tip. This experiment clearly shows that the reversible deposition of magnesium with an aqueous Mg-alginate-based electrolyte is possible.

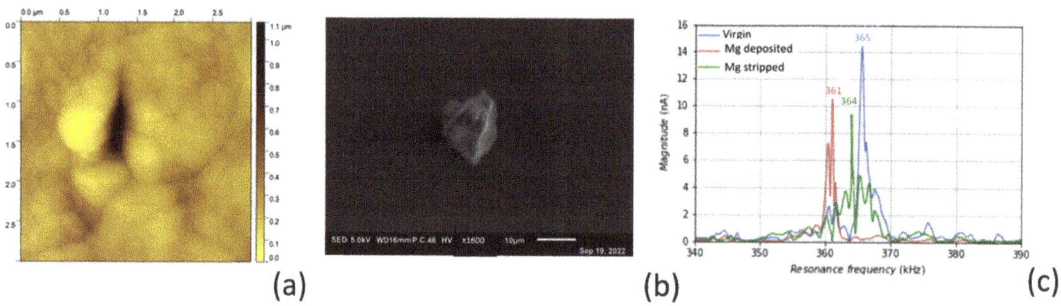

Figure 8. Deposition and stripping experiment: (**a**) AFM image; (**b**) SEM picture; (**c**) shift in resonance frequency.

3.5. *Electro-Chemical Magnesium Deposition at Bulk Level*

The results of the AFM experiments on a submicron scale gave us a reason to perform magnesium deposition at the macro scale, as described in Figure 2. Figure 9 shows SEM pictures of the deposition of Mg on substrates of aluminum (Figure 9a) and silicon (Figure 9b). The experiment with aluminum (Figure 9a) was performed using a glass fiber separator between the substrate and stripping target. Due to the relatively high energy of the SEM electron beam used (20 keV), the seed on which the magnesium was deposited was visible inside the particle. To avoid the preferred deposition on a seed-like feature, further experiments were performed using the cell configuration in Figure 2a, where the separator material was replaced with a Teflon spacer. Figure 2b shows the apparently random deposition on a silicon substrate, where magnesium appears to have a slight preference for clumping or accumulating on previously deposited material.

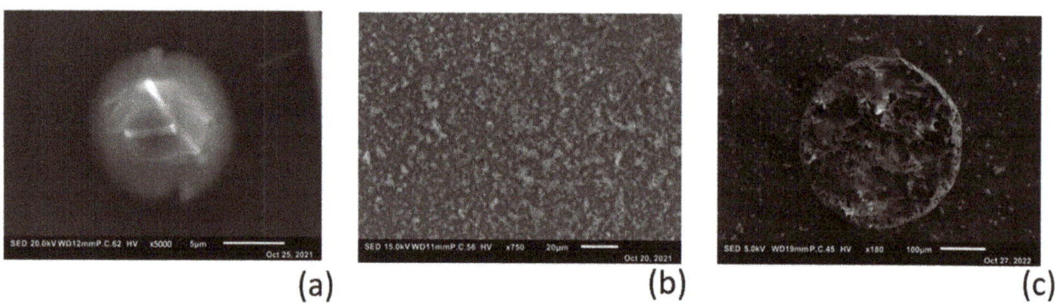

Figure 9. Deposition of Mg on different substrates with use of aqueous Mg-alginate electrolyte: (**a**) SEM picture of Mg on polished aluminum substrate. Inside the spherical particle, a small piece of glass fiber from the separator is visible, which acted as a seed for deposition. The inset image shows the EDX scanning result for Mg; (**b**) SEM picture of Mg on silicon wafer. Deposition without a separator shows random-oriented deposition on the substrate; (**c**) Mg deposited on sputtered Mg substrate using a mask with a pinhole of 350 μm.

To verify the sub-micron-scale findings in Figure 8b, where the ion current through the electrolyte appears to follow a very straight path, the macro-scale deposition on a flat magnesium substrate was performed using a mask (as schematically shown in Figure 2b). The SEM image in Figure 9c shows the result of this experiment, where the magnesium forms a cylindrical projection of the mask with the exact size of the pinhole (350 μm). The amount of magnesium that ended up next to the circular deposition was negligibly small.

4. General Conclusions

In the study described here, we focused on the basic and critical phenomena that are needed to develop an alginate-based aqueous secondary magnesium battery. The synthesized alginate powder was characterized by XRD and FTIR (Figure 4c,d), after which different concentrations of Mg-alginate were dissolved in water and analyzed via EIS and CV. In general, we concluded that the aqueous Mg-alginate solution shows electrolyte activity. The 2 wt% Mg-alginate in water with an ion conductivity of ~1 $mS.cm^{-1}$ was used further for testing.

A black layer formed on the interface between the aqueous Mg-alginate and the magnesium metal anode surface. The results indicated that this layer had a different composition than the $MgOH_2$ normally formed at the interphase of magnesium and water because it did not seem to block Mg ions to and from the metal surface. However, the specific composition of the black layer was not fully understood because it was an amorphous layer. The layer may have had a composition consisting of magnesium, oxygen, carbon, and hydrogen. Further research using specific measuring techniques is required to characterize the precise composition of the layer. The black layer did not appear to have any adverse effect and may have acted as a passivation layer for the magnesium surface. Further research into this ion-permeable passivation layer and the contribution and necessity of the alginate in its formation is recommended.

We successfully created an active AFM probe by providing the probe tip with metallic magnesium. With this active probe, it is possible to apply controlled magnesium to a substrate by means of electrochemical deposition. The electrochemical deposition of magnesium using a Mg-alginate electrolyte was demonstrated. Because the AFM probe can be considered an electrode point source with respect to the infinite 2D and plate-shaped electrode substrate, it was possible to analyze the deposition of magnesium both on the tip and on the substrate. These measurements provided insight into the way in which magnesium ions were transported through the alginate electrolyte. It appeared that the magnesium had a strong preference for taking the shortest path, as evidenced by the shapes on both from stripping and deposition on the substrate (Figure 8a,b). The magnesium formed solid structures, unlike lithium, where unwanted dendrite formation occurs.

The strong preference for a nondendritic morphology of magnesium provides good opportunities to use this deposition/stripping method for measuring ion current, which we achieved in this work with a qualitative current signal (Figure 6f). It also opens the way for future researchers to perform nonfaradaic impedance measurements at a local scale. Unlike with the use of lithium, as we investigated in our earlier work [9], a small amount magnesium can be deposited on an AFM probe with a relatively simple electrodeposition process. This makes it technically less complicated to measure the ion current while scanning over a surface with a magnesium-containing probe tip and thus to create an active-probe AFM image. The development of this 2D ion current scanning probe method creates a new and powerful analysis technique in the palette of AFM -based measuring techniques. With advanced AFM equipment, cathode materials can be investigated in this way in shading mode, providing insight into the relationship between the ion current and morphology of the electrode material.

Author Contributions: Methodology, W.J.L.; Formal analysis, W.J.L., L.K. and M.K.; Investigation, W.J.L.; Writing—original draft, W.J.L.; Writing—review & editing, L.K. and E.M.K.; Visualization, W.J.L. and M.K.; Supervision, E.M.K. All authors have read and agreed to the published version of the manuscript.

Funding: This research received no external funding.

Data Availability Statement: Data are contained within the article.

Conflicts of Interest: The authors declare no conflict of interest.

References

1. Shen, X.; Liu, H.; Cheng, X.B.; Yan, C.; Huang, J.Q. Beyond lithium ion batteries: Higher energy density battery systems based on lithium metal anodes. *Energy Storage Mater.* **2018**, *12*, 161–175. [CrossRef]
2. Zhang, X.; Wang, A.; Liu, X.; Luo, J. Dendrites in lithium metal anodes: Suppression, regulation, and elimination. *Acc. Chem. Res.* **2019**, *52*, 3223–3232. [CrossRef]
3. Biemolt, J.; Jungbacker, P.; van Teijlingen, T.; Yan, N.; Rothenberg, G. Beyond lithium-based batteries. *Materials* **2020**, *13*, 425. [CrossRef]
4. Lin, M.C.; Gong, M.; Lu, B.; Wu, Y.; Wang, D.Y.; Guan, M.; Angell, M.; Chen, C.; Yang, J.; Hwang, B.-J.; et al. An ultrafast rechargeable aluminium-ion battery. *Nature* **2015**, *520*, 324–328. [CrossRef]
5. Matsui, M. Study on electrochemically deposited Mg metal. *J. Power Sources* **2011**, *196*, 7048–7055. [CrossRef]
6. Aurbach, D.; Gizbar, H.; Schechter, A.; Chusid, O.; Gottlieb, H.E.; Gofer, Y.; Goldberg, I. Electrolyte solutions for rechargeable magnesium batteries based on organomagnesium chloroaluminate complexes. *J. Electrochem. Soc.* **2001**, *149*, A115. [CrossRef]
7. Salama, M.; Attias, R.; Hirsch, B.; Yemini, R.; Gofer, Y.; Noked, M.; Aurbach, D. On the feasibility of practical Mg–S batteries: Practical limitations associated with metallic magnesium anodes. *ACS Appl. Mater. Interfaces* **2018**, *10*, 36910–36917. [CrossRef]
8. Wang, F.; Fan, X.; Gao, T.; Sun, W.; Ma, Z.; Yang, C.; Han, F.; Xu, K.; Wang, C. High-voltage aqueous magnesium ion batteries. *ACS Cent. Sci.* **2017**, *3*, 1121–1128. [CrossRef] [PubMed]
9. Legerstee, W.J.; Boekel, M.; Boonstra, S.; Kelder, E.M. Scanning Probe Microscopy Facility for Operando Study of Redox Processes on Lithium ion Battery Electrodes. *Front. Chem.* **2021**, *9*, 505876. [CrossRef] [PubMed]
10. Grant, G.T.; Morris, E.R.; Rees, D.A.; Smith, P.J.C.; Thom, D. Biological interactions between polysaccharides and divalent cations: The Egg−Box model. *FEBS Lett.* **1973**, *32*, 195–198. [CrossRef]
11. Haug, A.; Smidsrod, O.A.; Larsen, B.; Gronowitz, S.; Hoffman, R.A.; Westerdahl, A. The Effect of Divalent Metals on the Properties of Alginate Solutions. II. Comparison of Different Metal Ions. *Acta Chem. Scand.* **1965**, *19*, 341–351. [CrossRef]
12. Brus, J.; Urbanova, M.; Czernek, J.; Pavelkova, M.; Kubova, K.; Vyslouzil, J.; Abbrent, S.; Konefal, R.; Horský, J.; Vetchy, D.; et al. Structure and Dynamics of Alginate Gels Cross−Linked by Polyvalent Ions Probed via Solid State NMR Spectroscopy. *Biomacromolecules* **2017**, *18*, 2478–2488. [CrossRef] [PubMed]
13. Topuz, F.; Henke, A.; Richtering, W.; Groll, J. Magnesium ions and alginate do form hydrogels: A rheological study. *Soft Matter* **2012**, *8*, 4877–4881. [CrossRef]
14. Gore, P.; Raja, V.S.; Birbilis, N. Use of sodium bicarbonate as a chloride-free aqueous electrolyte to explore film formation and the negative difference effect on pure magnesium. *J. Electrochem. Soc.* **2018**, *165*, C849. [CrossRef]
15. Salleh, S.H.; Thomas, S.; Yuwono, J.A.; Venkatesan, K.; Birbilis, N. Enhanced hydrogen evolution on $MgOH_2$ covered Mg surfaces. *Electrochim. Acta* **2015**, *161*, 144–152. [CrossRef]
16. Jäckle, M.; Helmbrecht, K.; Smits, M.; Stottmeister, D.; Groß, A. Self-diffusion barriers: Possible descriptors for dendrite growth in batteries? *Energy Environ. Sci.* **2018**, *11*, 3400–3407. [CrossRef]
17. Jäckle, M.; Groß, A. Microscopic properties of lithium, sodium, and magnesium battery anode materials related to possible dendrite growth. *J. Chem. Phys.* **2014**, *141*, 174710. [CrossRef] [PubMed]

polymers

MDPI

Article

Sustainable Transformation of Cellulose-Containing Textile Waste into Multifunctional Panels with Tailored FR-Lignocellulosic Fibres

Hamid Lamoudan [1,*], Lahbib Abenghal [2], Dan Belosinschi [2,*], François Brouillette [2], Patricia Dolez [3], Raymond Panneton [4] and Cécile Fonrouge [5]

1 CTT Group, Saint-Hyacinthe, QC J2S 1H9, Canada
2 Innovations Institute in Ecomaterials, Ecoproducts and Ecoenergies (I2E3), Université du Québec à Trois-Rivières, Trois-Rivières, QC G8Z 4M3, Canada; lahbib.abenghal@uqtr.ca (L.A.); francois.brouillette@uqtr.ca (F.B.)
3 Department of Human Ecology, University of Alberta, Edmonton, AB T6G 2N1, Canada; pdolez@ualberta.ca
4 Centre de Recherche en Acoustique-Signal-Humain, Université de Sherbrooke, Sherbrooke, QC J1K 2R1, Canada; raymond.panneton@usherbrooke.ca
5 Institut de Recherche sur les PME, Université du Québec à Trois-Rivières, Trois-Rivières, QC G8Z 4M3, Canada; cecile.fonrouge@uqtr.ca
* Correspondence: hlamoudan@gcttg.com (H.L.); dan.belosinschi@uqtr.ca (D.B.)

Abstract: The fashion industry significantly impacts the environment, mainly through the substantial generation of waste textiles fostered by fast fashion business models. This study introduces an innovative approach to textile waste management by recycling waste textiles without the use of chemical or mechanical treatments. Herein, we developed a method adhering to the principles of circular economy to transform these textile wastes into high-quality construction panels using a papermaking process. This method not only provides a sustainable solution to reduce landfill dependency but also enhances resource efficiency in the construction industry. The fabricated panels, composed of a blend of 45% textile waste microfibres and 55% fire-retardant fibres, exhibit several advantageous properties. They feature a low apparent density ranging between 170–180 kg/m^3 and a low thermal conductivity coefficient of 0.047 W/m∗K at 50 kPa. It revealed that phosphorylated fibres not only provide flame-retardant properties, but they also significantly improve the mechanical properties of the panels. For example, load at break increases from 12.4 to 81.1 N, stress at break from 0.44 to 3.59 MPa, and E-modulus from 29.2 to 198.8 MPa after the addition of these 55% fibres. Moreover, these panels successfully met the criteria set by international standards for construction products satisfying the fire test, EN ISO 11925-2. These characteristics make the panels superior options for sustainable construction materials, offering enhanced fire resistance and insulation properties, which are critical to meet modern building standards. They mark a pivotal step towards sustainable construction and waste reduction in the fashion industry.

Keywords: textile waste; management; circular economy; sustainable construction; papermaking process; fire-resistant fibres

1. Introduction

Textile manufacturing has become one of the most important industry sectors today, playing a vital part in the economies of many countries [1]. According to forecasts, the textile market's revenues are expected to exceed 1.36 trillion dollars in 2024, and reach 1.78 trillion dollars by 2029 [2–4]. This industry produces over 100 billion garments per year, a number that continues to rise year after year, as people are influenced by fashion and most garments are only worn a few times (seven to ten) before being discarded [5–7]. As a result, 92 million tonnes of textile waste end up in recycling facilities every year, some of which are still in good condition and can be reused after minor cleaning and repairs [8,9]. However, only 1% of textile waste is recycled after use to make new garments, while the rest ends up in incinerators or landfills [8]. This presents a serious problem for municipal waste managers, and contributes to the high environmental footprint of textiles. Indeed, studies have shown that this industry alone is responsible for 8–10% of total greenhouse gas emissions and 20% of the world's wastewater [10,11]. Recycling and reuse of textile waste are good ways to solve this problem. This transition can decrease reliance on natural fibres like cotton, which currently makes up approximately 22% of the world's annual fibre production [12]. Reducing the use of cotton, which is heavily utilized in garment manufacturing, could help mitigate environmental impacts such as high water usage, pesticide dependency, and extensive land requirements. It is estimated that the production of one kilogram of cotton can require up to 20,000 L of water, which means that the manufacture of a single T-shirt can take 2700 L of water [5].

In general terms, textile wastes are divided into two categories: pre-consumer wastes and post-consumer wastes [13]. The first category is generated during the manufacturing process, while the second is produced by consumers after use. The recycling of these textile wastes can be carried out according to three procedures: biological recycling, mechanical recycling and chemical recycling [11,14]. The appropriate procedure is selected based on the waste fibre content: cellulose-based fibres such as cotton, hemp, viscose, and lyocell; synthetic fibres such as polyamide and polyester; and finally, protein fibres such as silk and wool [7].

Biological recycling involves the transformation of textile wastes into simple molecules, namely water, carbon dioxide, methane, ethanol and ammonia, by enzymes or micro-organisms [15]. However, this process is not widely applicable to textile wastes, as in most cases they contain a blend of synthetic and natural fibres, and micro-organisms and enzymes would need several years to achieve complete decomposition. The decomposition period of synthetic polymers is lengthy, and plastic pollution is a serious environmental challenge in this century [16]. Mechanical recycling is considered the most widespread process, and involves the use of mechanical force to reduce textile wastes to fibres, which are then mixed with virgin fibres to produce yarns and then fabrics. Unfortunately, mechanical recycling tends to damage the structure of the fibres and reduce their length, limiting their use for the manufacture of high-value products [17,18]. Finally, chemical recycling is based on fibre depolymerization using strong acids such as phosphoric acid or sulfuric acid or polymer chain dissolution using organic solvents [19,20]. For instance, S. Miguel et al. [20] have used a two-step procedure to depolymerize cotton fibres contained in textile wastes. This process relies on the use of concentrated sulfuric acid in the first step, followed by a second hydrolysis step using a dilute solution from the same acid for total depolymerization of the cotton cellulose to produce glucose. Glucose solutions with concentrations of about 40 g/L were obtained with a yield of 90%. Palme et al. [21] used solutions of 5–15% sodium hydroxide at 70–90 °C to dissolve polyester cotton blended textiles in order to separate the cotton from the PET (polyethylene terephthalate). Ionic liquids or solvents such as N-methylmorpholine-N-oxide (NMMO), which are more environmentally friendly than what is employed in the viscose process, can also be used to dissolve cellulose fibres and produce cellulose solutions for the production of regenerated cellulose fibres [19,22]. However, the chemical process also has certain disadvantages, such as the use of strong acids which can rapidly damage process equipment, and the difficult recovery of acids and solvents [23]. In addition, they may lead to the severe degradation of the fibres.

To fill this gap, this study presents a simple and efficient method for reusing textile wastes to manufacture new value-added products that are in demand in today's market. This method is based on the use of a blend of textile waste microfibres and chemically modified lignocellulosic fibres to manufacture fire-resistant panels that can be used in the construction industry to improve the building safety.

2. Materials and Methods

2.1. Materials and Synthesis

2.1.1. Materials

Textile waste microfibres were collected from domestic laundry appliances (washer and dryer combo) by the researchers. They contained both synthetic (mainly polyester) and cellulose-based fibres (cotton, lyocell, and viscose). In this research, flame-retardant lignocellulose fibres were produced by phosphorylating unbeaten bleached softwood kraft pulp fibres (referred to as KF), which were provided by Kruger Wayagamack Inc. (Trois-Rivières, QC, Canada). The phosphorylation process utilized 1-decanol, polyphosphoric acid, phosphorus pentoxide (from Sigma-Aldrich, St. Louis, MI, USA), and urea (from Alfa Aesar, Ward Hill, MA, USA).

2.1.2. Synthesis

The KF were phosphorylated using the previously synthesized phosphate ester (EPs) (Figure 1). This modification was carried out in molten urea, with a molar ratio of urea to phosphate ester and anhydroglucose unit from cellulose polymer set at 17:3:1. The phosphorylation reaction is a commonly used process in our research group and is exhaustively described in several previous papers [24,25]. The phosphorylated fibres (PKF) used in this study had a phosphorus content of 11.2%.

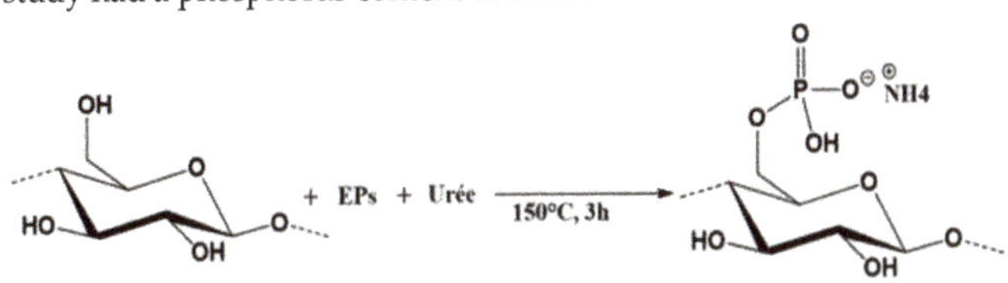

Figure 1. Phosphorylation reaction scheme of lignocellulosic fibres using the phosphate ester/urea system.

2.2. Textile Waste Panel Preparation

An empiric method was used for the production of prototypes (Figure 2). The panels were prepared by adding 16 g of textile waste microfibres (45%) and 19 g of SF fibres (55%) in 2 L of tap water, followed by dispersion in a standard laboratory disintegrator operating at 3000 revolutions per minute for a duration of 15 min. Subsequently, the fibre suspension was poured into a Buchner funnel containing a paper-making wire mesh with a diameter of 150 mm. A wet fibre pad was formed by applying a 500 mm Hg vacuum to the system. Finally, the wet structure was taken and dried in an oven at 110 °C for 2 h. A series of 10 panels was manufactured. For panels #2 to #10, 90% of the white water was recovered from the previous test and reused in the process. For this study, only the last five panels were kept for further analyses and investigation. The first five panels were excluded to allow the dissolved and colloidal substances to accumulate in the process water by recirculation [26].

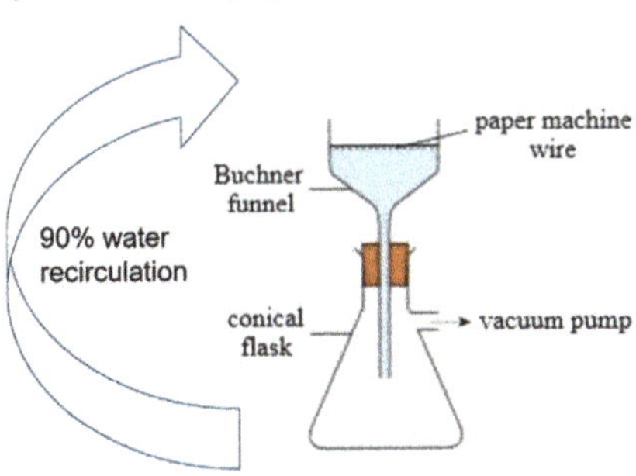

Figure 2. Wet fibre pad formation.

2.3. Fibre Length Distribution

An L&W Fibre Tester optical device (ABB, Brampton, ON, Canada) was used to determine the mean arithmetic fibre length and width, and the length weighted percentage of fine fibres (less than 0.2 mm in length). A fibre sample of approximately 0.2 g was disintegrated in 200 mL of water with a mixer until the fibres were completely individualized. The device requires a total of 3000 to 5000 fibres to achieve a consistent statistic for each test. Five measurements were conducted for each fibre sample and used to calculate the mean and standard deviation.

2.4. Zeta Potential

The Zeta potential of the aqueous textile waste and phosphorylated kraft fibres (PKF) suspensions was determined by simultaneously measuring their conductivity, pressure and streaming potential with a Mutek SZP-06 analyzer (BTG Americas, Norcross, GA, USA). Samples were drawn into the suction tube under vacuum, forming a fibre plug within the measuring cell. Water flow passing through the fibre plug displaced mobile charges from the shear plane, generating a streaming current and thereby establishing a measurable potential difference across two electrodes. The conductivity of the suspensions was adjusted to 0.1–7 mS/cm with 0.1 M KCl to generate a readable streaming potential and ensure reproducibility.

2.5. Surface Charge

The surface charge of the textile waste microfibres was evaluated using a Mutek PCD-03 analyzer (BTG Americas, Norcross, GA, USA) equipped with an automatic titrator [27]. Polyelectrolyte titration involves the use of standard polydiallyldimethylammonium chloride (poly-DADMAC, positively charged) and poly(vinyl sulfate) potassium salt (PVSK, negatively charged). Initially, 0.2 g of fibres were added to 50 mL of distilled water. The suspension was then stirred using a magnetic stirrer. Next, an excess of 0.001 N poly-DADMAC was introduced (10 mL for textile waste microfibres and 100 mL for phosphorylated fibres), and the mixture was stirred for 30 min at room temperature. The suspension underwent gravitational filtration using a 200 mL volumetric flask, funnel, and a 202 grade filter paper. Finally, 10 mL of the resulting filtrate were extracted from the volumetric flask and poured into the measuring cell of the PCD-03. The excess of poly-DADMAC in the filtrate was titrated with the negatively charged PVSK standard 0.001 N solution. The surface charge density of the fibre samples was calculated as the difference between the total amount of poly-DADMAC added and the excess amount found after adsorption on the fibre sample.

2.6. Nitrogen Content

The Kjeldahl method [28] is frequently employed to quantitatively assess the nitrogen content of organic compounds. Samples are initially digested at elevated temperatures in a strong acid and salts, such as concentrated sulfuric acid and potassium sulfate, with a catalyst like copper sulfate, to convert organic nitrogen into ammonia. Subsequently, the solution is cooled down and neutralized with an alkaline solution, releasing ammonia in gaseous form. The ammonia is then captured in a weak acid solution like boric acid through distillation. The nitrogen amount in the solution is determined by titration with a standard solution of strong acid. This method enables the evaluation of nitrogen content in both textile waste microfibres and SF fibre samples.

2.7. Fibre Composition

As the textile waste microfibres contained a blend of synthetic and cellulose based fibres, an empirical method was developed to assess their chemical composition. Cupriethylenediamine (Cuen) was used as a solvent for the cellulosic fibres. Initially, a sample of 0.3 g of fibre was inserted into a 150 mL glass bottle; 100 mL of 1M Cuen were then added to the fibres and the solution was well mixed for 60 min. Finally, the undissolved fibres associated with the synthetic fraction were separated through filtration with a silicone filter paper, thoroughly washed with water, dried, and weighed. The synthetic fraction was calculated by dividing the weight of the undissolved fibres by the total weight of the sample. The result was expressed as the average of three measurements.

2.8. Morphological Examination

Scanning electron microscopy (15 kV, variable pressure) coupled with energy dispersive Xray spectroscopy (SEM/EDX, Hitachi High-Tech Corporation, Tokyo, Japan, SU1510 with Oxford X-max 20 mm^2) was used to track structural changes occurring at the fibre surface during the phosphorylation reaction, and to evaluate the phosphorus content and its distribution on the surface of treated fibres.

2.9. FT-IR Spectroscopy

Transmission mode FT-IR spectra were recorded to identify the functional groups of the fibre composing our panel. The analysis was performed using a Thermo Scientific is10 FTIR spectrometer, (Thermo Fisher Scientific, Waltham, MA, USA) over a spectral range of 4000 to 400 cm^{-1}, with a resolution of 4 cm^{-1} and 32 scans per sample.

2.10. Thermal Conductivity

The thermal conductivity of the textile waste panels was assessed with a MTPS effusivity tester (C-Therm, Fredericton, NB, Canada), following the standard testing method ASTM D7984-21 [29]. The samples were conditioned at 20 °C and 65% humidity for at least 24 h prior to analysis. The tests were conducted in the same environment. Pressure of 50 kPa, the maximum value recommended by the ASTM D7984-21 testing method, was applied to the samples using a pressure foot to ensure good contact between the sample and the temperature sensor. The result was expressed as the average of six replicates.

2.11. Open Porosity and Bulk Density

Open porosity is defined as the fraction of volume that is occupied by the fluid in the interconnected porous network. The bulk density is the in-vacuum density of the porous aggregate. Both properties are measured using a bulk volume of 196 cm^3 with a Porosity/Density meter (Mecanum Inc., Sherbrooke, QC, Canada) according to the procedure outlined in [30]. The average of three measurements on each panel was reported for subsequent interpretation.

2.12. Tortuosity

Tortuosity is here defined as the square of the ratio of effective acoustic path length over direct path length through the porous medium. It was assessed using an ultrasonic Transmission Tortuosity Meter (Mecanum Inc.), according to a previously developed procedure [31]. The testing frequency range spanned from 100 kHz to 1000 kHz, with a sampling interval of 50 kHz. The average of three measurements on each panel was reported for subsequent interpretation.

2.13. Airflow Resistivity and Permeability

Airflow resistivity (R) was determined using the Airflow Resistance Meter (Mecanum Inc.) according to the ISO 9053-91 standard [32]. A panel sample with a 100 mm diameter was inserted into the instrument channel, and the edges were sealed with petroleum jelly. The average of three measurements on each panel was reported for subsequent interpretation. Note that the corresponding airflow permeability (P) of the material is obtained from the relation $P = \mu/R$, where μ is the dynamic viscosity of air.

2.14. Viscous and Thermal Characteristic Lengths

The viscous and thermal characteristic lengths are the average macroscopic dimensions of the cells related to the viscous and thermal losses, respectively, of acoustic waves propagating in the material. The former may be seen as an average radius of the smaller pore channels, and the latter as the average radius of the larger pore channels. Together with open porosity, tortuosity, and airflow resistivity, they are used in acoustical models to predict the sound absorption coefficient and the sound transmission loss of open-cell porous media [33]. The identification of these two lengths was performed using the inverse procedure described in [34]. The average of five samples of 44.44 mm in diameter was reported for subsequent interpretation.

2.15. Resistance to Ignition

The flammability of the textile waste panels was assessed using a vertical flammability test setup following EN ISO 11952-2 [35]. A propane gas flame with a height of 20 mm was applied on the specimen surface at a 45-degree angle for 15 s. Following the removal of the flame, the panels were inspected to determine the length of flame spread from the point of flame application. Based on EN 13501-1 [36], if the flame spread is less than 150 mm, the sample is categorized as class E material.

2.16. Tensile Strength Properties

Paper handsheets made from 100% phosphorylated fibres, 100% textile waste, and 55/45% mix of phosphorylated fibres and textile waste were prepared according to the TAPPI T 205 method "Forming handsheets for physical tests of pulp". Two major changes were made to adapt the method to our specific requirements: (1) The basis weight of paper was set at 500 g/m^2. (2) The drying was carried out at 110 °C in a laboratory dryer for paper.

The samples of paper were cut and tested to tensile strength according to the TAPPI T 220 method "Physical testing of pulp handsheets". The samples were preconditioned for 24 h at a temperature of 23 °C and a relative humidity of 50% before analysis. 10 tests for each sample were performed and the result is expressed as their average. The tensile index was measured with an Instron 4201 (Instron, Norwood, MA, USA) (TAPPI/ANSI T 494 om-22) [37].

3. Results and Discussion

3.1. Characterisation of Fibres

The fibre length, width and fines content are shown in Table 1. The PKF are on average at least twice as long as the textile waste microfibres. The PKF is produced from softwood kraft pulp fibres, so a length between 1 and 3 mm was expected [38]. The average diameter of PKF fibre is also significantly larger, almost 30% higher than the microfibres from textile waste. As expected, the fines content of PKF is very low. The fibre is also associated with a narrow size distribution and long and thick fibres. On the other hand, the microfibres from textile waste are characterised by a high fines content and short fibres. From a papermaking perspective, mixing the long and thick fibres of PKF with the short and thin textile waste microfibres creates an opportunity to develop uniform and dense panel structures, with the short and thin textile waste fibres filling the voids between the longer PKF. However, as papermaking is a wet-web formation process, a high amount of fines might increase the risk of fines accumulation in the process water. In this case, the high fines content in the textile waste microfibres did not present a significant problem as most of the fines were physically retained into the panel structure; a closed loop process was made possible, where more than 80% of the process water was reused.

Table 1 also shows the electrostatic properties of the fibres in water. Both types of fibres were electronegatively charged. The PKF fibres exhibited a surface charge density at least 10 times higher than the textile waste microfibres. This can be attributed to the high density of grafted phosphate moieties on the surface of PKF that dissociate and ionize very easily in water. Despite the large difference in charge density between the PKF and the textile waste fibres, the zeta potential of the PKF fibres was only slightly more negative than that of the textile waste microfibres. The charge on the surface of a particle is concentrated in two physical zones [39]: a compact or fixed layer and the diffuse layer. The Zeta potential is the measurement of the potential at the slippage plan that separates these two layers [40]. A high Zeta potential value means the charges are concentrated in the diffuse layer while a low value means they tend to accumulate in the fixed layer. In the case of the PKF fibres, the high density of charges on the surface and relatively low Zeta potential indicate that the charges tend to accumulate in the fixed layer. On the other hand, the fact that the textile waste microfibres are characterised by low values of both surface charge density and Zeta potential points to surface charges mainly concentrating in the diffuse layer. The negative Zeta potential values measured for both types of fibres suggest a certain degree of electrostatic repulsion between the fibres suspended in water, with a greater stability and a lower tendency to aggregate in the case of PKF. In papermaking, dissolved and colloidal substances tend to accumulate as anionic contaminants in the process water. To monitor and semi-quantitatively assess this "anionic trash", papermakers measure the Zeta potential and the cationic charge demand of the white water. A high cationic charge demand coupled with a strong negative Zeta potential signals a significant buildup of anionic trash, indicating an important contamination of the white water. In addition, the low values of Zeta potential and surface charge density of the textile waste microfibres suggest minimal dissociation and solubilization in water. As the electrostatic properties of fibres are caused by the dissociation or solubilization of some substances/fractions [41], this is a clear indication the microfibres from textile waste are relatively clean and pose a low risk to water contamination despite their heterogenous composition of synthetic and natural fibres. In textile waste microfibres, the nitrogen (N) content is a clear indicator of a heterogeneous composition, suggesting the presence of various fibre types, including synthetic, cellulosic, wool, hair and even certain dyes. This diverse makeup is reflected in the relatively high nitrogen content of about 1.6%, as measured by the Kjeldahl method. In specialty fibres (SF), nitrogen is a fundamental component of their chemical structure, often combined with grafted phosphate groups. This combination of phosphorus (P) and nitrogen in SF creates a synergistic effect that significantly enhances their fire-retardant properties.

Table 1. Characteristics of fibres.

	Fibre Size			Electrostatic Properties in Water		Composition		
	Length (mm)	Width (μm)	Fines (%)	Surface Charge (μeq/g)	Zeta Potential (mV)	Kjeldahl Nitrogen (%)	Synthetic/ Cellulosic (%)	Ash at 525 °C (%)
PKF	1.99 ± 0.03	31.9 ± 0.6	0.2	350 ± 15	−55 to −30	5.3	0/100	45 *
Textile waste fibres	0.81 ± 0.02	24.6 ± 0.1	5.6	30 ± 5	−25 to −10	1.6	33/67	0.8

* Char yield due to fire retardant properties of PKF fibres.

3.2. *Properties of Panels Made with Textile Waste Microfibres*

3.2.1. Structural Properties

The panels made from the textile waste microfibres and PKF fibres (Figure 3) showed good rigidity when assessed qualitatively and were evenly colored despite their heterogeneous composition. Table 2 shows that the prototype panels reached typical values for building insulation panels of mass per unit area (about 1800 g/m^2) and thickness (about 10 mm). However, the standard deviation for these parameters was quite high. This can be attributed to the manual nature of the manufacturing method used. The structures were lightweight, with a bulk density of about 180 kg/m^3, and were highly permeable, with about 86% open voids. The pores were quite small, about 5 μm in diameter. Additionally, the path that the fibres follow within the panels was quite convoluted, with a tortuosity of 2.

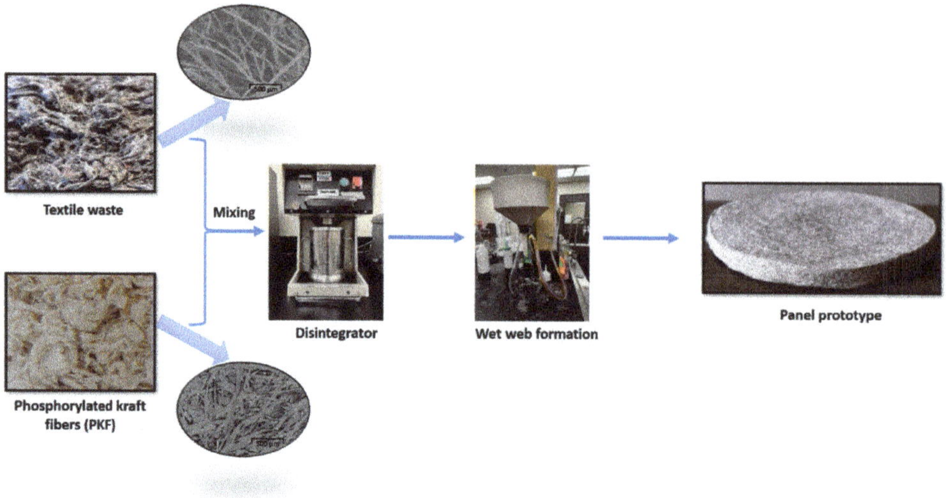

Figure 3. Manufacturing process of panels from textile waste microfibres.

Table 2. Physical characteristics of panels.

Criteria	Value
Composition, % textile waste microfibres/% PKF	45/55
Mass per unit area, g/m^2	1864 ± 103
Thickness, mm	9.62 ± 1.02
Density, kg/m^3	183.3 ± 4.9
Open porosity, %	86 ± 1
Viscous characteristic length, μm	5.4 ± 0.3
Tortuosity	2 ± 1
Thermal conductivity, W/m∗K	0.047 ± 0.003
Thermal characteristic length, μm	60.8 ± 4.8
Airflow resistivity at 0.5 mm/s, N∗s/m^4	896,260 ± 57,266

The panels made with the blend of textile waste microfibres and PKF fibres exhibited a thermal conductivity of 0.047 W/m∗K, which is almost 50% higher than the values reported in the literature for similar structures made from wood fibres and/or synthetic fibres [42]. This can be attributed to the application of 50 kPa of pressure during the test, which was required to ensure appropriate contact between the MTPS sensor and the surface of the panel. This applied pressure compressed the panel structure and removed the air from the panel voids, thus reducing the overall thermal insulation of the panel. As a result, the value recorded for the thermal conductivity corresponds to the constitutive materials of the panel rather than the structure of the panel itself [43].

3.2.2. Fibre Morphology—SEM/EDX

The morphology of the fibres inside the prototype panel was compared to that of the constitutive fibres in their original condition using SEM (Figure 4). In Figure 4a, the textile waste microfibres appear as short and thin flattened strands with a smooth non-fibrillated surface. The PKF fibres were observed to be longer and thicker than the textile waste microfibres from Figure 4b, in agreement with the length and diameter values reported in Table 1. Their surface was also smooth without a significant presence of fibrils. Figure 4c depicts the microstructure of the composite panel. The fibres here are visibly more twisted and heterogenous in size, with a mix of long and short units. Despite the large polydispersity in fibre sizes, the resulting panel shows a homogenous texture with a relatively uniform distribution of fibres and voids throughout the structure. This may be caused by the entanglement of long and short fibres during the panel formation process. Overall, the SEM images reveal slight differences in the morphology and structure of the two types of fibres. These differences allow for a good mixing and interconnection of fibres into a more homogenous panel structure.

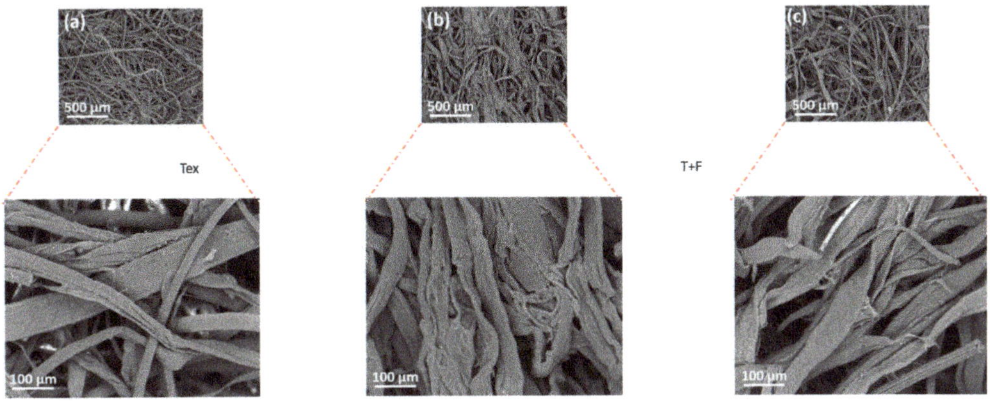

Figure 4. SEM images of (**a**) textile waste microfibres, (**b**) PKF and (**c**) Panel made from a mixture of textile waste microfibres and PKF.

The EDX results are presented in Table 3 below.

Table 3. Elemental distribution (atomic weight) as determined by EDX.

Samples	%C	%O	%N	%P
Phosphorylated fibres (PKF)	43.5	44.3	7.8	4.4
Textile waste	65.8	33.4	0.5	N.D *
Mixture of textile waste microfibres and PKF (45/55%)	53.9	40	3.1	2.7

* Not detected.

The PKF sample contains significant amounts of oxygen (44.3%), phosphorus (4.4%), and nitrogen (7.8%). The presence of phosphorus in such quantities is expected, as the PKF sample is obtained through grafting of phosphate moieties on cellulosic fibres. The presence of nitrogen is also beneficial, as it works synergistically with phosphorus to induce flame-resistant properties in the fibres. The high content of phosphorus and nitrogen in PKF confirms that the phosphorylation treatment was effective. The textile waste sample, on the other hand, has a high carbon content and a residual nitrogen content, which is consistent with its composition. Textiles are typically made from carbon-rich organic polymers, such as polyester and cotton, which was further confirmed through Cuen analysis. The low nitrogen content in the textile waste suggests the presence of some contaminants such as wool, hair, or certain nitrogen-based dyes like azo dyes. A low nitrogen content in the textile waste has also been confirmed through Kjeldahl analysis. As expected, the value distribution of elements in the panel prototype made from the mixture of textile waste and PKF lies between those of textile waste and the PKF samples. The carbon content (53.9%) is lower than that of the textile waste but higher than PKF, according to the composition of the mixture (45/55%, w/w). The presence of nitrogen and phosphorus in larger quantities than in textile waste but smaller than in the PKF sample suggests a potential fireproof behavior induced in the panel prototypes.

3.2.3. FTIR Spectroscopy

Figure 5 shows the FT-IR spectra of four samples: cellulosic fibres (as standard), polyester fibres (as standard), and textile waste before and after Cuen treatment.

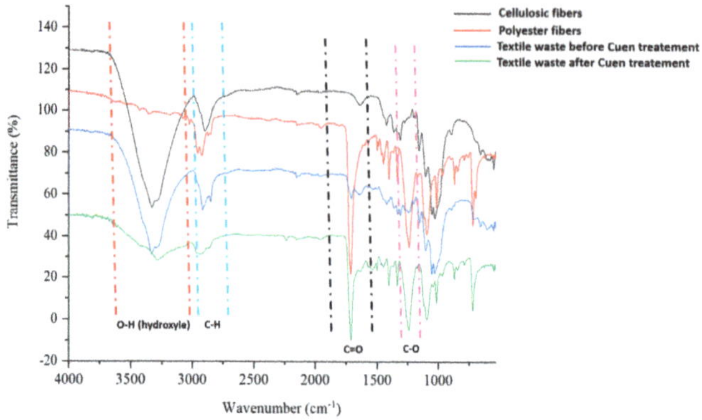

Figure 5. FT-IR spectra of cellulosic fibres, polyester fibres, and textile waste before and after Cuen treatment.

The FT-IR spectra reveal several characteristic peaks. There is a broad, strong peak around 3500–3300 cm^{-1} corresponding to hydroxyl (O–H) groups, which are prevalent in cellulosic fibres and textile waste before Cuen treatment, but it is missing in polyester fibres and vanishes almost completely in the sample of textile waste after Cuen treatment. An important peak is observed in the 1300–1000 cm^{-1} region due to the C–O stretching vibrations. This peak is strong in intensity in case of ester bonds present in polyester fibres and textile waste sample after Cuen treatment but rather weak in intensity in case of cellulose fibres and textile waste sample before Cuen treatment. In synthetic polymers, like polyester fibres and textile waste sample after Cuen treatment, strong peaks appear around 2900 cm^{-1} (C–H stretch) and 1700 cm^{-1} (C = O stretch). These peaks are weak or undefined in case of spectra of cellulose fibres and textile waste before Cuen treatment. In conclusion, Cuen treatment of textile waste causes an almost complete disappearance of O–H stretching peak due to cellulose dissolution and a huge increase of C = O stretching peak due to polyester fibre isolation. The treatment with Cuen solution proves to be effective in determining the composition of the textile waste.

3.2.4. Acoustical Properties

Both the sound absorption coefficient and the sound transmission loss of the manufactured panels were assessed. Figure 6 shows the acoustic performance of a panel of 9.62 mm mean thickness in terms of sound absorption and transmission loss at normal incidence over the 100–4300 Hz frequency range. Figure 6a shows the sound absorption coefficient when the panel was placed directly on a hard wall. The absorption coefficient increased from 0.2 to 0.5 when the frequency increased from 500 to 4300 Hz. The curve shows a peak between 2000 and 3000 Hz which is most likely due to a first plate-like resonance (circular plate clamped on its edge). In fact, when a test specimen of 44.44 mm in diameter is installed in the tube of the same size, the axial motion on its edge is limited due to friction. To proof this, a simulation on a 44.44-mm sample with bonded edge (dashed line in Figure 6a), using a poroelastic model [33], provided a similar profile with a peak located around 2000 Hz. In this case, the elastic properties were inversely identified to best fit the results. On the other hand, this peak is not visible in the case of a simulation with a large-size sample (continuous line in Figure 6a). Overall, the panel prototype has low performance in terms of sound absorption (less than 50% of sound absorption). This low performance is attributed to the low airflow permeability (0.2×10^{-10} m^2—see Table 2) due the high compaction of the material. As a result, the panel produced a shielding phenomenon for the sound. The same type of behavior was obtained when a 20-mm air cavity was present between the panel and the hard wall (Figure 6b). Here, the absorption peak is more pronounced, suggesting an even stronger resonance effect likely caused by the air cavity. Figure 6c illustrates the sound transmission loss in decibels (dB) of the panel prototype. The value of the sound transmission loss is close to 20 dB for the entire frequency range tested. As expected, the resonance effect is also present in the sound transmission loss curve and is attributed to the size of the tested specimen.

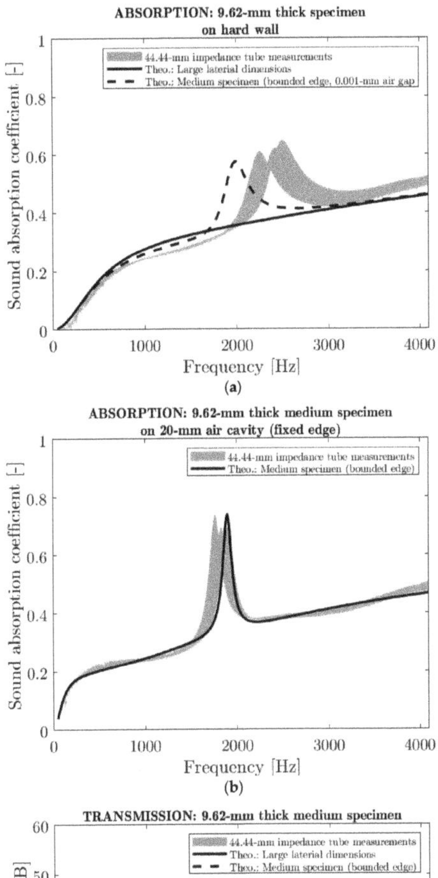

Figure 6. Sound absorption coefficient and transmission loss of panel prototypes (**a**) Normal incidence sound absorption coefficient on hard wall. (**b**) Normal incidence sound absorption coefficient on 20-mm air cavity backed by hard wall. (**c**) Normal incidence sound transmission.

In summary, the prototype panel excelled in sound insulation but not in sound absorption due to its low airflow permeability. This low airflow permeability is a consequence of the high compression rate of the material that results in small pore channels, which are restrictive to airflow. A lower compression rate would improve sound absorption at the expense of sound insulation. The impact of the compression rate on the physical and acoustical properties of a fibre assembly is discussed in detail elsewhere [44].

3.2.5. Ignitability

The standard specification EN 13501-1 [36] assigns construction products and floor coverings to a specific class based on their propensity to ignite, assessed according to the EN ISO 11925-2 ignitability test method [35]. This test method was used here to evaluate the fire resistance of the panels made from a blend of textile waste microfibres and PKF using the test conditions for class E products, i.e., with a flame exposure of 15 s. Two parameters were assessed: the extent of flame spread during the test and the ability of the panel to maintain its structural integrity.

The tendency of the panels to ignite depends on several factors, but the most important is the chemical composition of the material. In the case of these panels, the textile waste microfibres contain highly combustible materials like cellulose and polyester, while PKF is a fire-resistant phosphorylated fibre. The Cuen analysis revealed that 66% of the textile waste microfibres were cellulose-based; the remaining 33% were most likely polyester. Studies have shown that polyester is less flammable than cotton (cellulose) [45], but cotton/polyester blends can generate more heat and burn faster than fabrics composed solely of cotton [46]. Preliminary trials showed that specimen panels made solely from textile waste microfibres burned and smoked continuously after the 15 s of flame exposure. On the other hand, when 55% PKF fibres were blended with the textile waste microfibres to manufacture the prototype panel, satisfactory results were obtained. When exposed to direct flame, the flame did not spread; it self-extinguished instantly after the withdrawal of the applied flame. Moreover, the panel maintained its structural integrity (Figure 7), the smoke was suppressed in less than a minute and no droplets or incandescent particles fell from the sample. The flame spread was 115 ± 5 mm, significantly less than the 150 mm threshold, indicating that such panels could meet the fire-resistance requirements of class E materials according to EN 13501-1 [36].

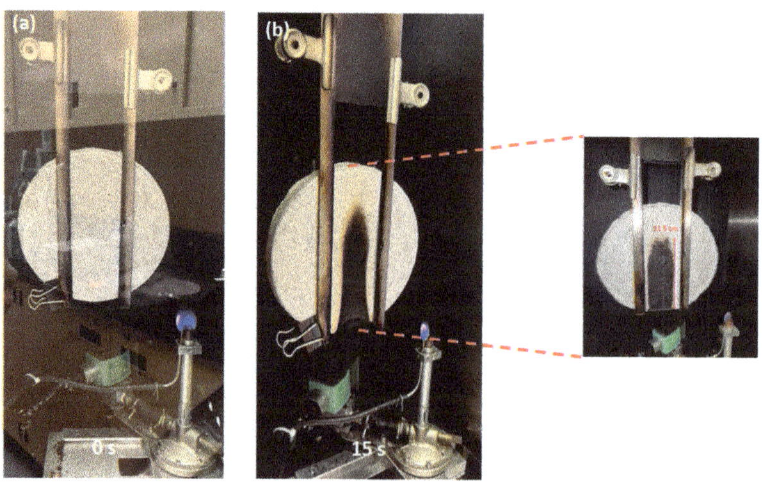

Figure 7. Photographs of prototype panel during the ignitability test (**a**) before and (**b**) after flame exposure.

3.2.6. Tensile Strength Properties

The tensile strength results are shown in Figure 8 and Table 4. As can be easily observed, the tensile strength of paper samples made from 100% phosphorylated fibres is at least one order of magnitude higher than that obtained from 100% textile waste. This can be attributed to the capacity of phosphorylated fibres to create strong bonds in the paper structure due to hydrogen bridges, electrostatic interactions, and even covalent bonds through the crosslinking potential of phosphate moiety. On the other hand, the microfibres from textile waste have an ultralow bonding potential.

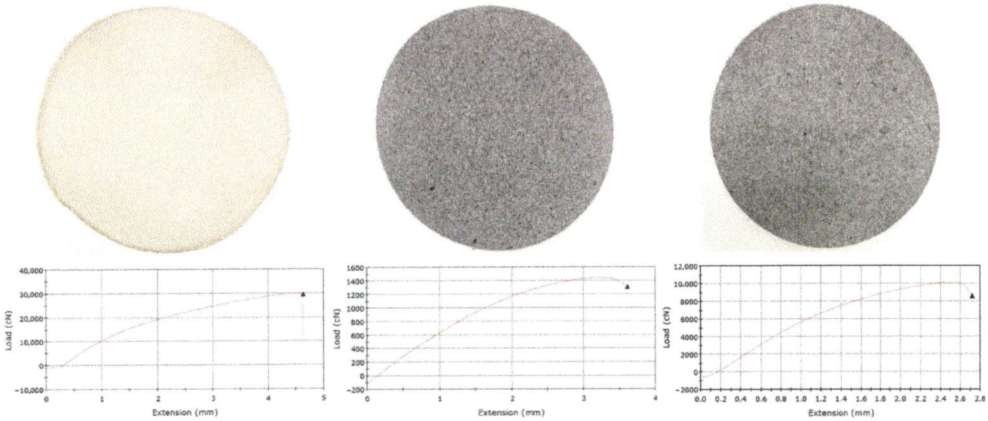

Figure 8. Paper samples and the corresponding stress–strain curves for 100% phosphorylated fibres (**left**), 100% textile waste (**middle**), and 55/45% mix of phosphorylated fibres and textile waste (**right**).

Table 4. Tensile strength properties of papers made from phosphorylated fibres and textile waste.

Sample	Basis Weight, g/m^2	Thickness, mm	Load at Break, N	Energy at Break, J	Tensile Strain at Break, %	Tensile Stress at Break, MPa	E-Modulus, MPa
Phosphorylated fibres (100%)	500 ± 10	1.15 ± 0.02	292.3 ± 14.6	0.95 ± 0.04	4.67 ± 0.2	26.38 ± 1.31	922.3 ± 46.1
Textile waste (100%)	500 ± 10	2.81 ± 0.11	12.4 ± 0.9	0.03 ± 0.002	3.27 ± 0.21	0.44 ± 0.03	29.2 ± 2.3
Mix of phosphorylated fibres and textile waste (55/45%)	500 ± 10	2.26 ± 0.06	81.1 ± 4.8	0.19 ± 0.01	3.18 ± 0.18	3.59 ± 0.28	198.8 ± 15.9

The low values recorded for tensile strength suggested a fibre–fibre interaction in the 100% textile waste sample exclusively based on a mechanical interlocking mechanism. Obviously, the paper samples made from a mixture of phosphorylated fibres and textile waste microfibres show tensile strength that places them between the two extremes. In conclusion, the phosphorylated fibres not only confer fire resistance to the panels, but they also bring mechanical resistance to these composite structures.

4. Conclusions

The study successfully demonstrates the transformation of microfibres derived from textile waste into multifunctional panels suitable for the construction industry. The proposed method is straightforward and uses technological processes commonly found in the papermaking industry, and provides a very promising alternative to approaches based on complex mechanical or chemical sorting of waste materials. The incorporation of flame-resistant PKF fibres into the panel composition not only imparts fire-resistant properties but also gives stiffness for structural integrity without compromising the lightweight characteristics. These fibres also improve the mechanical properties of the panels, such as load at break, strain at break, stress at break and E-modulus, due to strong hydrogen bonds and electrostatic interactions.

This simple, efficient and low-cost approach makes the method particularly appealing for rapid and large-scale implementation, contributing to environmental protection and more sustainable management of textile resources. Furthermore, the study underscores its significance within a circular-based economy by providing a viable solution for upcycling textile waste into value added products. The study also demonstrates the true power of cooperation. The innovation described here was made possible by bringing together scientists and engineers from various disciplines including materials, textile and papermaking. Finally, this study bolsters the commitment to more ecological and sustainable practices, which are crucial for addressing current and future environmental challenges.

Author Contributions: Data curation: L.A. and H.L.; Formal analysis: H.L., P.D., L.A. and H.L.; Funding acquisition: D.B.; Investigation: F.B. and D.B.; Methodology: H.L. and D.B.; Project administration: D.B.; Resources: L.A. and H.L.; Software: H.L. and L.A.; Supervision: F.B.; Validation: H.L., R.P., C.F., P.D., F.B. and D.B.; Writing—original draft: H.L.; Writing—review and editing: H.L., L.A., P.D., D.B., F.B., H.L., R.P. and C.F. All authors have read and agreed to the published version of the manuscript.

Funding: This work was supported by the Lecturers Research Scholarship Award 06042023 granted to Dan Belosinschi by Quebec University at Trois-Rivieres Foundation and UQTR Union of Lecturers (3351, boul. des Forges, C.P. 500, Trois-Rivières, G8Z 4M3, Québec, Canada, Foundation registration number: 119001725 RR0002).

Institutional Review Board Statement: Not applicable.

Data Availability Statement: The original contributions presented in the study are included in the article, further inquiries can be directed to the corresponding authors.

Acknowledgments: The authors wish to thank Rashedul Islam (University of Alberta) for performing the thermal conductivity measurements.

Conflicts of Interest: Author Hamid Lamoudan was employed by the company CTT Group. The remaining authors declare that the research was conducted in the absence of any commercial or financial relationships that could be construed as a potential conflict of interest.

References

1. Wang, C.; Su, J.; Liu, T.; Ge, S.; Liew, R.K.; Zhang, H.; Naushad, M.; Lam, S.S.; Ng, H.S.; Sonne, C.; et al. A sustainable strategy to transform cotton waste into renewable cellulose fiber self-reinforcing composite paper. *J. Clean. Prod.* **2023**, *429*, 139567. [CrossRef]
2. Mordor Intelligence. Apparel Market (2020–2025) | Growth | Trends | Forecasts. Available online: https://www.mordorintelligence.com/industry-reports/apparel-market (accessed on 15 June 2024).
3. Global Apparel Industry Statistics Statistics: Market Data Report 2024. Available online: https://www.gitnux.org/global-apparel-industry-statistics/ (accessed on 30 September 2024).
4. Oberlo. Apparel Industry Statistics (2012–2025) | Oberlo. Available online: https://www.oberlo.com/statistics/apparel-industry-statistics (accessed on 15 June 2024).
5. Igini, M. 10 concerning Fast Fashion Waste Statistics. Available online: https://earth.org/statistics-about-fast-fashion-waste/ (accessed on 20 September 2024).
6. Huang, X.; Tan, Y.; Huang, J.; Zhu, G.; Yin, R.; Tao, X.; Tian, X. Industrialization of Open-and Closed-Loop Waste Textiles Recycling towards Sustainability: A Review. *J. Clean. Prod.* **2024**, *436*, 140676. [CrossRef]
7. Ütebay, B.; Çelik, P.; Çay, A. Textile Wastes: Status and Perspectives. In *Waste in Textile and Leather Sectors*; IntechOpen Limited: Londan, UK, 2020. [CrossRef]
8. Ruiz, A. 17 Most Worrying Textile Waste Statistics & Facts. Available online: https://theroundup.org/textile-waste-statistics/ (accessed on 30 September 2024).
9. Bartl, A. *End-of-Life Textiles*; Waste: Somerset, UK, 2019; pp. 323–336. [CrossRef]
10. Teixeira França Alves, P.H.; Clarke-Sather, A.; Carlson, S.; Martini, A. *Theoretical Method for Characterizing Textile Failure Mechanics in Mechanical Recycling with Carded Drums*; American Society of Mechanical Engineers: New York, NY, USA, 2023. [CrossRef]
11. Ribul, M.; Lanot, A.; Tommencioni Pisapia, C.; Purnell, P.; McQueen-Mason, S.J.; Baurley, S. Mechanical, Chemical, Biological: Moving towards Closed-Loop Bio-Based Recycling in a Circular Economy of Sustainable Textiles. *J. Clean. Prod.* **2021**, *326*, 129325. [CrossRef]

12. Lawson, L.; Degenstein, L.M.; Bates, B.; Chute, W.; King, D.; Dolez, P.I. Cellulose Textiles from Hemp Biomass: Opportunities and Challenges. Sustainability. *Sustainability* **2022**, *14*, 15337. [CrossRef]
13. Arafat, Y.; Uddin, A.J. Recycled Fibers from Pre- and Post-Consumer Textile Waste as Blend Constituents in Manufacturing 100% Cotton Yarns in Ring Spinning: A Sustainable and Eco-Friendly Approach. *Heliyon* **2022**, *8*, e11275. [CrossRef]
14. Subramanian, K.; Chopra, S.S.; Cakin, E.; Li, X.; Lin, C.S.K. Environmental Life Cycle Assessment of Textile Bio-Recycling—Valorizing Cotton-Polyester Textile Waste to Pet Fiber and Glucose Syrup. *Resour. Conserv. Recycl.* **2020**, *161*, 104989. [CrossRef]
15. Wojnowska-Baryła, I.; Bernat, K.; Zaborowska, M. Strategies of Recovery and Organic Recycling Used in Textile Waste Management. *Int. J. Environ. Res. Public Health* **2022**, *19*, 5859. [CrossRef]
16. Thomas, J.; Patil, R.S.; Patil, M.; John, J. Navigating the Labyrinth of Polymer Sustainability in the Context of Carbon Footprint. *Coatings*. **2024**, *14*, 774. [CrossRef]
17. Eppinger, E. Recycling Technologies for Enabling Sustainability Transitions of the Fashion Industry: Status Quo and Avenues for Increasing Post-Consumer Waste Recycling. *Sustain.Sci. Pract. Policy* **2022**, *18*, 114–128. [CrossRef]
18. Harmsen, P.; Scheffer, M.; Bos, H. Textiles for Circular Fashion: The Logic behind Recycling Options. *Sustainability* **2021**, *13*, 9714. [CrossRef]
19. Costa, C.; Viana, A.; Silva, C.; Marques, E.F.; Azoia, N.G. Recycling of Textile Wastes, by Acid Hydrolysis, into New Cellulosic Raw Materials. *Waste Manag.* **2022**, *153*, 99–109. [CrossRef] [PubMed]
20. Sanchis-Sebastiá, M.; Ruuth, E.; Stigsson, L.; Galbe, M.; Wallberg, O. Novel Sustainable Alternatives for the Fashion Industry: A Method of Chemically Recycling Waste Textiles via Acid Hydrolysis. *Waste Manag.* **2021**, *121*, 248–254. [CrossRef] [PubMed]
21. Palme, A.; Peterson, A.; de la Motte, H.; Theliander, H.; Brelid, H. Development of an Efficient Route for Combined Recycling of PET and Cotton from Mixed Fabrics. *Text. Cloth. Sustain.* **2017**, *3*, 4. [CrossRef]
22. Piribauer, B.; Bartl, A. Textile Recycling Processes, State of the Art and Current Developments: A Mini Review. *Waste Manag. Res.* **2019**, *37*, 112–119. [CrossRef]
23. Clarke, C.J.; Tu, W.-C.; Levers, O.; Bröhl, A.; Hallett, J.P. Green and Sustainable Solvents in Chemical Processes. *Chem. Rev.* **2018**, *118*, 747–800. [CrossRef] [PubMed]
24. Lamoudan, H.; Brouillette, F. fabrication of high strength paper from different types of phosphorylated fibers using hot pressing and forming agents. *Cellul. Chem. Technol.* **2024**, *58*, 313–322. [CrossRef]
25. Shi, Y.; Belosinschi, D.; Brouillette, F.; Belfkira, A.; Chabot, B. The Properties of Phosphorylated Kraft Fibers. *BioResources* **2015**, *10*, 4375–4390. [CrossRef]
26. Bobu, E.; Belosinschi, D. Effects of coagulants on the dcs accumulation in process water of papermaking. *Environ. Eng. Manag. J.* **2008**, *7*, 269–276. [CrossRef]
27. Lamoudan, H.; Abenghal, L.; Brouillette, F. Improving the Mechanical Strength of Paper Sheets Made from Phosphorylated Fibers through the Use of Forming Agents. *J. Nat. Fibers* **2024**, *21*, 2332705. [CrossRef]
28. Bradstreet, R.B. Kjeldahl Method for Organic Nitrogen. *Anal. Chem.* **1954**, *26*, 185–187. [CrossRef]
29. *ASTM D7984-21*; Standard Test Method for Measurement of Thermal Effusivity of Fabrics Using a Modified Transient Plane Source (MTPS) Instrument. ASTM: West Conshohocken, PA, USA, 2021.
30. Salissou, Y.; Panneton, R. Pressure/Mass Method to Measure Open Porosity of Porous Solids. *J. Appl. Phys.* **2007**, *101*, 124913. [CrossRef]
31. Allard, J.F.; Castagnède, B.; Henry, M.; Lauriks, W. Evaluation of Tortuosity in Acoustic Porous Materials Saturated by Air. *Rev. Sci. Instrum.* **1994**, *65*, 754–755. [CrossRef]
32. *ISO 9053-1*; Acoustics—Determination of Airflow Resistance. Part 1: Static Airflow Method. International Organization for Standardization: Geneva, Switzerland, 2018.
33. Allard, J.F.; Atalla, N. *Propagation of Sound in Porous Media*; John Wiley and Sons: Hoboken, NJ, USA, 2009. [CrossRef]
34. Verdiere, K.; Panneton, R.; Atalla, N.; Elkoun, S. *Inverse Poroelastic Characterization of Open-Cell Porous Materials Using an Impedance Tube*; SAE Technical Paper Series; SAE International: Warrendale, PA, USA, 2017. [CrossRef]
35. *CEN ISO 11925-2*; Reaction to Fire Tests. Ignitability of Products Subjected to Direct Impingement of Flame. Single-Flame Source Test. European Committee for Standardization: Brussels, Belgium; International Organization for Standardization: Geneva, Switzerland, 2010.
36. *CEN 13501-1*; Fire Classification of Construction Products and Building Elements. Classification Using Test Data from Reaction to Fire Tests. Standard European Committee for Standardization: Brussels, Belgium, 2007.
37. *TAPPI/ANSI T 494*; Tensile Properties of Paper and Paperboard (Using Constant Rate of Elongation Apparatus). TAPPI Press: Atlanta, GA, USA, 2022.
38. Smook, G.A. Chapter 2—Characteristics of Wood and Wood Pulp Fibers. In *Handbook for Pulp & Paper Technologists*, 3rd ed.; TAPPI PRESS: Atlanta, GA, USA, 2002.
39. Behrens, S.H.; Borkovec, M. Electrostatic Interaction of Colloidal Surfaces with Variable Charge. *J. Phys. Chem. B* **1999**, *103*, 2918–2928. [CrossRef]
40. Hunter, R.J. *Zeta Potential in Colloid Science: Principles and Applications*; Academic Press: London, UK, 1981.

41. Lindman, B.; Medronho, B.; Alves, L.; Costa, C.; Edlund, H.; Norgren, M. The Relevance of Structural Features of Cellulose and Its Interactions to Dissolution, Regeneration, Gelation and Plasticization Phenomena. *Phys. Chem. Chem. Phys.* **2017**, *19*, 23704–23718. [CrossRef]
42. Belosinschi, D.; Brouillette, F.; Shi, Y.; Paradis, J.; Doucet, J. Phosphorylated Lignocellulosic Fibers, Uses and Processes of Preparation Thereof. International Patent Application WO 2017/214719A1, 21 December 2017.
43. Islam, R.; Golovin, K.; Dolez, P.I. Thermal Effusivity Assessment of Sportswear Fabrics in the Dry State: Stacked and Air-Hoop Methods. *J. Test. Eval.* **2024**, *52*, 2468–2482. [CrossRef]
44. Tran, Q.V.; Perrot, C.; Panneton, R.; Hoang, M.T.; Dejaeger, L.; Marcel, V.; Jouve, M. Effect of Polydispersity on the Transport and Sound Absorbing Properties of Three-Dimensional Random Fibrous Structures. *Int. J. Solids Struct.* **2024**, *296*, 112840. [CrossRef]
45. Dipietro, J.; Stepniczka, H.; Nametz, R.C. A Study of Flammability of Cotton, Polyester, and Their Blends. *Text. Res. J.* **1971**, *41*, 593–599. [CrossRef]
46. Jarvis, C.W.; Barker, R.H. Flammability of Cotton-Polyester Blend Fabrics. In *Flame—Retardant Polymeric Materials*; Lewin, M., Atlas, S.M., Pearce, E.M., Eds.; Springer: Boston, MA, USA, 1978. [CrossRef]

polymers

MDPI

Article

Modified Cellulose-Based Waste for Enhanced Adsorption of Selected Heavy Metals from Wastewater

Katarina Trivunac [1,*], Snežana Mihajlović [1], Marija Vukčević [1], Marina Maletić [2], Biljana Pejić [1,3], Ana Kalijadis [4] and Aleksandra Perić Grujić [1]

1 Faculty of Technology and Metallurgy, University of Belgrade, Karnegijeva 4, 11000 Belgrade, Serbia; marijab@tmf.bg.ac.rs (M.V.); biljanap@tmf.bg.ac.rs (B.P.); alexp@tmf.bg.ac.rs (A.P.G.)
2 Innovation Center, Faculty of Technology and Metallurgy, Karnegijeva 4, 11000 Belgrade, Serbia; mvukasinovic@tmf.bg.ac.rs
3 Textile School for Design, Technology and Management, Academy of Technical and Art Applied Studies Belgrade, StarineNovaka 24, 11000 Belgrade, Serbia
4 Department of Materials, "VINCA" Institute of Nuclear Sciences—National Institute of the Republic of Serbia, University of Belgrade, Mike Petrovica Alasa 12-14, 11000 Belgrade, Serbia; ana.kalijadis@vin.bg.ac.rs
* Correspondence: trivunac@tmf.bg.ac.rs

Abstract: Due to industrial growth and its impact on the environment, the increasing amount of industrial waste requires a comprehensive approach aligned with the principles of sustainable development. The main goals are not only to preserve natural resources but also to encourage innovation in the reuse of waste materials. In an attempt to reduce the problems regarding waste disposal and wastewater treatment in the textile industry, fibrous textile waste was used as a starting material to obtain carbon adsorbents for the removal of pollutants from wastewater. Waste cotton and mixed yarns, mainly consisting of polysaccharide cellulose, were hydrothermally carbonized and activated with KOH to convert them into efficient carbon adsorbents for heavy metal removal from water. Characterization of carbonized material showed that after activation, an increase in specific surface area (up to 872 m^2/g) and content of surface oxygen groups (6.04 mmol/g) leads to a higher affinity towards heavy metal ions, especially lead ions, and high adsorption capacity of 19.98 mg/g obtained for activated cotton yarns. The results of this research represent a contribution to the reduction of waste materials by modifying them into adsorbents, while the regeneration of adsorbents is an example of the practical application of polysaccharide-based materials in the purification of wastewater containing various heavy metal ions.

Keywords: waste; polysaccharide-based material; cotton; hydrothermal carbonization; heavy metal; adsorption; wastewater

1. Introduction

During different stages of industrial production (textile, paper, decorative fabrics), various cellulose-based waste materials, in the form of fibers, yarns, and woven, are generated and disposed of, endangering human health or harming the environment. Furthermore, the textile industry generates large amounts of wastewater, polluted with various hazardous and harmful substances [1,2]. The long-term goals of the European Union's waste policy are not only to reduce the amount of waste generated but to promote the use of waste as a potential resource, as well as to achieve a higher level of recycling and safe disposal [3]. Waste policy aims to protect the environment and human health while facilitating the transition to a circular economy. Cellulosic materials represent the most widespread natural, renewable, biodegradable, and low-cost raw material. The main component of these materials is the polysaccharide cellulose, a linear homopolymer consisting of β-D-anhydroglucopyranose repeating units linked by β-1,4-glycosidic bonds [4]. It can be obtained from various sources: higher plants, trees, algae, bacteria, and even some marine animals. Cotton fibers are mainly made of cellulose. Polysaccharide-based materials are widespread and widely used (biomedicine, electronics and printing, food, for the production of packaging, biofuels, sorbents, etc.) [5,6] and, therefore, accumulate in large quantities in the form of residues, unused and waste materials. Considering the forecasts that the global production and consumption of fibers will increase until 2030, an increase in the amount of waste cotton and polyester fibers and their blends must be expected [7]. Textile waste is recognized as a significant polluter of the environment due to the large quantities generated during production and after use. Today's textile and fashion industry implies mass and cheap production using inferior materials, chemicals, and dyes. That is why, in addition to natural materials such as cotton, linen, and hemp, artificial, synthetic materials are also used, which are mixed with natural ones to improve certain properties such as strength, color fastness, and behavior during washing and drying. Polyester fibers are the most popular synthetic textile material due to their physical properties, low cost, and recyclability, but they are not biodegradable, so they stay longer in the environment. Considering that the majority of waste cotton and polyester is mostly burned and disposed of in landfills, it is very important to implement their more massive recycling and reuse in order to reduce negative impacts on water, soil, air, survival of the living world, and human health [8–11]. Materials produced from biomass or industrial waste can be used as fuels [12,13], catalysts [14], flame retardancy [15], composites, raw material for the production of 5′-hydroxymethylfurfural [16], electrocatalytic carbon material [13], etc. Also, its use in the biomedical field and environmental treatment is significant [14,17–20]. One possibility is to use these waste materials as adsorbents for purifying water contaminated with different pollutants [8,21]. In the case of raw waste cellulosic materials, the adsorption processes take place mostly slowly, and they have low selectivity and relatively low adsorption capacities. Therefore, for the application of these materials, it is necessary to perform chemical or thermal modifications to obtain efficient, easily applicable materials with increased affinity for pollutant binding thanks to their specific chemical structure, i.e., the presence of different functional groups. Cotton stalk, cotton fibers, denim, household and industrial textile waste, and cotton cloth can be used as raw materials for the production of cotton-based adsorbents. Niu et al. prepared functional cotton fiber with tetraethylenepentamine and chitosan and investigated their adsorption capacity towards Cu(II), Pb(II), and Cr(III) ions. The maximum adsorption capacity obtained by the Langmuir model was highest for Cu(II) ions. Chemical modification of waste textile fibers was performed by the carboxymethylation process [22]. Obtained materials were used for the adsorption of Cd(II) ions and showed great potential for the removal of heavy metals in comparison to commercial resin. The graft copolymerization of household and industrial textile waste by polyacrylic acid was used to prepare adsorbents for the removal of Pb(II) and Cr(VI) ions from water [23]. The maximum adsorption capacity for Pb(II) ions was ten times higher. Modified cotton-based materials showed an increase in affinity towards pollutants than the initial, unmodified ones. Chemical modification has advantages due to a large number of potential reagents and the possibility of obtaining selective adsorbents, but a major disadvantage is the creation of new waste after the adsorption process and secondary pollution. Also, the procurement and storage of chemicals are not economically profitable. The high carbon content in the composition of waste biomass makes them a good starting material for obtaining carbon adsorbents with a large specific surface area and porous structure [24]. The thermal modification of cotton is most often performed using carbonization with pyrolysis and hydrothermal treatment. Pyrolysis of cotton stalks was performed in the temperature range of 250 to 650 °C to obtain biochar, and it was found that materials obtained at temperatures higher than 450 °C had better adsorption characteristics [25]. Cotton waste was transformed into a carbon microtube by direct pyrolysis at 900, 1100, 1300, and 1500 °C and used for sorption of tannic acid [14]. Recently, research has been directed at hydrothermal modification of cotton from different sources and process conditions for obtaining carbon materials. Through the process of hydrothermal carbonization, the high carbon content of cellulosic materials can be converted into hydrothermal carbon materials without releasing CH_4 and CO_2 into the atmosphere. Hydrothermal carbonization, which takes place at lower temperatures, is a more energy-efficient process than pyrolysis at high temperatures of 600–1000 °C. Besides temperature, which is the main factor in the carbonization process, pH values can have a significant role in the structure and hydrothermal behavior of cellulose molecules. This consumption was investigated by Cui et al. by treatment of cotton fibers in subcritical water [26]. The authors found that pH has an influence on the morphology of the resulting spherical particles, carbon content, and calorific value. In order to improve the properties of the final activated carbon, the cotton was impregnated with urea [27] and then subjected to the processes of pyrolysis at 800 °C and hydrothermal carbonization. After activation with KOH, it was determined that there was an increase in pores and specific surface area, as well as that this material has a high affinity for iodine adsorption up to 2254.8 mg/g.

The concentrations of heavy metals in ground and surface waters are constantly increasing. The presence of most metals, some potentially harmful, in the environment is a consequence of human activity. There are different mechanisms of metals entering the water, washing streets, fields, unorganized landfills, coal and ore mining, and discharge of wastewater from various industrial plants [28,29]. If the measured concentrations of heavy metals are above the maximum allowed and desired limit, the toxic effects are manifested in a way that is harmful to people. The removal of metal ions from water, therefore, becomes inevitable to maintain the balance in the ecosystem. A very effective adsorbent for many pollutants is activated carbon, and much scientific research is directed toward obtaining carbon materials from various natural and waste sources [30–36]. To the best of our knowledge, the application of hydrothermally treated cotton, and specially cotton/polyester, for heavy metal removal in wastewater treatment has not been examined yet. Thus, in this study, hydrothermal carbonization and subsequent activation of waste cotton and cotton/polyester mix and characterization of obtained carbon materials as potential adsorbents of selected heavy metal ions were performed. Special attention is paid to the examination of the influence of further chemical activation on the porosity, specific surface area, the type and amount of surface oxygen groups, as well as the adsorption characteristics towards heavy metal ions of obtained materials.

2. Materials and Methods

2.1. Materials Preparation

Polysaccharide-based waste (cotton, C and mixed cotton/polyester ((50% cotton–50% polyester) yarn, C/P) generated at the factory SIMPO Dekor (Vranje, Serbia) was used as a precursor to obtain carbon materials. Hydrothermal carbonization of the yarn was carried out in a stainless steel autoclave with a Teflon insert at a temperature of 180 °C, under self-generated pressure, for 24 h. The reaction mixture consisted of 6 g of yarn, 40 cm^3 of distilled water, and 0.015 g of citric acid as a catalyst. After hydrothermal carbonization, the solid product was filtered and washed with ethanol and distilled water. Two hydrothermally carbonized materials, C_{HTC} and C/P_{HTC}, were obtained from cotton and mixed cotton/polyester yarn, respectively. Subsequent activation was performed with potassium hydroxide as an activating agent. Carbonized materials were mixed with KOH in the mass ratio 1:2 and heated in an electric furnace under a nitrogen atmosphere, up to 900 °C with a heating rate of 5 °C/min. The obtained samples were washed with distilled water until neutral reaction, dried at 100 °C overnight, and labeled C_{aHTC} and C/P_{aHTC}.

2.2. Material Characterization

The morphological characteristics of the obtained materials were determined by scanning electron microscopy (Mira Tescan 3X, Tescan Orsay Holding, Brno, Czech Republic).

X-ray diffraction (XRD) patterns of all examined materials were recorded by Ultima IV Rigaku diffractometer, equipped with CuKα1,2 radiations. XRD spectra were recorded in the 2θ range of 10–60° (hydrothermally treated samples) and 15–60° (activated samples) in a continuous scan mode with a scanning step size of 0.02° and at a scan rate of 2° min^{-1}.

Nitrogen adsorption and desorption isotherms were measured at the temperature of liquid nitrogen on a Micromeritics ASAP 2020 instrument (Micromeritics N.V./S.A., Brussels, Belgium) for all hydrothermally carbonized and activated samples. Using instrument ASAP 2020 software, the results of specific surface area S_{BET}, external (S_{ext}), and microporous (S_{micro}) surface area, as well as total pore volume (V_{total}) and volume of micropores (V_{micro}), were obtained. Attenuated total reflectance Fourier transforms infrared spectroscopy (ATR-FTIR) was used to qualitatively analyze carbon materials' surface oxygen groups. FTIR spectra were recorded in the 500–4000 cm^{-1} range, using ATR-FTIR Bomem MB-Series, Hartmann Braun (Quebeck, QC, Canada). For comparison, FTIR spectra of starting materials, cotton and cotton/polyester yarn, were also recorded. The amount of surface oxygen groups on examined carbon materials was determined by the acid-base titration method and this experiment was performed in triplicate for each material sample. For determination of the acidic and basic sites, small quantities (0.1 g) of each carbon material were mixed with 10 cm^3 of 0.1 M NaOH and 0.1 M HCl, respectively, in 25 cm^3 beakers. The beakers were sealed and shaken under a nitrogen atmosphere for 24 h. Afterward, filtered solutions were titrated with 0.05 M H_2SO_4 and 0.1 M NaOH to determine the amount of acidic and basic oxygen groups, respectively.

2.3. Adsorption Experiments

Adsorption of lead and cadmium ions on C_{HTC}, C_{aHTC}, C/P_{HTC}, and C/P_{aHTC} was performed in a batch system by immersing 0.025 g of carbon adsorbent sample in 25 cm^3 of adsorbate solution, initial concentration of 20 mg/dm^3. The concentrations of Pb(II) and Cd(II) in the samples were determined using atomic absorption spectroscopy (AAS) at certain time intervals (5, 10, 15, 30, 60, 120 and 180 min). Experimental data were analyzed using pseudo-first (Equation (1)) [37] and pseudo-second-order (Equation (2)) [38] models for the optimization of the process and a better insight into the adsorption mechanism:

$$q_t = q_e \times \left(1 - e^{-k_1 t}\right), \tag{1}$$

$$q_t = q_e - \left(\frac{1}{q_e} + k_2 t\right)^{-1}, \tag{2}$$

where q_e and q_t (mg/g) are the adsorption capacity at equilibrium and at time t (min), k_1 (1/min) and k_2 (g/mg min) are the pseudo-first-order and pseudo-second-order rate constants.

Adsorption data obtained at different initial concentrations of lead ions (5, 15, 25, 35, 50, 75, and 100 mg/dm^3) and cadmium ions (5, 7.5, 10, and 15 mg/dm^3) were used to construct adsorption isotherms for all thermally modified samples. Langmuir [39] and Freundlich [40] isotherm models (Equations (3) and (4), respectively) were used to evaluate the affinity or capacity of the C_{HTC}, C_{aHTC}, C/P_{HTC}, and C/P_{aHTC}:

$$q_e = \frac{Q_0 K_L C_e}{1 + K_L C_e} \tag{3}$$

$$q_e = K_f \times C_e^{\frac{1}{n}} \tag{4}$$

where C_e is the concentration of metal at equilibrium (mg/dm^3), Q_0 is the maximum adsorption capacity (mg/g), K_L is a Langmuir adsorption equilibrium constant (dm^3/mg) that is related to the apparent energy of sorption, K_f ($mg\ g^{-1}(mg\ dm^{-3})^{-1/n}$) is Freundlich constant, and $1/n$ is the heterogeneity factor.

The possibility of reusing hydrothermally carbonized and subsequently activated waste yarn C_{aHTC} for the removal of Pb ions was investigated in a flow system. The material (0.03 g) was placed between two polyethylene frits in a cartridge. Under vacuum, a solution of Pb(II) ions (10 cm^3 solution, initial concentration 5 mg/dm^3) was passed through the cartridge at a flow rate of 1 cm^3/min. After the adsorption cycle, the desorption of metal ions was performed by passing 5 cm^3 of 2% HNO_3 solution through the cartridge. The concentration of ions in the eluents (obtained after adsorption or desorption) was determined by atomic absorption spectroscopy. The adsorption/desorption procedure in the flow system was carried out in three cycles. The efficiency and capacity of adsorption and desorption processes (E_{ads}, q_a, E_{des}, and q_d, respectively) were calculated by Equations (5)–(8):

$$E_{ads} = \frac{(c_0 - c_e)}{c_0} \times 100, \quad (5)$$

$$q_a = \frac{(c_0 - c_e)}{m} \times 100, \quad (6)$$

$$E_{des} = \frac{q_d}{q_a} \times 100, \quad (7)$$

$$q_d = \frac{c_d V}{m}, \quad (8)$$

where c_0 (mg/dm^3), c_e (mg/dm^3), and c_d (μg/dm^3) are the initial equilibrium and concentration of metal ions in the desorption solution, respectively, V (dm^3) is the solution volume and m (g) is the adsorbent mass.

3. Results and Discussion

3.1. Material Characterization

Scanning electron microscopy (SEM) was used to study the surface structure, topography, and morphology of hydrothermally modified and activated cotton and mixed cotton/polyester yarns (Figure 1).

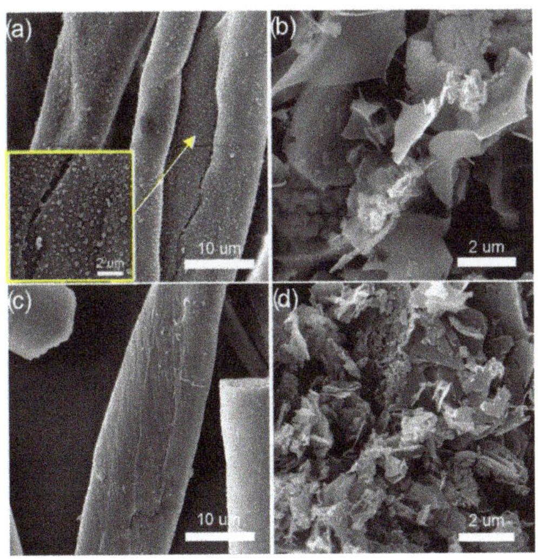

Figure 1. SEM photographs of samples: **(a)** C_{HTC}, **(b)** C_{aHTC}, **(c)** C/P_{HTC}, and **(d)** C/P_{aHTC}.

Hydrothermally carbonized samples C_{HTC} and C/P_{HTC} (Figure 1a,c) retain the structure of the starting material: hydrothermally carbonized cotton and polyester components are characterized by spirally twisted fibers and smooth filament, respectively. In the case of sample C_{HTC}, condensed smooth spherical particles, which are characteristic of hydrothermally carbonized materials, are observed on the surface. After the activation of the hydrothermally carbonized samples, with potassium hydroxide as an activating agent, the destruction of the fiber structure, both cotton and polyester components, in the samples is noticeable (Figure 1b,d).

To examine the structure of hydrothermally treated and activated yarn XRD analysis was performed (Figure 2). XRD spectrum of C_{HTC} exhibits the typical diffraction peaks of the crystalline structure of cellulose I at 2θ = 14.7°, 16.4° and 22.6° [41], while XRD spectrum of C/P_{HTC} along with the peaks from cellulose, exhibits diffraction peaks characteristic for polyester at 2θ = 17.2°, 25.7° and 22.8° (superimposed with cellulose peak at 22.6°) [42].

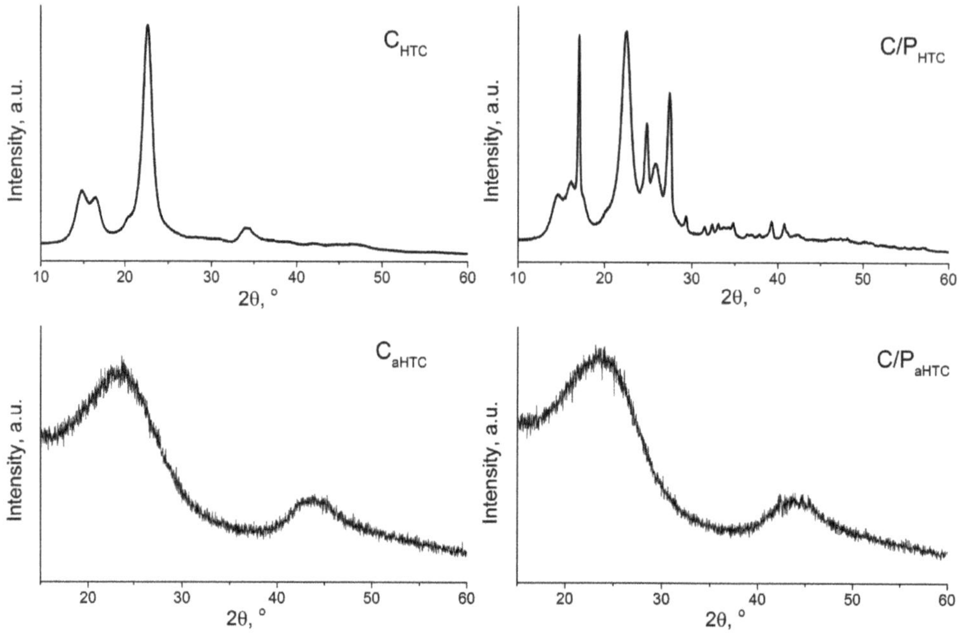

Figure 2. XRD spectra of hydrothermally treated and activated cotton and cotton/polyester yarn.

As starting materials (cotton and cotton/polyester yarn) consist of pure cellulose and polyester, the XRD patterns of the hydrothermally treated samples, C_{HTC} and C/P_{HTC}, confirm that these samples are not fully carbonized and that hydrothermal treatment leads to the partial decomposition of cellulose and its conversion to aromatic carbon network [43].

XRD spectra of chemically activated samples, C_{aHTC} and C/P_{aHTC}, exhibit two typical broad peaks around 23.5° and 43.4° that indicate the disordered carbon structure and confirm that samples C_{aHTC} and C/P_{aHTC} are fully carbonized by applied activation.

The observed changes in the morphology and structure of the activated samples can influence the specific surface area and affect the ability of the material to adsorb selected pollutants. Starting cotton and cotton/polyester yarns have no developed porosity, with the specific surface area immeasurably small and close to their geometric surface. Specific surface area and porosity of examined materials were determined from the adsorption-desorption isotherm of N_2, shown in Figure 3.

The nitrogen adsorption-desorption curves for samples C_{aHTC}, C/P_{aHTC}, and C/P_{HTC} exhibit a hysteresis loop corresponding to type IV (C_{aHTC}) and type V (C/P_{aHTC}, C/P_{HTC}) isotherms in the relative pressure range of 0.4–0.95, 0.5–0.95, and 0.8–0.95, respectively. This behavior is characteristic of mesoporous materials, which is in agreement with the average pore diameter (D_{mean}) and the data presented in Table 1. The obtained results (Table 1) show that modification by hydrothermal carbonization does not lead to the development of the specific surface area of the material (S_{BET}). A very low S_{BET} value for sample C_{HTC} makes it impossible to determine the pore volume and the mean pore diameter (D_{mean}). The lack of porosity for sample C_{HTC} may be the consequence of the surface coverage by smooth condensed particles (Figure 1a). Sample C/P_{HTC} is a mesoporous material with an average pore diameter of 28.75 nm and a preserved fibrous structure of the starting precursor (Figure 1c). By applying the activation process, carbonized samples' specific surface area and porosity increase while the average pore diameter decreases. After activation with KOH, there is a significant increase in the specific surface area for sample C_{aHTC} (872.2 m^2/g). During activation, potassium is incorporated into the structure of the carbon material, forming various compounds that are washed after the process, leaving behind a free porous surface [44]. Based on mean pore diameter, samples C/P_{HTC} and C/P_{aHTC} can be classified as mesoporous, while sample C_{aHTC} shows a higher proportion of microporous surface area in the total specific surface area.

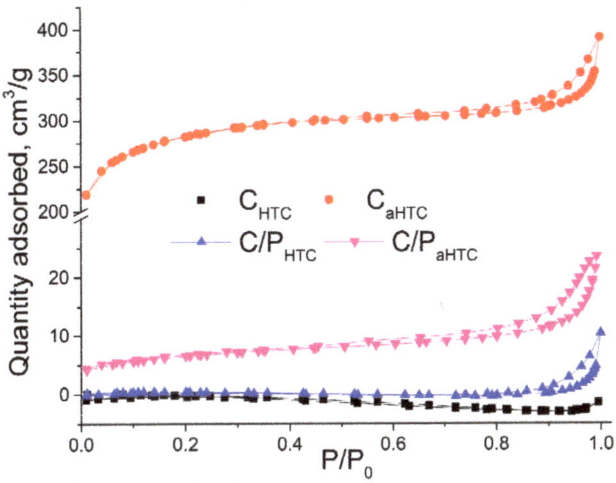

Figure 3. Adsorption–desorption isotherm of N_2 on different adsorbents.

Table 1. Surface properties of hydrothermally carbonized and activated cotton and mixed cotton/polyester yarns.

Sample	S_{BET}, m^2/g	S_{ext}, m^2/g	S_{micro}, m^2/g	V_{total}, cm^3/g	V_{micro}, cm^3/g	D_m, nm	Amount of Surface Oxygen Groups, mmol/g		
							Basic	Acidic	Total
C_{HTC}	0.003	-	-	-	-	-	0.983 ± 0.020	3.782 ± 0.017	4.765 ± 0.037
C/P_{HTC}	0.64	0.122	0.516	0.01	4×10^{-4}	28.75	0.345 ± 0.015	4.866 ± 0.012	5.211 ± 0.027
C_{aHTC}	872.2	386.1	486	0.532	0.269	4.82	2.134 ± 0.008	3.940 ± 0.025	6.074 ± 0.033
C/P_{aHTC}	21.75	19.03	2.72	0.035	0.002	8.93	2.333 ± 0.010	4.811 ± 0.015	7.144 ± 0.025

In addition to the specific surface and mean diameters of pores, the sorptive behavior of the material depends on the presence and type of surface oxygen groups. The quantitative composition of oxygenic acid and base groups present on the surface of carbonaceous materials was determined by the acid-base titration method (Table 1). Total acidic groups (carboxyl, lactone, phenol) are more abundant than basic ones for all samples of carbonaceous materials and are in the range of 3.763–4.866 mmol/g.

Determination of functional groups and examination of changes occurring during the modification process was performed by recording FTIR spectra of precursors and modified samples. As seen in Figure 4, a broad band in the region of 3450–3150 cm^{-1} for precursors and 3650–3450 cm^{-1} for thermally treated samples may originate from the stretching vibrations of the O-H bond at carboxyl (COOH) or hydroxyl (OH) groups [45]. After activation, the increased temperature (900 °C) may lead to dehydration and loss of hydroxyl and carboxyl groups [46], which is visible from the decrease in the intensity of corresponding bands on FTIR spectra of activated samples. The peak at 2920 cm^{-1} is attributed to asymmetric, and at 2850 cm^{-1} to symmetric stretching vibrations of the C–H bond (carboxylic, methyl, or methylene groups) [47]. Also, a peak at 1640 cm^{-1} can be attributed to the bending vibrations of the O–H bond from carboxyl functional groups. The stretching vibration of the aliphatic C=C bond can also contribute to this peak. The deformational vibrations of the C–O bond in the carboxyl group can be observed as a peak at 1384 cm^{-1} [48].

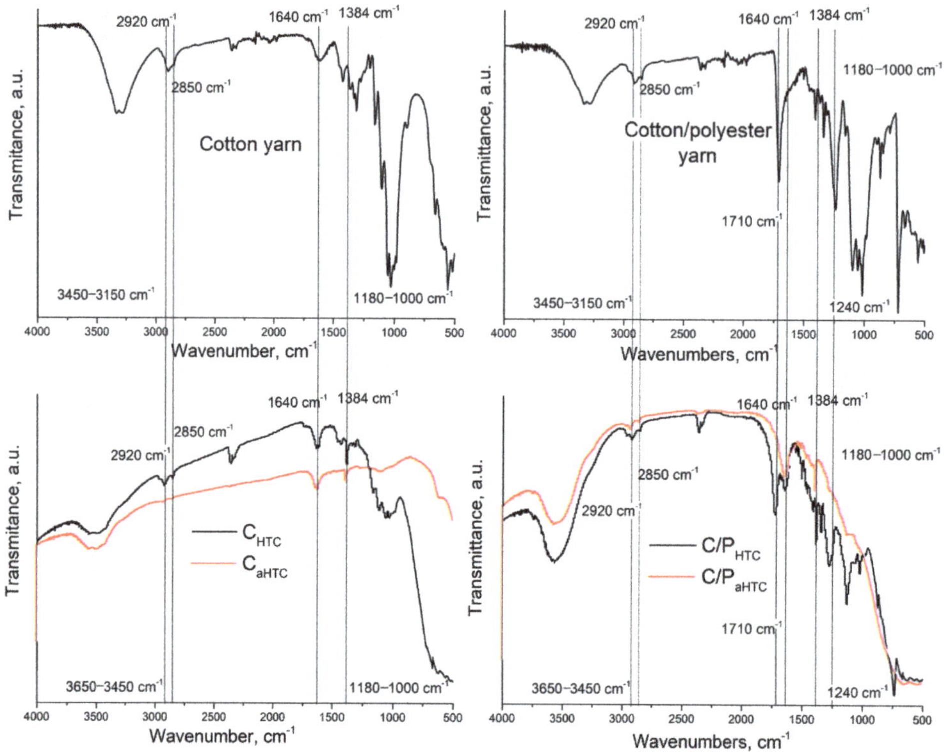

Figure 4. FTIR spectra of untreated cotton and mixed cotton/polyester yarns, and hydrothermally carbonized and activated samples.

On the FTIR spectra of the hydrothermally carbonized samples, C_{HTC} and C/P_{HTC}, a significantly higher number of bands in the range 1800–1400 cm^{-1} is observed compared to the hydrothermally carbonized activated samples C_{aHTC}, C/P_{aHTC}. As the hydrothermal carbonization is performed under milder conditions and the precursors are not completely converted into carbon material, the FTIR spectra of these samples may also show some characteristic bands of the precursors but with the lower intensity compared with the FTIR spectra of cotton and mixed cotton/polyester yarns. Thus, the bands in the region 1180–1000 cm^{-1}, observed in the samples C_{HTC} and C/P_{HTC}, may originate from the stretching of the C-O bond (carboxyl groups, esters, ethers). The peaks at 1710 cm^{-1} and 1240 cm^{-1} indicate the presence of ester groups, and the peaks at 870 cm^{-1} and 720 cm^{-1} correspond to the bending vibrations of the C-H bond of the benzene ring in the polyester chain. After the activation of the C/P_{HTC} sample, the bands characteristic for the polyester component disappear (Figure 4).

3.2. Adsorption Experiments

Dependence of adsorption on time, as well as agreement of experimental data with pseudo-first- and pseudo-second-order models for the adsorption of (a) lead and (b) cadmium ions on thermally modified and activated cotton and cotton/polyester yarns (C_{HTC} and C_{aHTC}, C/P_{HTC} and C/P_{aHTC}) is shown in Figure 5. All examined samples showed higher capacities and rates of adsorption for the adsorption of lead ions, reaching the equilibrium after 30 min (Figure 5a). The highest adsorption capacity was observed for sample C_{aHTC}, which also showed the fastest adsorption, reaching the removal efficiency of 99.9% after 15 min. Although the adsorption of cadmium ions onto examined samples can also be described as a fast process, examined samples showed much lower adsorption capacities for the removal of Cd(II) ions. These observations follow the kinetic parameters shown in Table 2. Based on correlation coefficients (R^2) and comparison between experimentally obtained ($q_{e,exp}$) and calculated ($q_{e,mod}$) values of maximal adsorption capacities, adsorption of lead and cadmium ions onto examined samples can be described by the pseudo-second-order kinetic model, which indicates that heavy metal ions adsorption on carbonaceous materials takes place by a complex mechanism, and that chemisorption can play the most significant role in determining the overall reaction rate. The experimentally obtained values of the equilibrium adsorption capacities ranged from 2.06 to 19.98 mg/g for the adsorption of Pb(II) and from 1.52 to 4.05 mg/g for the adsorption of Cd(II) ions, following the sequence C_{aHTC} > C/P_{aHTC} > C_{HTC} > C/P_{HTC}. Higher values of equilibrium adsorption capacities for Pb(II) ions may be a consequence of larger ionic radii and electronegativity compared to Cd(II) ions.

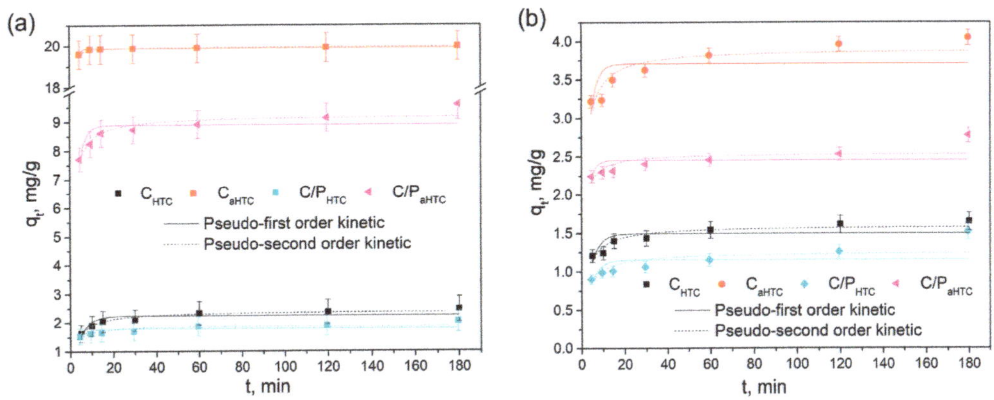

Figure 5. The influence of contact time on adsorption of (**a**) Pb and (**b**) Cd onto hydrothermally carbonized and activated cotton and mixed cotton/polyester.

Table 2. Kinetic parameters for adsorption of Pb(II) and Cd(II) ions onto hydrothermally carbonized and activated cotton and mixed cotton/polyester yarns.

Metal Ion	Sample	Pseudo-First-Order Model			Pseudo-Second-Order Model			
		R^2	k_1 (1/min)	$q_{e,mod}$ (mg/g)	R^2	k_2 (g/(mg min))	$q_{e,mod}$ (mg/g)	$q_{e,exp}$ (mg/g)
Pb(II)	C_{HTC}	0.71124	0.216	2.28	0.94000	0.159	2.42	2.46
	C/P_{HTC}	0.28953	0.325	1.83	0.70164	0.325	1.93	2.06
	C_{aHTC}	0.85576	0.842	19.91	0.91098	0.607	19.96	19.98
	C/P_{aHTC}	0.54732	0.372	8.95	0.86901	0.100	9.26	9.59
Cd(II)	C_{HTC}	0.47925	0.255	1.52	0.85098	0.295	1.61	1.65
	C/P_{HTC}	0.23811	0.212	1.22	0.56231	0.243	1.31	1.52
	C_{aHTC}	0.31121	0.337	3.75	0.78406	0.175	3.93	4.05
	C/P_{aHTC}	0.10068	0.446	2.48	0.50107	0.417	2.57	2.78

The highest values of equilibrium adsorption capacities obtained for the C_{aHTC} sample can be a result of the highest specific surface area (Table 1). However, a comparison between the values of specific surface area (Table 1) and the adsorption capacities of examined samples (Table 2) indicates that the specific surface area is not the key factor determining the adsorption capacities. The higher adsorption capacities of activated samples can also be influenced by the higher total amount of surface oxygen groups, which can serve as the main adsorption sites for metal ions. The achieved results are in agreement with literature data and scientific research [49–53].

To obtain additional information about the system, carbon material—lead and cadmium ions in an equilibrium state at a constant temperature, adsorption isotherms were used. The obtained equilibrium experimental results were fitted by Langmuir and Freundlich adsorption isotherms and shown in Figure 6, while the isotherm parameters are presented in Table 3. For all examined samples, adsorption capacities increase with the initial concentration of heavy metal ions. In the examined concentration range, the characteristic plot on the isotherm curve that indicates surface saturation was observed only for the adsorption of lead ions on C_{HTC} and C/P_{HTC} (Figure 6a). The increase in adsorption capacity is particularly pronounced for the C_{aHTC} sample during the removal of Pb(II) ions (Figure 6a). The maximum adsorption capacity (q_{max}) obtained by the Langmuir model is very high, which can be attributed to the lack of equilibrium surface saturation in the examined concentration range and large specific surface area (Table 1). The process of adsorption on all thermally modified and activated yarns for Pb(II) ions, according to the obtained values of the correlation coefficient R^2, is better described by the Langmuir model, which is most often the case during the removal of heavy metals from carbon materials whose precursors are cellulose materials [54,55]. A high value of the Freundlich constant indicates a high adsorption capacity, that is, a high affinity of the C_{aHTC} sample towards Pb(II) compared to other samples. Considering that the chemisorption process applies to Langmuir's model, and the obtained values of $1/n < 1$ indicate a heterogeneous adsorbent surface, it can be assumed that the adsorption mechanism is probably a combination of physical and chemical adsorption. In thermally modified cotton/polyester samples (C/P_{HTC}, C/P_{aHTC}), during the removal of Pb(II) ions, much lower values of adsorption capacity were obtained compared to thermally modified and activated cotton.

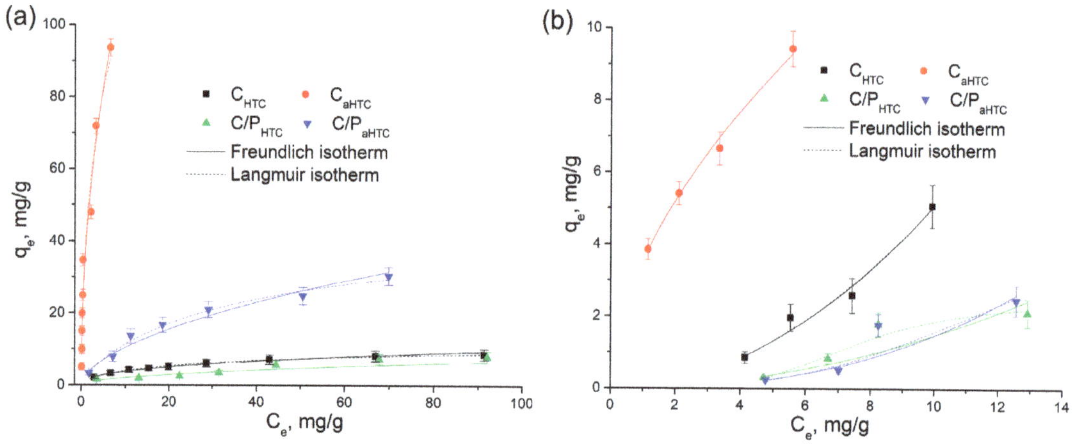

Figure 6. The influence of initial adsorbate concentration on adsorption of (**a**) Pb and (**b**) Cd onto hydrothermally carbonized and activated cotton and mixed cotton/polyester yarns.

Table 3. Langmuir and Freundlich parameters for adsorption of Pb(II) and Cd(II) ions onto hydrothermally carbonized and activated cotton and mixed cotton/polyester yarns.

Metal Ion	Sample	Langmuir Isotherm			Freundlich Isotherm		
		q_{max} (mg/g)	b (dm^3/mg)	R^2	K_f ((mg/g)/(mg/dm$_3$)$^{1/n}$)	1/n	R^2
Pb(II)	C_{HTC}	260.8	0.006	0.99999	1.683	0.372	0.98237
	C/P_{HTC}	81.6	0.004	0.94043	0.361	0.694	0.92571
	C_{aHTC}	3345.0	0.012	0.99999	40.596	0.448	0.95892
	C/P_{aHTC}	49.7	0.046	0.98189	3.634	0.501	0.97498
Cd(II)	C_{HTC}	1357.0	4.62×10^{-5}	0.93807	0.063	1.906	0.96911
	C/P_{HTC}	2.2	1.01×10^{-6}	0.92372	0.086	1.275	0.71281
	C_{aHTC}	843.4	0.00414	0.97853	3.480	0.573	0.98945
	C/P_{aHTC}	2.4	4.81×10^{-13}	0.94990	0.026	1.807	0.76007

The adsorption process on C_{HTC} and C_{aHTC} for Cd(II) ions (Figure 6b) can be equally well described by both Langmuir and Freundlich isotherm models. The value of the heterogeneity factor less than 1 for the C_{aHTC} sample implies a relatively homogeneous surface, while adsorption occurs through the chemical process on the surface functional groups as active sites [56]. On the other hand, equilibrium adsorption data obtained for Cd(II) adsorption onto C/P_{HTC} and C/P_{aHTC} fits better with the Langmuir isotherm model, while the dimensionless heterogeneity factor $1/n$ has values approximately and greater than 1, indicating that with the increase in the concentration of Cd(II) ions, the free energy for further adsorption decreases [57]. Applied chemical activation improved the adsorption capacity of hydrothermally carbonized cotton, while in the case of cotton/polyester, activation did not bring the expected increase in adsorption capacity for the adsorption of Cd(II) ions.

Very few researchers have investigated the use of thermally modified cotton and cotton/polyester blends for metal ion adsorption. A comparison of results obtained in this study with the one found in the literature is given in Table 4.

Table 4. Comparison of maximal adsorption capacities of carbon adsorbents based on cotton and cotton/polyester for removal of heavy metal.

Material	Modification	Initial Concentration, mg/dm^3	Heavy Metal	Adsorption Capacity q_{max}, mg/g	Reference
Cotton fabric	Pyrolysis + Activation H_3PO_4	80–500	Pb(II)	361.54	[20]
Cotton/polyester fabric 75:25		80–500		385.77	
Cotton fibre	Microwave-assisted carbonization	0.05–0.4	Hg(II)	169.2	[58]
Cotton stalk	Pyrolysis	100–500	Pb(II)	146.78	[25]
Cotton stalk	Hydrothermal carbonization + Activation KOH	5–300	Cd(II)	30.40	[17]
Cotton stalk	Pyrolysis	10–80	Pb(II)	42.55	[59]
		0.1–1.0	Cd(II)	0.53	
		1.0–10	Ni(II)	5.25	
		0.1–1.0	Co(II)	0.54	

Table 4. *Cont.*

Material	Modification	Initial Concentration, mg/dm^3	Heavy Metal	Adsorption Capacity q_{max}, mg/g	Reference
	Pyrolysis + H_2SO_4	10–80	Pb(II)	38.76	
		0.1–1.0	Cd(II)	0.53	
		1.0–10	Ni(II)	2.21	
		0.1–1.0	Co(II)	0.52	
	Pyrolysis + NaOH	10–80	Pb(II)	37.59	
		0.1–1.0	Cd(II)	0.51	
		1.0–10	Ni(II)	2.05	
		0.1–1.0	Co(II)	0.50	
	Pyrolysis + $H_2C_2O_4$	10–80	Pb(II)	44.64	
		0.1–1.0	Cd(II)	0.65	
		1.0–10	Ni(II)	6.20	
		0.1–1.0	Co(II)	0.52	
Waste cotton yarns	Hydrothermal carbonization	5–100	Pb(II)	260.8	This study
Waste cotton/polyester yarns		5–100		81.6	
Waste cotton yarns		5–15	Cd(II)	3345.0	
Waste cotton/polyester yarns		5–15		49.7	
Waste cotton yarns	Hydrothermal carbonization + Activation KOH	5–100	Pb(II)	1357.0	
Waste cotton/polyester yarns 50:50		5–100		2.2	
Waste cotton yarns		5–15	Cd(II)	843.4	
Waste cotton/polyester yarns 50:50		5–15		2.4	

According to Table 4, the adsorption capacities of hydrothermally treated and activated cotton and mixed cotton/polyester yarns investigated in this study are comparable or better than in cited studies. These results indicate that this way of modifying textile waste can produce an effective adsorbent for the removal of heavy metals, which corresponds to the principles of circular economy and sustainability.

3.3. Adsorption/Desorption

The possibility of reusing thermally modified yarns was tested by the adsorption of lead ions on the C_{aHTC} sample, which showed the highest adsorption capacity. The adsorption/desorption procedure in the flow system was carried out in three cycles, where HNO_3 was used as the desorption reagent (Figure 7a). The results of adsorption efficiency (%) and desorption efficiency (%) for all three cycles are shown in Figure 7b.

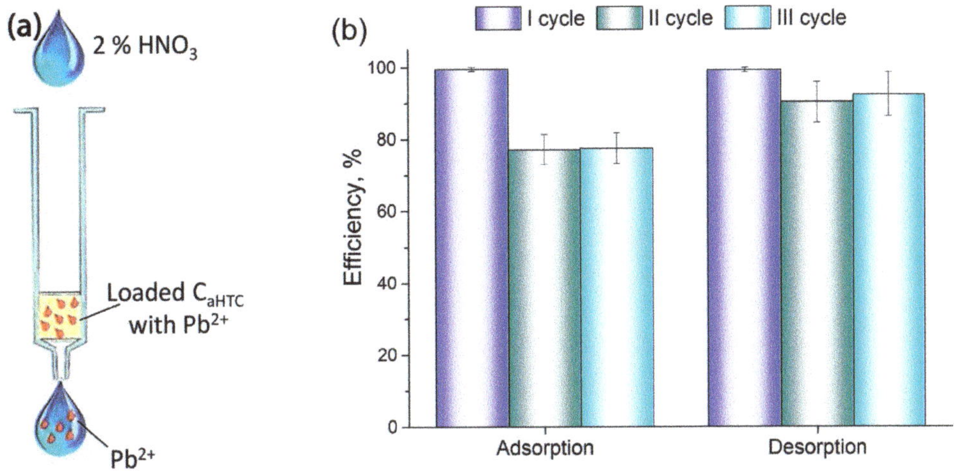

Figure 7. Scheme of lead ions desorption (**a**) and adsorption and desorption efficiency (%) of the C_{aHTC} sample (**b**).

As can be seen from Figure 7b, the adsorption efficiency of Pb(II) ions in the first cycle on the C_{aHTC} sample is close to 100%, with a very high desorption efficiency of 99.23%. In the second and third cycles, lower adsorption efficiency values were obtained (77.16% and 77.53%), but desorption efficiencies were still relatively high (90.33% and 92.43%). The high efficiency of diluted nitric acid to desorb Pb(II) ions indicates the weak bound between adsorbate and adsorbent surface, leading to an assumption that Pb(II) ions are adsorbed through the mechanism of ion exchange or physisorption. Similar results on adsorption and desorption efficiency throughout several consecutive cycles were shown in the literature for the adsorption of Pb(II) ions onto carbonized cotton stalks [25] and adsorption of Pb(II), Cd(II), and Ni(II) ions onto carbonized cherry pits [35].

The results obtained from adsorption/desorption experiments have shown that hydrothermally carbonized and activated cotton yarn can be used as a highly efficient adsorbent for lead ions removal from water in at least three consecutive cycles of adsorption/desorption, enabling its practical application for wastewater treatment.

4. Conclusions

Following the increasing demand for cheap, available, and ecological adsorbents, waste materials with a large portion of natural cellulose polysaccharide were used in this work as starting raw materials to obtain effective carbon adsorbents for the removal of Pb(II) and Cd(II) ions from water. Hydrothermal carbonization was used to convert cotton and mixed cotton/polyester yarn to carbon adsorbents. Material characterization showed that additional chemical activation changed the morphology of materials and led to an increase in the specific surface area and the number of surface oxygen groups, consequently improving the affinity of obtained carbon materials for removing cations. Carbon adsorbent showing the highest values of Freundlich constant and Langmuir maximum adsorption capacity (3345.0 mg/g for Pb(II) ions adsorption) were obtained by hydrothermal carbonization and subsequent activation of cotton yarn. The adsorption of Pb(II) and Cd(II) ions onto hydrothermally carbonized and activated cotton and the mixed yarn is a relatively fast process following the pseudo-second-order kinetic, while the equilibrium adsorption process can be described by Langmuir isotherm model. The adsorption process most likely takes place by a complex mechanism, including physisorption and ion exchange on the surface oxygen groups. It was shown that hydrothermally carbonized and activated cotton yarns can be successfully reused in a flow system as a highly efficient adsorbent for lead ions in three consecutive adsorption/desorption cycles, with adsorption efficiency ranging from 100 to 77% throughout three cycles. The results of this research represent a significant contribution to solving the problem of removing heavy metal ions from water, using waste materials as adsorbents, and can serve as an example of the practical application of these materials in the purification of wastewater containing various heavy metals.

Author Contributions: Conceptualization, M.V. and M.M.; methodology, M.V. and B.P.; investigation, M.M. and S.M.; resources, S.M. and A.K.; data curation, M.M.; writing—original draft preparation, K.T.; writing—review and editing, M.V. and B.P.; visualization, K.T., M.V. and M.M.; supervision, A.P.G.; funding acquisition, A.P.G. All authors have read and agreed to the published version of the manuscript.

Funding: This research was funded by the Science Fund of the Republic of Serbia, Program Ideas, GRANT No 7743343, Serbian Industrial Waste towards Sustainable Environment: Resource of Strategic Elements and Removal Agent for Pollutants—SIW4SE.

Institutional Review Board Statement: Not applicable.

Data Availability Statement: The original contributions presented in the study are included in the article, further inquiries can be directed to the corresponding author.

Conflicts of Interest: Author Marina Maletić is employed by the Innovation Center of the Faculty of Technology and Metallurgy. The remaining authors declare that the research was conducted in the absence of any commercial or financial relationships that could be construed as a potential conflict of interest.

References

1. Panigrahi, T.; Santhoskumar, A.U. Adsorption process for reducing heavy metals in Textile Industrial Effluent with low cost adsorbents. *Prog. Chem. Biochem. Res.* **2020**, *3*, 135–139. [CrossRef]
2. Al-Asadi, S.T.; Al-Qaim, F.F.; Al-Saedi, H.F.S.; Deyab, I.F.; Kamyab, H.; Chelliapan, S. Adsorption of methylene blue dye from aqueous solution using low-cost adsorbent: Kinetic, isotherm adsorption, and thermodynamic studies. *Environ. Monit. Assess.* **2023**, *195*, 676. [CrossRef] [PubMed]
3. Minelgaitė, A.; Liobikienė, G. Waste problem in European Union and its influence on waste management behaviours. *Sci. Total Environ.* **2019**, *667*, 86–93. [CrossRef]
4. John, M.J.; Anandjiwala, R.D. Recent developments in chemical modification and characterization of natural fiber-reinforced composites. *Polym. Composite* **2008**, *29*, 187–207. [CrossRef]
5. Mohammed, A.S.A.; Naveed, M.; Jost, N. Polysaccharides; Classifcation, Chemical Properties, and Future Perspective Applications in Fields of Pharmacology and Biological Medicine (A Review of Current Applications and Upcoming Potentialities). *J. Polym. Environ.* **2021**, *29*, 2359–2371. [CrossRef] [PubMed]
6. Peltzer, M.A.; Salvay, A.G.; Delgado, J.F.; Wagner, J.R. Use of edible films and coatings for functional foods developments: A review. In *Functional Foods: Sources, Health Effects and Future Perspectives*; Nelson, D.L., Ed.; Nova Science Publishing: Hauppauge, NY, USA, 2016; pp. 1–26.
7. Textile Exchange. Available online: https://textileexchange.org/app/uploads/2021/08/Textile-Exchange_Preferred-Fiber-and-Materials-Market-Report_2021.pdf (accessed on 5 June 2024).
8. Bediako, J.K.; Apalangya, V.; Hodgson, I.O.A.; Anugwom, I.; Repo, E. Adsorbents for water decontamination: A recycling alternative for fiber precursors and textile fiber wastes. *Sci. Total Environ.* **2024**, *919*, 171000. [CrossRef]
9. Stefan, D.S.; Bosomoiu, M.; Stefan, M. Methods for Natural and Synthetic Polymers Recovery from Textile Waste. *Polymers* **2022**, *1419*, 3939. [CrossRef]
10. Fazli, A.; Rodrigue, D. Sustainable Reuse of Waste Tire Textile Fibers WTTF, as Reinforcements. *Polymers* **2022**, *1419*, 3933. [CrossRef]
11. Srichola, P.; Witthayolankowit, K.; Sukyai, P.; Sampoompuang, C.; Lobyam, K.; Kampakun, P.; Toomtong, R. Recycling of Nanocellulose from Polyester–Cotton Textile Waste for Modification of Film Composites. *Polymers* **2023**, *1515*, 3324. [CrossRef]
12. Xu, Z.; Qi, R.; Zhang, D.; Gao, Y.; Xiong, M.; Chen, W. Co-hydrothermal carbonization of cotton textile waste and polyvinyl chloride waste for the production of solid fuel: Interaction mechanisms and combustion behaviors. *J. Clean. Prod.* **2021**, *316*, 128306. [CrossRef]
13. Zhao, Y.; Chen, W.; Liu, F.; Zhao, P. Hydrothermal pretreatment of cotton textile wastes: Biofuel characteristics and biochar electrocatalytic performance. *Fuel* **2022**, *316*, 123327. [CrossRef]
14. Shirvanimoghaddama, K.; Czech, B.; Wiącek, A.E.; Ćwikła-Bundyra, W.; Naebe, M. Sustainable carbon microtube derived from cotton waste for environmental applications. *Chem. Eng. J.* **2019**, *361*, 1605–1616. [CrossRef]
15. Adolfsson, K.H.; Yadav, N.; Hakkarainen, M. Cellulose-derived hydrothermally carbonized materials and their emerging applications. *Curr. Opin. Green Sustain. Chem.* **2020**, *23*, 18–24. [CrossRef]

16. Kawamura, K.; Sako, K.; Ogata, T.; Tanabe, K. Environmentally friendly, hydrothermal treatment of mixed fabric wastes containing polyester, cotton, and wool fibers: Application for HMF production. *Bioresour. Technol. Rep.* **2020**, *11*, 100478. [CrossRef]
17. Sun, K.; Tang, J.; Gong, Y.; Zhang, H. Characterization of potassium hydroxide (KOH) modified hydrochars from different feedstocks for enhanced removal of heavy metals from water. *Environ. Sci. Pollut. Res.* **2015**, *22*, 16640. [CrossRef]
18. Shi, S.; Feng, X.; Gao, L.; Tang, J.; Guo, H.; Wang, S. Hydrolysis and carbonization of reactive dyes/cotton fiber in hydrothermal environment. *Waste Manag.* **2020**, *103*, 370–377. [CrossRef] [PubMed]
19. Soffian, M.S.; Halim, F.Z.A.; Aziz, F.; Rahman, M.A.; Amin, M.A.M.; Chee, D.N.A. Carbon-based material derived from biomass waste for wastewater treatment. *Environ. Adv.* **2022**, *9*, 100259. [CrossRef]
20. Carraro, P.S.; Spessato, L.; Crespo, L.H.S.; Yokoyama, J.T.C.; Fonseca, J.M.; Bedin, K.C.; Ronix, A.; Cazetta, A.L.; Silva, T.L.; Almeida, V.C. Activated carbon fibers prepared from cellulose and polyester–derived residues and their application on removal of Pb^{2+} ions from aqueous solution. *J. Mol. Liq.* **2019**, *289*, 111150. [CrossRef]
21. Parmakoğlu, E.Ü.; Cxay, A.; Yanik, J. Valorization of Solid Wastes from Textile Industry as an Adsorbent Through Activated Carbon Production. *AATCC J. Res.* **2023**, *10*, 133–143. [CrossRef]
22. Bediako, J.K.; Wei, W.; Yun, Y.-S. Conversion of waste textile cellulose fibers into heavy metal adsorbents. *J. Ind. Eng. Chem.* **2016**, *43*, 61–68. [CrossRef]
23. Racho, P.; Waiwong, W. Modified textile waste for heavy metals removal. *Energy Rep.* **2020**, *6*, 927–932. [CrossRef]
24. Ambaye, T.G.; Vaccari, M.; van Hullebusch, E.D.; Amrane, A.; Rtimi, S. Mechanisms and adsorption capacities of biochar for the removal of organic and inorganic pollutants from industrial wastewater. *Int. J. Environ. Sci. Technol.* **2021**, *18*, 3273–3294. [CrossRef]
25. Gao, L.; Li, Z.; Yi, W.; Li, Y.; Zhang, P.; Zhang, A.; Wang, L. Impacts of pyrolysis temperature on lead adsorption by cotton stalk-derived biochar and related mechanisms. *J. Environ. Chem. Eng.* **2021**, *9*, 105602. [CrossRef]
26. Cui, L.-P.; Shi, S.; Hou, W.-S.; Yan, Z.-F.; Dan, J.-M. Hydrolysis and carbonization mechanism of cotton fibers in subcritical water. *New Carbon Mater.* **2018**, *33*, 245–251. [CrossRef]
27. Luo, M.; Chen, J.; Li, Q.; Wang, J. Cotton-Based Activated Carbon Fiber with High Specific Surface Area Prepared by Low-Temperature Hydrothermal Carbonization with Urea Enhancement. *Ind. Eng. Chem. Res.* **2023**, *62*, 8744–8753. [CrossRef]
28. Briffa, J.; Sinagra, E.; Blundell, R. Heavy metal pollution in the environment and their toxicological effects on humans. *Heliyon* **2020**, *6*, e04691. [CrossRef]
29. Singh, V.; Ahmed, G.; Vedika, S.; Kumar, P.; Chaturvedi, S.K.; Rai, S.N.; Vamanu, E.; Kumar, A. Toxic heavy metal ions contamination in water and their sustainable reduction by eco-friendly methods: Isotherms, thermodynamics and kinetics study. *Sci. Rep.* **2024**, *14*, 7595. [CrossRef] [PubMed]
30. Bayuo, J.; Rwiza, M.J.; Choi, J.W.; Mtei, K.M.; Bandegharaei, A.H.; Sillanpa, M. Adsorption and desorption processes of toxic heavy metals, regeneration and reusability of spent adsorbents: Economic and environmental sustainability approach. *Adv. Colloid. Interface Sci.* **2024**, *329*, 103196. [CrossRef] [PubMed]
31. Husien, S.; El-taweel, R.M.; Salim, A.I.; Fahim, I.S.; Said, L.A.; Radwan, A.G. Review of activated carbon adsorbent material for textile dyes removal: Preparation, and modelling. *Curr. Res. Green Sustain. Chem.* **2022**, *5*, 100325. [CrossRef]
32. Shukla, S.K.; Al Mushaiqri, N.R.S.; Al Subhi, H.M.; Yoo, K.; Al Sadeq, H. Low-cost activated carbon production from organic waste and its utilization for wastewater treatment. *Appl. Water Sci.* **2020**, *10*, 62. [CrossRef]
33. Azharul Islam, M.; Benhouria, A.; Asif, M.; Hameed, B.H. Methylene blue adsorption on factory-rejected tea activated carbon prepared by conjunction of hydrothermal carbonization and sodium hydroxide activation processes. *J. Taiwan Inst. Chem. Eng.* **2015**, *52*, 57–64. [CrossRef]
34. Babeker, T.M.A.; Chen, Q. Heavy Metal Removal from Wastewater by Adsorption with Hydrochar Derived from Biomass: Current Applications and Research Trends. *Curr. Pollut. Rep.* **2021**, *7*, 54–71. [CrossRef]
35. Pap, S.; Radonić, J.; Trifunović, S.; Adamović, D.; Mihajlović, I.; Vojinović Miloradov, M.; Turk Sekulić, M. Evaluation of the adsorption potential of eco-friendly activated carbon prepared from cherry kernels for the removal of Pb^{2+}, Cd^{2+} and Ni^{2+} from aqueous wastes. *J. Environ. Manag.* **2016**, *184*, 297–306. [CrossRef]
36. Saini, R.; Pandey, M.; Mishra, R.K.; Kumar, P. Adsorption potential of hydrochar derived from hydrothermal carbonization of waste biomass towards the removal of methylene blue dye from wastewater. *Biomass. Convers. Biorefin.* 2024. [CrossRef]
37. Lagergren, S. Zur the oriedersogennanten adsorption geloesterstoffe, Kungliga Svenska Vetenskapsakademiens. *Handlingar* **1898**, *24*, 1–39.
38. Ho, Y.S.; Mckay, G. Pseudo-second order model for sorption processes. *Process Bio-Chem.* **1999**, *34*, 451–465. [CrossRef]
39. Langmuir, I. The adsorption of gases on plane surfaces of glass, mica and platinum. *J. Am. Chem. Soc.* **1918**, *40*, 1361–1403. [CrossRef]
40. Freundlich, H. Adsorption in solutions. *J. Phys. Chem.* **1906**, *57*, 384–410.
41. Gong, J.; Li, J.; Xu, J.; Xiang, Z.; Mo, L. Research on cellulose nanocrystals produced from cellulose sources with various polymorphs. *RSC Adv.* **2017**, *7*, 33486. [CrossRef]
42. Ren, Y.; Zhao, Z.; Jiang, W.; Zhang, G.; Tan, Y.; Guan, Y.; Zhou, L.; Cui, L.; Choi, S.W.; Li, M.-X. Preparation of Y_2O_3/TiO_2-Loaded Polyester Fabric and Its Photocatalytic Properties under Visible Light Irradiation. *Polymers* **2022**, *14*, 2760. [CrossRef]
43. Titirici, M.-M.; White, R.J.; Falco, C.; Sevilla, M. Black perspectives for a green future: Hydrothermal carbons for environment protection and energy storage. *Energy Environ. Sci.* **2012**, *5*, 6796. [CrossRef]

44. Amen, R.; Bashir, H.; Bibi, I.; Shaheen, S.M.; Niazi, N.K.; Shahid, M.; Hussain, M.M.; Antoniadis, V.; Shakoor, M.B.; Al-Solaimani, S.G.; et al. A critical review on arsenic removal from water using biochar-based sorbents: The significance of modification and redox reactions. *Chem. Eng. J.* **2020**, *396*, 125195. [CrossRef]
45. Liu, J.; Zhang, S.; Jin, C.E.S.; Sheng, K.; Zhang, X. Effect of swelling pretreatment on properties of cellulose-based hydrochar. *ACS Sustain. Chem. Eng.* **2019**, *7*, 10821–10829. [CrossRef]
46. Chen, J.; Li, S. Characterization of biofuel production from hydrothermal treatment of hyperaccumulator waste (*Pteris vittata* L.) in sub- and supercritical water. *RSC Adv.* **2020**, *10*, 2160–2169. [CrossRef] [PubMed]
47. Kharrazi, S.M.; Mirghaffari, N.; Dastgerdi, M.M.; Soleimani, M. A novel post-modification of powdered activated carbon prepared from lignocellulosic waste through thermal tension treatment to enhance the porosity and heavy metals adsorption. *Powder Technol.* **2020**, *366*, 358–368. [CrossRef]
48. Kalijadis, A.; Đorđević, J.; Trtić-Petrović, T.; Vukčević, M.; Popović, M.; Maksimović, V.; Rakočević, Z.; Laušević, Z. Preparation of boron-doped hydrothermal carbon from glucose for carbon paste electrode. *Carbon* **2015**, *95*, 42–50. [CrossRef]
49. Wang, H.; Gao, B.; Wang, S.; Fang, J.; Xue, Y.; Yang, K. Removal of Pb(II), Cu(II), and Cd(II) from aqueous solutions by biochar derived from $KMnO_4$ treated hickory wood. *Bioresour. Technol.* **2015**, *197*, 356–362. [CrossRef]
50. Tyagi, U. Enhanced adsorption of metal ions onto Vetiveriazizanioides biochar via batch and fixed bed studies. *Bioresour. Technol.* **2022**, *345*, 126475. [CrossRef]
51. Xia, Y.; Yang, T.; Zhu, N.; Li, D.; Wang, Z.; Lang, Q.; Liu, Z.; Jiao, W. Enhanced adsorption of Pb(II) onto modified hydrochar: Modeling and mechanism analysis. *Bioresour. Technol.* **2019**, *288*, 1–8. [CrossRef]
52. Khoshbouy, R.; Lejiu, R.; Takahashi, F.; Yoshikawa, K. Effect of acid-assisted hydrothermal carbonization (HTC) process of tree branches using nitric acid on cadmium adsorption. *E3S Web Conf.* **2018**, *67*, 03033. [CrossRef]
53. Zhou, Z.; Xu, Z.; Feng, Q.; Yao, D.; Yu, J.; Wang, D.; Lv, S.; Liu, Y.; Zhou, N.; Zhong, M.E. Effect of pyrolysis condition on the adsorption mechanism of lead, cadmium and copper on tobacco stem biochar. *J. Clean. Prod.* **2018**, *187*, 996–1005. [CrossRef]
54. Duan, C.; Ma, T.; Wang, J.; Zhou, Y. Removal of heavy metals from aqueous solution using carbon-based adsorbents: A review. *J. Water Process Eng.* **2020**, *37*, 101339. [CrossRef]
55. Anastopoulos, I.; Pashalidis, I.; Hosseini-Bandegharaei, A.; Giannakoudakis, D.A.; Robalds, A.; Usman, M.; Escudero, L.B.; Zhou, Y.; Colmenares, J.C.; Núñez-Delgado, A.; et al. Agricultural biomass/waste as adsorbents for toxic metal decontamination of aqueous solutions. *J. Mol. Liq.* **2019**, *295*, 111684. [CrossRef]
56. Madaeni, S.S.; Salehi, E. Adsorption of cations on nanofiltration membrane: Separation mechanism, isotherm confirmation and thermodynamic analysis. *Chem. Eng. J.* **2009**, *150*, 114–121. [CrossRef]
57. Yin, W.; Zhao, C.; Xu, J. Enhanced adsorption of Cd (II) from aqueous solution by a shrimp bran modified Typha orientalis biochar. *Environ. Sci. Pollut. Res.* **2019**, *26*, 37092–37100. [CrossRef]
58. Bhakta, J.N.; Majumdar, P.B.; Munekage, Y. Development of Activated Carbon from Cotton Fibre Waste as Potential Mercury Adsorbent: Kinetic and Equilibrium Studies. *Int. J. Chem. Eng.* **2014**, *1*, 176483. [CrossRef]
59. Mosa, A.A.; El-Ghamry, A.; Al-Zahrani, H.; Selim, E.M.; El-Khateeb, A. Chemically Modified Biochar Derived from Cotton Stalks: Characterization And Assessing Its Potential for Heavy Metals Removal from Wastewater. *Environ. Biodivers. Soil Secur.* **2017**, *1*, 33–45. [CrossRef]

polymers

MDPI

Article

Optimizing Soil Stabilization with Chitosan: Investigating Acid Concentration, Temperature, and Long-Term Strength

Runshen Wang [1,2], Dominic E. L. Ong [2,3,*], Hossein Sadighi [4], Mohammad Goli [5], Peng Xia [1], Hadi Fatehi [2,3,*] and Tianchi Yao [6]

1 Key Laboratory of Geomechanics and Embankment Engineering of Ministry of Education, Hohai University, Nanjing 210024, China; panda.jasonwang@gmail.com (R.W.); xiapeng_1994@163.com (P.X.)
2 Cities Research Institute, Griffith University, Southport, QLD 4215, Australia
3 School of Engineering and Built Environment, Griffith University, Nathan, QLD 4111, Australia
4 Department of Civil Engineering, Isfahan University of Technology, Isfahan 84156-83111, Iran; h.sadighi@alumni.iut.ac.ir
5 Department of Civil, Ocean, Environmental Engineering, Stevens Institute of Technology, Hoboken, NJ 07030, USA; mgoli1@stevens.edu
6 College of Science, Australia National University, Canberra, ACT 2600, Australia; u7886714@anu.edu.au
* Correspondence: d.ong@griffith.edu.au (D.E.L.O.); hadi.fatehieglab@griffithuni.edu.au (H.F.)

Abstract: Civil and geotechnical researchers are searching for economical alternatives to replace traditional soil stabilizers such as cement, which have negative impacts on the environment. Chitosan biopolymer has shown its capacity to efficiently minimize soil erosion, reduce hydraulic conductivity, and adsorb heavy metals in soil that is contaminated. This research used unconfined compression strength (UCS) to investigate the impact of chitosan content, long-term strength assessment, acid concentration, and temperature on the improvement of soil strength. Static triaxial testing was employed to evaluate the shear strength of the treated soil. Overall, the goal was to identify the optimum values for the mentioned variables so that the highest potential for chitosan-treated soil can be obtained and applied in future research as well as large-scale applications in geotechnical engineering. The UCS results show that chitosan increased soil strength over time and at high temperatures. Depending on the soil type, a curing temperature between 45 to 65 °C can be considered optimal. Chitosan biopolymer is not soluble in water, and an acid solution is needed to dissolve the biopolymer. Different ranges of acid solution were investigated to find the appropriate amount. The strength of the treated soil increased when the acid concentration reached its optimal level, which is 0.5–1%. A detailed chemical model was developed to express how acid concentration and temperature affect the properties of the biopolymer-treated soil. The SEM examination findings demonstrate that chitosan efficiently covered the soil particles and filled the void spaces. The soil was strengthened by the formation of hydrogen bonds and electrostatic interactions with the soil particles.

Keywords: biopolymer; chitosan; environmentally friendly; cost-effective

1. Introduction

The rapid growth of the world's population has increased the socio-economic demand for the development and expansion of civil infrastructure at an unprecedented rate. The construction of civil foundations requires proper soil improvement methods; more than 40,000 soil improvement projects with a total cost of more than 6 billion dollars per year are implemented worldwide [1]. The main purpose of ground improvement is to enhance the bearing capacity, shear strength, soil stiffness, resistance improvement against surface erosion, and soil hydraulic conductivity and seepage [2–6]. Climate change and global warming are two of the most concerning trends of the last several years. Regarding the materials used in construction projects, both synthetic and conventional materials with a calcium basis, including cement and lime, have contributed to the production of greenhouse gases that are directly linked to global warming. The production of one unit of cement emits one ton of carbon dioxide, accounting for five to eight percent of global carbon dioxide emissions [7–10].

Soil stabilization methods are usually divided into three main categories: mechanical, biological, and chemical. Conventional chemical stabilizers such as cement and lime are still the most common materials in large-scale soil improvement projects. However, their excessive use leads to significant environmental issues, including CO_2 emissions during cement production, suppression of plant growth, groundwater pollution, and global warming. In contrast, biological approaches such as microbially induced calcite precipitation (MICP), enzyme-induced calcite precipitation (EICP), and biopolymer application in soil enhancement have attracted greater attention in the field of geotechnical engineering due to their minimal negative effects on the environment [11–14].

Biopolymers are natural polymeric products that are environmentally friendly and sustainable, with high cohesive properties, and can be widely found in the environment in different forms [15–18]. These properties make them a suitable choice for soil improvement applications. Biopolymers, using natural polymers such as polysaccharides, have several advantages for engineering purposes in the enhancement of soil structure, water retention, and the reduction in wind, water, and unwanted vegetation growth erosion [19,20]. However, the widespread use of biopolymers is not still very common in engineering fields due to their relatively high costs. The costs of biopolymer production and application have significantly reduced over the past years. Polysaccharides are known for their high potential and effectiveness in improving soil properties [21–24].

Chitosan is a natural polysaccharide derived from chitin as well as the exoskeletons of crustaceans such as crabs and shrimps [25]. Chitosan has soil-binding, water-retention, and biodegradable properties. Several studies have been conducted to examine the potential of chitosan in soil enhancement [26]. It was shown that adding chitosan improved the mechanical properties of different types of soils (clay and sand) by positively increasing the interparticle interaction of the soil particles and biopolymer. Chitosan was presented as an ecologically acceptable substitute for conventional additives, such as cement and lime, when Jamshidi et al. (2023) investigated its impact on marl soil qualities and found that the polymer greatly enhanced soil shear wave velocity and maximum shear modulus by 51 and 128% after 7 days curing even after 8F-T cycles compared to those of unstabilized samples [27]. Ilman et al. (2023) examined the impact of chitosan on earthen construction materials. Chitosan was shown to act as a reliable stabilizer, enhancing the mechanical properties of the soil and preventing the development of visible cracks. This is explained by the creation of a stable framework that connects the particles of the soil, therefore increasing its durability [28]. Additionally, it was found by Adamczuk et al. (2022) that chitosan behaves differently in different types of soils [29]. No secondary environmental impacts have been observed when using chitosan as an adhesive in soil improvement applications, aligning with the goal of transitioning from traditional to new-generation materials in geotechnical engineering [30].

The number of studies on the use of chitosan as a soil stabilizer is limited; however, there is still a significant gap in exploring all aspects of this research field. This paper aims to investigate the behavior of chitosan-treated soil under various curing temperatures, long-term curing periods, and different acidity concentrations of chitosan solution. Biopolymer content and soil particle size are additional parameters examined in this research.

These parameters were evaluated using the unconfined compressive strength (UCS) test. Consolidated drained (CD) static triaxial testing was employed to examine how optimal conditions of these parameters influence the shear strength of kaolinite and sand. SEM images were included to provide visual insights into chitosan-treated sand and kaolinite. Additionally, chemical analysis was conducted to assess how temperature and the acidity concentration of the biopolymer solution affected soil properties. Upon completing this study, we aim to deepen understanding of chitosan-treated soil and the effectiveness of different parameters in geotechnical engineering applications.

2. Materials

2.1. Kaolinite

The kaolinite silt used in this research was purchased from Kaolin Malaysia, a mining products specialist. The kaolinite was taken from depths of 4.0 to 6.0 m below the ground surface. Table 1 shows the results of the XRD analysis performed on the kaolinite silt. The composition of kaolinite is 78% silt, 21.12% clay, and 0.88% sand, as determined by geotechnical criteria derived from particle size distribution and hydrometer tests (Table 2). According to the Unified Soil Classification System (USCS), it is classified as high-plasticity kaolinite silt (MH) with a Skempton activity rating of 0.77%. This description illustrates its versatility, necessitating a thorough examination of its properties and behavior in order to optimize its utilization.

Table 1. MH silt XRD analysis.

Component	Percentage (%)
SiO_2	48.97
Al_2O_3	35.19
K_2O	2.51
Fe_2O_3	0.88
MgO	0.35
TiO_2	0.23

Table 2. The kaolinite silt's geotechnical characteristics.

Soil Type	Sand Fraction (%)	Silt Fraction (%)	Clay Fraction (%)	L_L (%)	P_L (%)	P_I (%)	USCS	Activity = P_I/Clay Content (%)
Kaolinite	0.88	78.00	21.12	62	46	16	MH	0.77

2.2. Sand

The sand employed in this study was obtained from a construction site located in Gold Coast, Australia. According to the USCS classification, it is classified as poorly graded sand (SP). The gradation analysis of the utilized kaolinite silt and sand is depicted in Figure 1. The values of the gradation coefficient (Cc) and uniformity coefficient (Cu) are 2.77 and 0.91, respectively. The void ratio of the sand varies between 0.61 and 0.76. The XRD analysis of the sand's chemical composition is detailed in Table 3, whereas the geotechnical properties are presented in Table 4.

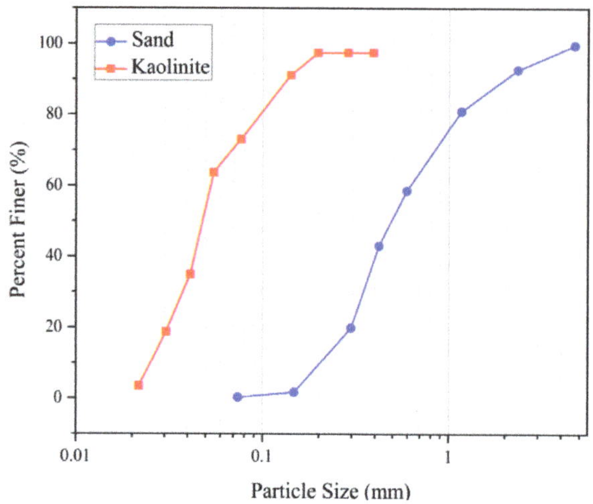

Figure 1. Particle size distribution of kaolinite silt and sand.

Table 3. XRD analysis of sand.

Component	Percentage (%)
Quartz	76.5
Calcite	3.4
Dolomite	0.1
Siderite	0.7
Siderite (Mg/Ca)	2.2
Andradite	1.0
Plagioclase (An0-15)	7.7
K-Feldspar	3.0
Kaolinite	0.9
Chlorite	1.4
Illite	3.2

Table 4. Geotechnical properties of the sand.

Soil Type	D_{50} (mm)	C_u	C_c	G_s	Shape	UCSC	e_{min}	e_{max}
Sand	0.49	2.77	0.91	2.63	Round	SP	0.61	0.76

2.3. Chitosan

For Chitosan manufacturing objectives, the United States, Japan, India, Canada, China, South Korea, Russia, and Norway frequently employ the byproducts of crustacean fisheries. The principal industrial source from which the biopolymer is extracted is fishery industry detritus, specifically the shells of prawns, crabs, and lobsters. An estimated 100 billion kilograms of chitin and its derivatives are produced on an annual basis. Because of its strong hydrogen bonds and polymeric structure, chitin is insoluble in most organic acids and water. With CAS No. 9012-76-4, the Chemsupply firm provided the chitosan that was used. Table 5 and Figure 2 provide specifics on the chitosan biopolymer. The molecular weight

(MW) of the chitosan was determined to be 120–140 kDa, and the degree of acetylation (DA) was approximately 15%.

Table 5. Properties of the chitosan biopolymer.

Product Name	Chitosan
Form	Solid
Chemical formula	{C6H11NO4}n
Appearance	Faint Beige to Beige
Melting point	>290 °C
Solubility in water	Insoluble in water and organic solvent, soluble in dilute aqueous acidic solution (pH < 6.5)
pH	6.5–7.5

Figure 2. Chitosan biopolymer used in this study.

A standard compaction test was conducted to determine the maximum dry density and optimum moisture content of the soil. This information was important for calculating the appropriate amounts of soil and water needed to prepare the samples for both untreated and treated soils. The experiment was performed based on the ASTM D698 [31] standard and the results are shown in Figure 3.

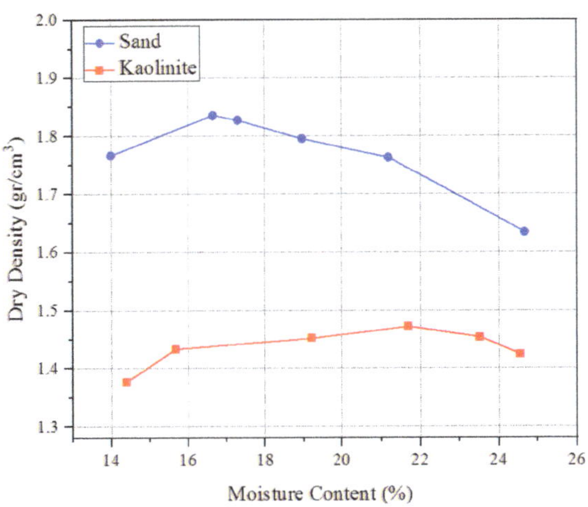

Figure 3. Compaction curves for sand–kaolinite mixtures.

3. Research Method

3.1. Sample Preparation

To attain maximum efficiency, the wet mixing method was implemented, in which, first, the biopolymer dissolves in the solvent until reaching a desirable concentration and then the biopolymer solution is mixed with soil. To prepare the biopolymer solution, the following steps were taken: (a) the solution was heated to 80 °C after dissolving the specified amount of glacial acetic acid in distilled water to obtain a 1% concentration; (b) chitosan powder was added to the acetic acid solution as required; and (c) constant stirring continued until a homogeneous solution was obtained, ensuring that the concentration remained moderate enough to dissolve all components.

The soil was subsequently mixed with the solution for 10 min after being dehydrated in an oven set at 105 °C for 24 h. The mixture was then packed in a bag for a day to ensure even moisture distribution. A small sample was taken and placed in the oven to determine its moisture content.

A specially designed mold was fabricated in compliance with ASTM D2850 [32] and ASTM D4609 [33] guidelines to ensure consistency in the sample preparation. A cylindrical stainless-steel tube with two wings attached to the sides served as the mold's main component. A square stainless-steel plate, a plug, short and long hammers, and other supporting components were included. The cylinder's fixed dimensions were 50 mm in diameter and 200 mm in height. During the sample fabrication process, lubrication was provided to the inside walls of the tube, the plug, and the short hammer's surfaces. The square plate made it easier to modify the mold's location once the mixture was put inside. Using a hydraulic jack, compaction was carried out in a single layer until the maximum density determined by compaction tests was exceeded by 90%. Following compaction, the samples were extruded and cured in a controlled setting at a temperature of 25 °C and a relative humidity of 50–60%. The cylindrical samples that were selected for triaxial and UCS testing had dimensions of 100 mm in height and 50 mm in diameter.

3.2. Experimental Program

This study involved various experiments, including compaction, UCS, static triaxial testing, and SEM imaging, to assess the impact of the biopolymer on the engineering properties of the soil. UCS and triaxial tests were used to evaluate the mechanical properties of the biopolymer-treated soil, and SEM photography was used to provide a microstructural representation of the biopolymer–soil interaction. Table 6 shows the experimental procedures for this study.

Table 6. Experimental program.

Test Type	Biopolymer Content (%)	Long Term Strength Assessment (Days)	Soil Type	Acid Concentration (g/L)	Temperature (°C)
UCS	0, 0.25, 0.5, 1, 1.5, 2	14	Kaolinite, Sand	0.5	25
	0, 0.5	0, 1, 3, 7, 14, 28, 60, 90, 120	Kaolinite, Sand	0.5	25
	0, 0.5	14	Kaolinite, Sand	0, 1, 2, 3, 4, 5, 10, 20, 50	25
	0, 0.5	14	Kaolinite, Sand	0.5	15, 25, 45, 60, 80
Triaxial	0, 0.5	14	Kaolinite, Sand	0.5	25
SEM	0, 0.5	14	Kaolinite, Sand	0.5	25

Details of the different parameters investigated in this study can be found in Table 6. Biopolymer content, long-term curing assessment, acid concentration, and temperature were evaluated through the UCS test. Samples were kept for up to 120 days to see how the compressive strength of the biopolymer-treated soil changed over time. Acid concentrations are a crucial factor in preparing chitosan solutions; therefore, different concentrations were prepared to evaluate this parameter. For the temperature effect, samples were cured from 15 to 80 °C.

3.3. Unconfined Compression Test

A universal testing machine was used to evaluate the compressive strength in accordance with ASTM D2166 [34]. UCS experiments were utilized to determine the ideal values for the following factors: temperature, soil type impact, acid concentration, biopolymer content, and long-term strength assessment. The device monitored and recorded stress–strain behavior as the axial strain rate increased from 1%/min (1 mm/min) to 7% of the total strain. The average of the three specimens evaluated in each condition was used to calculate the unconfined compressive strength.

3.4. Infiltration Test

To study the impact of water flow on the UCS of chitosan-treated soil, a controlled infiltration test was designed and performed. The test included samples placed inside a polyethylene cylinder with inner diameter of 50 mm and height of 100 mm. Each sample was enclosed with a cap tightened at both the top and bottom. A 10 mm diameter hole was made in the center of each cap, into which a small tube was inserted to facilitate water entry and exit. The overall shape of the sample can be seen in Figure 4.

(a)

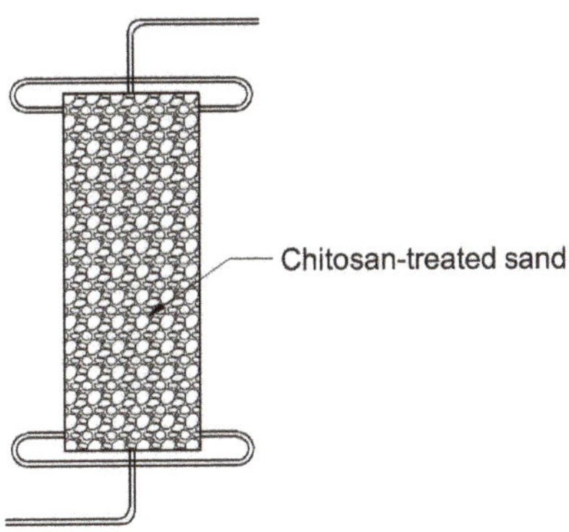

(b)

Figure 4. (**a**) Infiltration samples and (**b**) schematic form of the mold.

Water entered the samples through the upper cap, and a constant hydraulic head of 2 m was maintained to ensure a steady flow through the sample. Water exited through the lower cap, and the water flow was sustained for 8 h. Following the infiltration process, the samples were carefully extracted from the tubes and allowed to air dry for 14 days. After the drying period, UCS tests were performed to assess the extent of biopolymer loss due to water flow and its subsequent effect on the compressive strength of the treated samples.

3.5. Static Triaxial Test

The triaxial specimens were prepared under the same circumstances used for the UCS test, and they were then cured in a controlled atmosphere at 25 °C and around 60% relative humidity. Consolidated undrained (CU) triaxial testing was carried out in accordance with ASTM D7181 specifications [35].

In order to evaluate the impact of chitosan on sand and kaolinite, a series of static triaxial experiments were implemented. The initial confining pressure of each sample was kept at 100 kPa. Static triaxial testing was carried out on both saturated and dry samples. In order to facilitate the breakdown of air within the pore space into deaired water, saturation was achieved by maintaining an effective stress of 20 kPa while applying back pressure from the sample's bottom. Saturation persisted until a B value greater than 95% was obtained, indicating successful saturation. A small amount of consolidation was expected at this step due to the sample preparation using a hydraulic jack, so a quick consolidation under 50 kPa pressure was applied as a safety precaution. The applied strain rate was fixed at 0.1 mm/min during the shearing step. The overall shape of the system used for the triaxial testing is shown in Figure 5.

Figure 5. Triaxial testing system.

3.6. Scanning Electron Microscopy (SEM) Imaging

SEM images were obtained to investigate the microscopic interactions and structural characteristics of the biopolymer and soil. The motivation behind using SEM was to observe the detailed surface morphology and bonding between chitosan and the soil particles. For SEM characterization, soils treated with biopolymers as well as untreated soil (sand, kaolinite) were used. To ensure dryness, the samples were dried for 24 h at 25 °C in the oven. Carbon-conductive tabs were then used to attach them to a SEM mount. After applying a coating of carbon paint to guarantee adequate electric grounding, conductive tape was used to cover the samples. For specimen observation, the Zeiss Sigma VP Field Emission Scanning Electron Microscope (Oberkochen, Germany) was utilized.

4. Results

4.1. Unconfined Compression Strength (UCS) Test

4.1.1. Biopolymer Content Effect on the Compressive Strength of Chitosan-Treated Soil

Unconfined compression strength (UCS) testing plays an important role in evaluating biopolymer influence in soil treatment by measuring the maximum compressive load that a sample can withstand. Figure 6 shows the UCS and modulus of elasticity (EM) values for biopolymer-treated sand and kaolinite soils with varying amounts of chitosan (0, 0.25, 0.5, 1, 1.5, and 2%) at a curing temperature of 25 °C for a period of 14 days. From the compaction test, the optimal moisture content for this test for sand and kaolinite soil was determined as 17 and 35%, respectively. Figure 6 shows that EM represents the elasticity modulus, and UCS stands for the unconfined compressive strength.

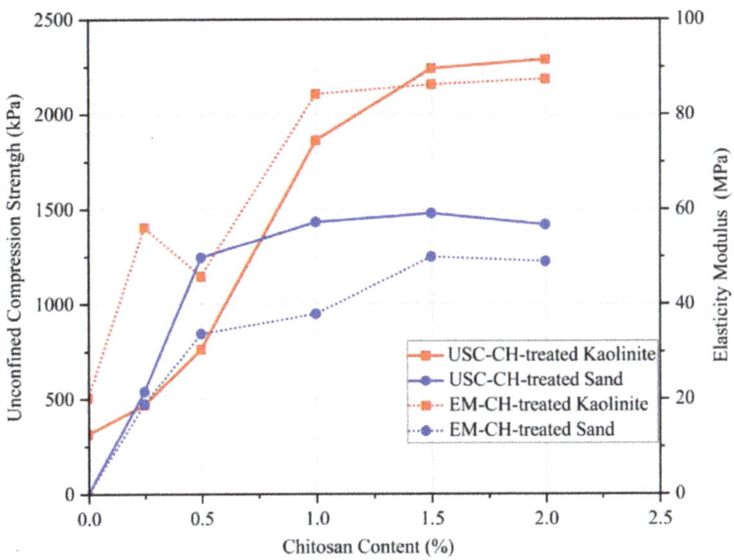

Figure 6. Effect of chitosan content on the mechanical behavior of CH-treated soils' compressive strength and elasticity modulus.

According to Figure 6, chitosan biopolymer increased the UCS of soils regardless of the soil type. The compressive strength of untreated sand and kaolinite was obtained as 0 and 316 kPa, respectively. Adding 0.25% chitosan increased the compressive strength to 540 kPa for the treated sand and 469 kPa for the treated kaolinite, indicating that sand outperformed kaolinite in the chitosan-treated soil. Up to 0.5% chitosan, sand still maintained a higher strength. The coarse-grained nature of sand facilitated more effective mixing with chitosan compared to kaolinite, which was primarily responsible for this improvement.

Kaolinite's compressive strength exceeded that of sand after adding 1% chitosan, highlighting its superior bonding and integration with the finer particles of kaolinite. The optimal compressive strength for stabilized sand was obtained as 1480 kPa with 1.5% chitosan, while for stabilized kaolinite, it was 2290 kPa with 2% chitosan. In conclusion, the addition of chitosan to sand and kaolinite soils forms bonds that enhance soil strength [16].

However, the UCS test indicates that, with increasing chitosan content, kaolinite soil exhibits greater strength compared to sand soil. This is due to various bonding mechanisms: (a) hydrogen bonding between chitosan and hydroxyl kaolinite surface; (b) electrostatic interaction between charged surfaces of kaolinite and chitosan; (c) hydrophobic bonding; and (d) covering kaolinite particles with chitosan chains to create a strong and integrated network in the particles. It prevents soil particles from shifting and increases soil strength. Sand particles' lack of significant electric charge makes electrostatic and hydrogen bonding phenomena unlikely.

Figure 6 shows that the addition of chitosan increases the soil's modulus of elasticity, which is similar to compressive strength. The modulus of elasticity of untreated sand and kaolinite was obtained as 0 and 20 MPa, respectively. Sandy soil lacks cohesion between its grains and therefore cannot withstand any significant strength. As a result, its modulus of elasticity was obtained as zero. On the other hand, kaolinite soil has a cohesive nature between its grains, which allows it to withstand some strength. This cohesive characteristic gives the kaolinite soil a non-zero modulus of elasticity. However, when 1.5% chitosan is added, the modulus of elasticity slightly decreases for the treated sand but stays constant for the treated kaolinite. Due to its lubricating effects, adding more chitosan to the soil results in a decrease in soil particle cohesion. Additionally, by distributing polymeric chains of biopolymers across the soil void space and covering the sand surfaces, a robust and well-integrated film network is created throughout the treated soil mass, improving the resistance forces and preventing soil grain movement. The optimal modulus of elasticity with 2% chitosan sand was 49 MPa, while for kaolinite, it was 86 MPa with 1.5% chitosan.

4.1.2. Long-Term Curing Time Effect on the Strength Behavior

The impact of chitosan treatment over time is investigated in this section. Samples were kept up to 120 days so that the effect of long-term curing could be studied using a UCS test. The samples were prepared at a temperature of 25 °C using a biopolymer solution with an acidity concentration of 10 g/L and a chitosan content of 0.5%. Various curing times were considered for the samples, including 0, 1, 3, 7, 14, 28, 60, 90, and 120 days.

Figure 7a shows the UCS and modulus of elasticity as well as the moisture content over time for sand and kaolinite soils treated with chitosan over time. As curing time progressed, the UCS of both treated soils increased regardless of the soil type. At the time right after sample preparation, the UCS of the treated kaolinite was around 334 kPa, while the treated sand showed zero strength due to the lack of cohesion between the soil grains. After one day of dehydration, the strength of the treated sand increased dramatically to 697 kPa, which was higher than that of the treated kaolinite. The reason is that sand has a more porous structure, making moisture loss easier compared to the treated kaolinite with much finer particles.

After seven days, approximately 90% of the UCS was achieved for both treated soils, indicating that the size of soil particles does not significantly impact the results after one week. Moisture content was reduced to less than 3%, as curing time has an inverse relationship with moisture content. Samples reached their final strength at 14 days after preparation, which was around 760 kPa for the treated kaolinite and 1250 for the treated sand. No significant changes in compressive strength were observed from 14 days to 120 days, demonstrating that the bonds between the biopolymer and soil particles maintained their integrity over four months. Additionally, the moisture content remained below 1.5%, indicating that the samples were almost completely dried, with only small amounts of moisture retained from the ambient humidity. The decline in humidity over a period of 14 days was a significant contributing factor to the rise in UCS. Moisture is crucial for dissolving chitosan in the solvent (a water and acetic acid solution). As the water content decreases over time, the formed gel becomes stronger and more brittle. This process enhances the polymeric chains of chitosan, creating a stronger soil–biopolymer mass, which results in increased compressive strength.

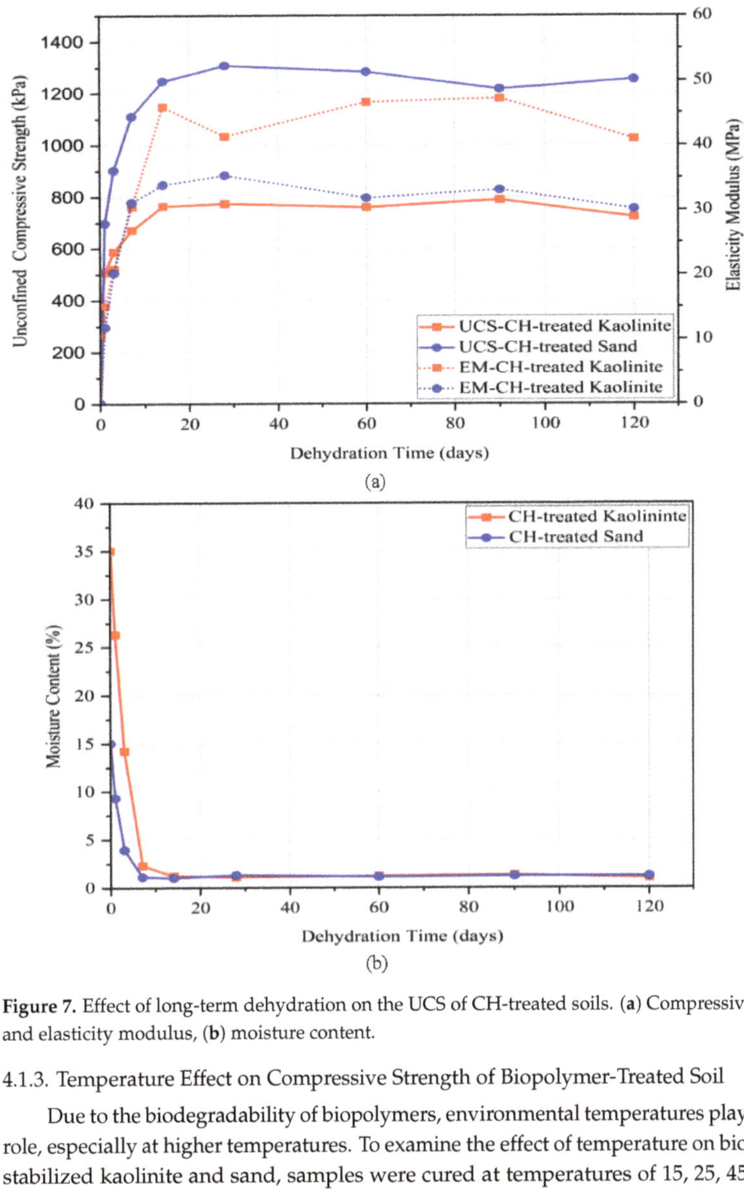

Figure 7. Effect of long-term dehydration on the UCS of CH-treated soils. (**a**) Compressive strength and elasticity modulus, (**b**) moisture content.

4.1.3. Temperature Effect on Compressive Strength of Biopolymer-Treated Soil

Due to the biodegradability of biopolymers, environmental temperatures play a crucial role, especially at higher temperatures. To examine the effect of temperature on biopolymer-stabilized kaolinite and sand, samples were cured at temperatures of 15, 25, 45, 60, and 80 °C and then tested using UCS tests. The samples were kept at the specified temperatures for 14 days, with a chitosan content of 0.5% and an acidity concentration of biopolymer solutions of 10 gr/L. Figure 8 shows the UCS for the biotreated sand and kaolinite soils versus temperature. As shown, the compressive strength of both treated soils increased with the elevation of temperature up to 45 °C. For the chitosan-treated kaolinite, 60 °C was the peak UCS, after which it started to decrease at 80 °C. For the chitosan-treated sand, the UCS decreased after 45 °C, indicating its optimal dehydration temperature. The increased soil compressive strength from 15 to 45 °C was due to the dehydration of biopolymer gel and forming stronger chemical bonds.

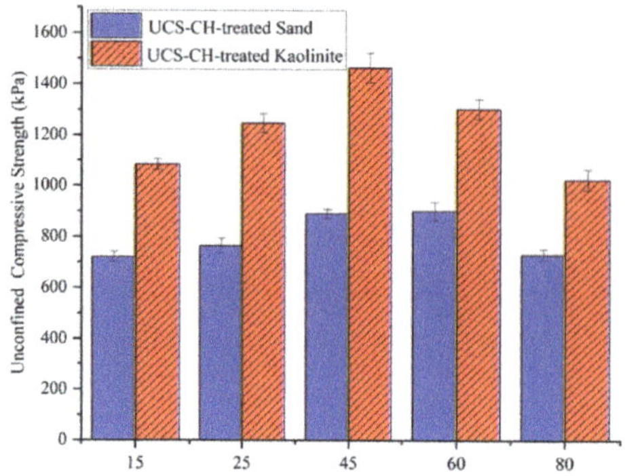

(a)

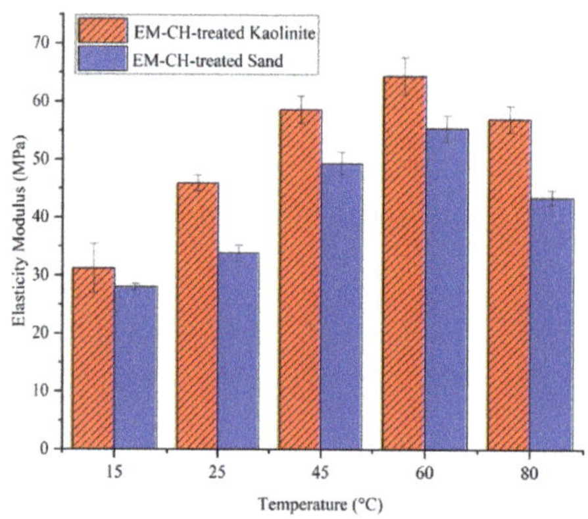

(b)

Figure 8. Effect of temperature on the UCS of CH-treated soils. (**a**) Compressive strength, (**b**) elasticity modulus.

From the elasticity modulus graph, it can be observed that very high temperatures (80 °C) led to a reduction in the stiffness of the samples. The reason is that the chitosan hydrogel shrinks due to dehydration, forming brittle chitosan fibers and involving a smaller surface area of soil particles in the bonding mechanisms. This effect is especially pronounced in treated sand, which has no electrical charge on its particle surfaces, resulting in smaller adhesion surface contact between the biopolymer film and soil grains and consequently lower compressive strength parameters. On the other hand, biopolymers have the ability to produce more gel at higher temperatures; therefore, increasing the temperature causes chitosan to generate more gel, which leads to soil improvement.

Figure 9 shows the variation of moisture content for samples cured at different temperatures. As seen, higher temperatures resulted in lower moisture content after 14 days of dehydration. An inverse correlation is observed between moisture content and UCS. Referring to the obtained results, it can be said that very low (around 15 °C) and very high (above 60 °C) temperatures result in relatively lower compressive strengths compared to the samples cured within this range. This is because lower temperatures retain higher moisture content, resulting in more flexible biopolymer films, whereas higher temperatures absorb almost all the moisture, leading to brittle biopolymer films.

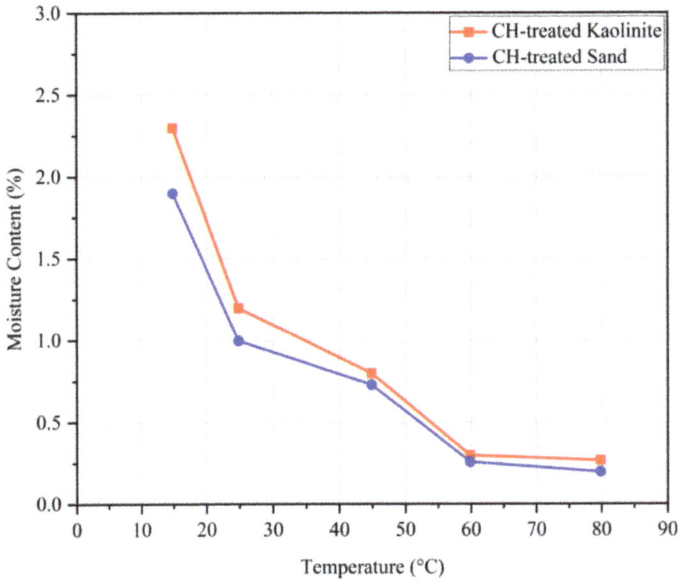

Figure 9. Effect of temperature on the UCS of CH-treated soils.

4.1.4. Acid Concentration Effect on the Behavior of the Treated Soil

Chitosan is a biopolymer that is insoluble in neutral water with a pH around 7. To prepare a chitosan solution, the pH needs to be lowered, as chitosan is soluble in diluted acid solutions. Acetic acid is commonly used for this purpose. The dissolution of chitosan in a diluted acid involves the protonation of its amine groups (-NH_2), making it soluble. The chemical reaction for dissolving chitosan in a diluted acid is as follows:

$$\text{Chitosan-}NH_2 + CH_3COOH \rightarrow \text{Chitosan-}NH_3^+ + CH_3COO^-$$

Chitosan needs a sufficient degree of protonation of its amino groups (-NH2) to dissolve completely in the acid solution. At lower concentrations, there may not be enough acetic acid molecules available to effectively protonate the chitosan molecules. Conversely, higher concentrations of acetic acid may overly protonate the chitosan molecules. An excessive positive charge can alter the structure and properties of chitosan and affect its solubility characteristics. Very high concentrations of acid can also lead to the precipitation of chitosan due to the strong acidic environment. Therefore, investigating how acid concentrations affect the behavior of chitosan-treated soil is of high importance. For this purpose, various concentrations of acid were prepared, including 0, 0.1, 0.2, 0.3, 0.4, 0.5, 1, 2, and 5%, with the biopolymer content set at 0.5% of the soil weight. Similar to the previous sections, kaolinite and sand were the soils used in this study. The treated samples were dehydrated for 14 days at a temperature of 25 °C.

The mechanical properties of the chitosan-treated soil can be seen in Figure 10. At 0% acid concentration, chitosan is approximately insoluble in neutral water, resulting in chitosan precipitating in the solution with only a small amount dissolving. Consequently, the sand samples were not able to maintain their structure, yielding zero strength. For the chitosan-treated kaolinite, only a slight improvement in the UCS was observed. Adding a small amount of acid to water (0.1%) significantly increased the compressive strength from 432 kPa to 546 kPa for the treated kaolinite and from zero to around 100 kPa for the treated sand. This increasing trend remained very promising up to a 0.5% acidity solution, indicating that a large percentage of chitosan dissolved into the solution. According to the literature, 0.5% to 1% acidity solution is commonly used for most chitosan applications, suggesting that this range provides enough acid molecules to effectively dissolve chitosan. However, at a 5% acid concentration, no significant change in the UCS was observed.

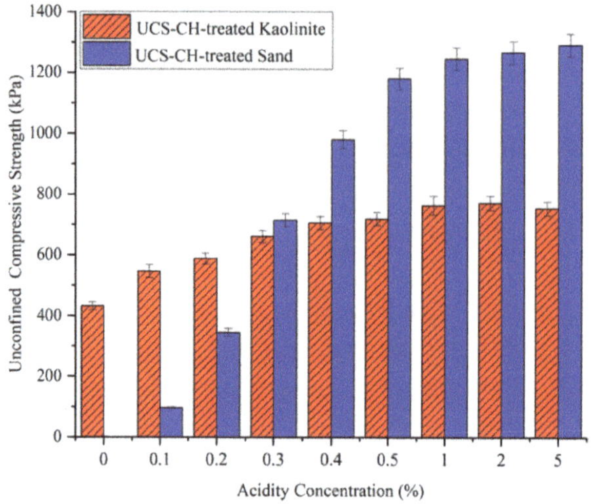

(a)

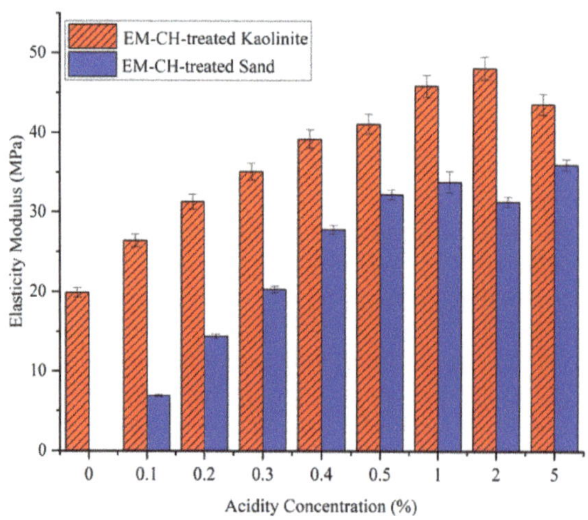

(b)

Figure 10. *Cont.*

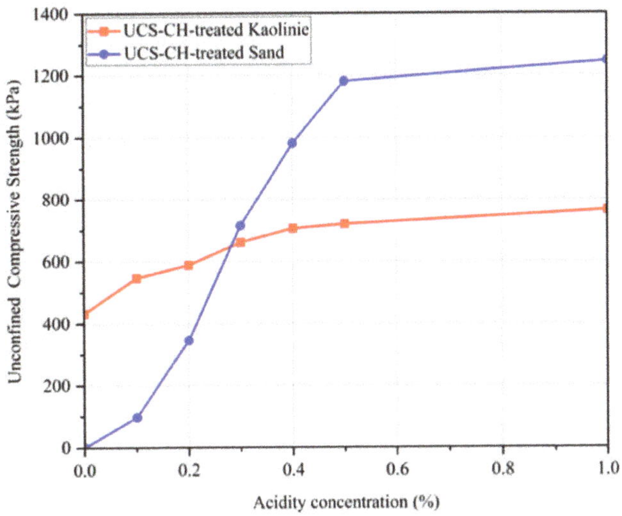

(c)

Figure 10. Effect of acid concentration on the UCS of CH-treated soils. (**a**) Compressive strength, (**b**) elasticity modulus, (**c**) variation of UCS vs. acidity concentration between 0–1%.

In real-world applications, whether temporary or long-term, acid rain is a potential concern regarding the leaching of biopolymer from chitosan-treated soil. According to available data, the typical acid concentration in acid rain ranges from 0.00002 to 0.00006 g/L. These concentrations are significantly lower than the levels required to solubilize chitosan, which necessitates much higher acid concentrations (1000 to 50,000 mg/L or 0.1% to 5%). Figure 11 shows a comparison between the solutions with and without enough acid concentration.

Figure 11. Chitosan in water vs. low concentration acid (0.5%): dissolution comparison.

4.2. Infiltration Test

Figure 12 shows the results obtained from the UCS tests before and after the infiltration process. The results prove that chitosan exhibits a remarkable resistance to a consistent water flow with minimal reduction in the compressive strength. For 0.25% chitosan, a decrease of 4.4% was observed in UCS, from 540 kPa to 516 kPa. Similarly, the 0.5% chitosan-treated sample saw a decrease from 1246 kPa to 1198 kPa, with a reduction percentage of just 3.85%. The higher chitosan content samples, 1% and 1.5%, experienced reductions in UCS from 1434 kPa to 1352 kPa and from 1480 kPa to 1402 kPa, corresponding to reductions of 5.72% and 5.27%, respectively.

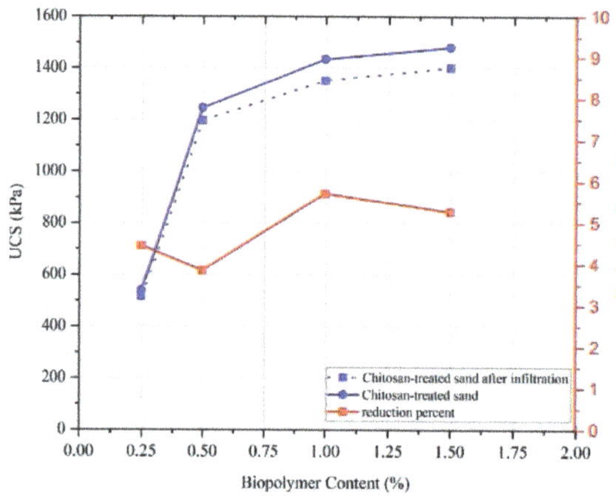

Figure 12. Compressive strength change for chitosan-treated sample after infiltration test.

These results clearly demonstrate that the infiltration process led to only slight decreases in compressive strength, which indicates that a small amount of the chitosan was washed away. This underscores the potential of chitosan as a reliable additive in civil and geotechnical applications, especially in environments where the soil may be subjected to water infiltration over time. Chitosan's insolubility in water ensures that its chemical structure remains intact, providing sustained stability and effectiveness as a soil stabilizer even under prolonged water exposure.

4.3. Static Triaxial Test

A set of consolidated drained (CD) triaxial tests were conducted on the untreated and chitosan-treated kaolinite samples, as well as on untreated and chitosan-treated sand, to investigate how chitosan biopolymer affects the shear strength properties of different soil types. The sample preparation followed the same procedure as the UCS samples, with compression applied using a hydraulic jack to achieve a dry density higher than 95% of the maximum dry density (MDD). Chitosan was used at a concentration of 0.5% (mb/ms) to treat both kaolinite and sand samples. The samples were allowed to dehydrate for 14 days at a temperature of 25 °C.

Three different confining pressures (50, 100, and 200 kPa) were applied consistently across all tests to compare the shear strength at failure for different samples. The results are illustrated in Figure 13 shows the stress–strain behavior of untreated and chitosan-treated kaolinite at confining pressures of 50, 100, and 200 kPa. The failure envelopes for the tested samples are presented in Figure 14.

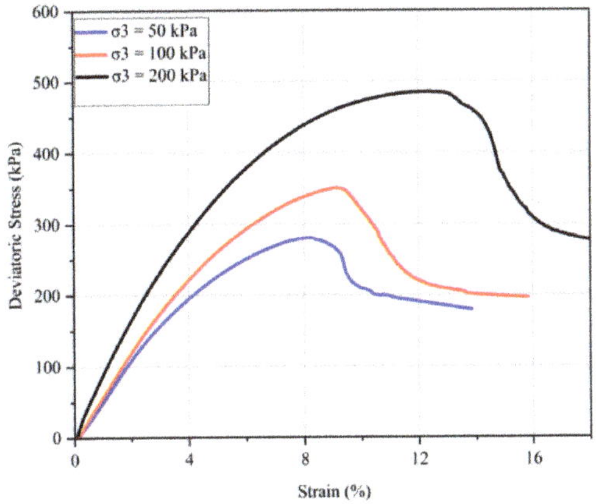

(a)

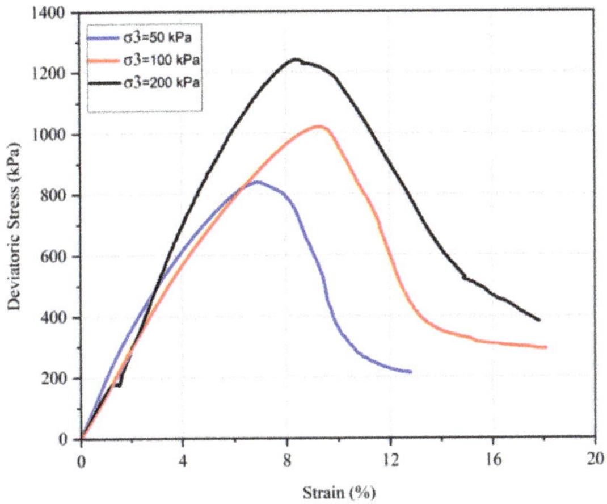

(b)

Figure 13. Stress–strain behavior of (**a**) untreated kaolinite and (**b**) chitosan-treated kaolinite.

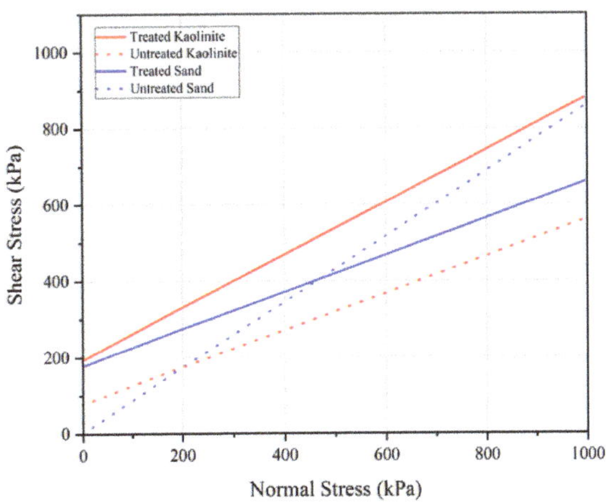

Figure 14. Failure envelope of treated and untreated kaolinite.

The data indicate that for both untreated and chitosan-treated kaolinite, an increase in confining pressure corresponds to an increase in strength. For instance, the maximum deviatoric stress for untreated kaolinite increased from 280 kPa to 350 kPa and further to 490 kPa when the confining pressure was increased from 50 kPa to 100 kPa and 200 kPa, respectively. In contrast, the chitosan-treated kaolinite samples exhibited a significant enhancement in shear strength, with deviatoric stresses rising from 840 kPa to 1020 kPa and 1240 kPa under the same confining pressures. Residual strengths for the chitosan-treated samples were also notably higher compared to the untreated kaolinite.

Similarly, for the chitosan-treated sand, the deviatoric stresses were recorded as 520 kPa, 651 kPa, and 785 kPa under confining pressures of 50, 100, and 200 kPa, respectively. The chitosan-treated sand also demonstrated substantial improvements in shear strength parameters, as shown by the cohesion value of 161 kPa and a friction angle of 39.5°, compared to untreated sand, which had a cohesion of 0 kPa and a friction angle of 36°.

Table 7 summarizes the Mohr–Coulomb shear strength parameters for the untreated and chitosan-treated samples.

Table 7. Summary of the results of the CD triaxial test.

Sample ID	Biopolymer Content (%)	C′	φ′
Untreated kaolinite	0	70	23.73
Chitosan-treated kaolinite	0.5	192	34.58
Untreated sand	0	0	36
Chitosan-treated sand	0.5	161	39.5

The results indicate a substantial improvement in the shear strength parameters for chitosan-treated kaolinite and sand. The effective cohesion (C′) of kaolinite increased from 70 kPa for untreated kaolinite to 192 kPa for chitosan-treated kaolinite, while the effective internal friction angle (φ′) increased from 23.76° to 34.58°. Similarly, for sand, the cohesion increased from 0 kPa to 161 kPa and the friction angle increased from 36.04° to 39.5° upon treatment with chitosan. This improvement can be attributed to the adhesive properties of chitosan, which enhances particle bonding and soil cohesion. The formation of chitosan–soil conglomerates during compaction further contributes to the increased interlocking and frictional resistance within the soil matrix.

In conclusion, the treatment of kaolinite and sand with 0.5% chitosan significantly enhances their shear strength, demonstrating the potential of chitosan as an effective soil stabilizer. The notable increase in both cohesion and friction angle underscores the efficacy of chitosan in improving soil mechanical properties, making it a viable option for various geotechnical applications.

4.4. Acid Concentration and Temperature Effects on the Interaction of Chitosan and Soil

Kaolinite, a hydrous aluminum phyllosilicate, is one of the most abundant clay minerals in Earth's crust and belongs to the dioctahedral 1:1 kaolin mineral group. In its structure, oxygen atoms bonded to silicon atoms form a tetrahedral sheet, referred to as the siloxane surface. Similarly, hydroxyl groups bonded to aluminum atoms form an octahedral alumina sheet, known as the aluminol surface. As depicted in Figure 15a, both sheets share the apical oxygen atoms. Higher MW polysaccharides may exhibit increased tensile strength but reduced solubility, while lower MW variants may lack sufficient bonding capabilities. Each kaolinite layer is considered as a strong dipole, where the aluminol surface is hydrophilic and dominated by positive charges, while the siloxane surface exhibits negative charges and is hydrophobic. Thus, the kaolinite layers are held together by strong hydrogen and dipolar interactions [36]. The edges of these layers consist of O atoms and OH groups.

In an acidic solution (pH < 3.6), the hydroxyl groups accept protons from the solution, forming protonated OH groups. Conversely, at pH > 3.6, they release protons into the solution, resulting in O^- ions. Therefore, the kaolinite surfaces exhibit two types of charges: a permanent negative charge on the tetrahedral face and a variable pH-dependent charge, either positive or negative, due to the protonation or deprotonation of hydroxyl groups at the amphoteric sites, including the edges and the octahedral faces [37,38]. Figure 15 shows scanning electron microscopic (SEM) images of untreated and treated soils.

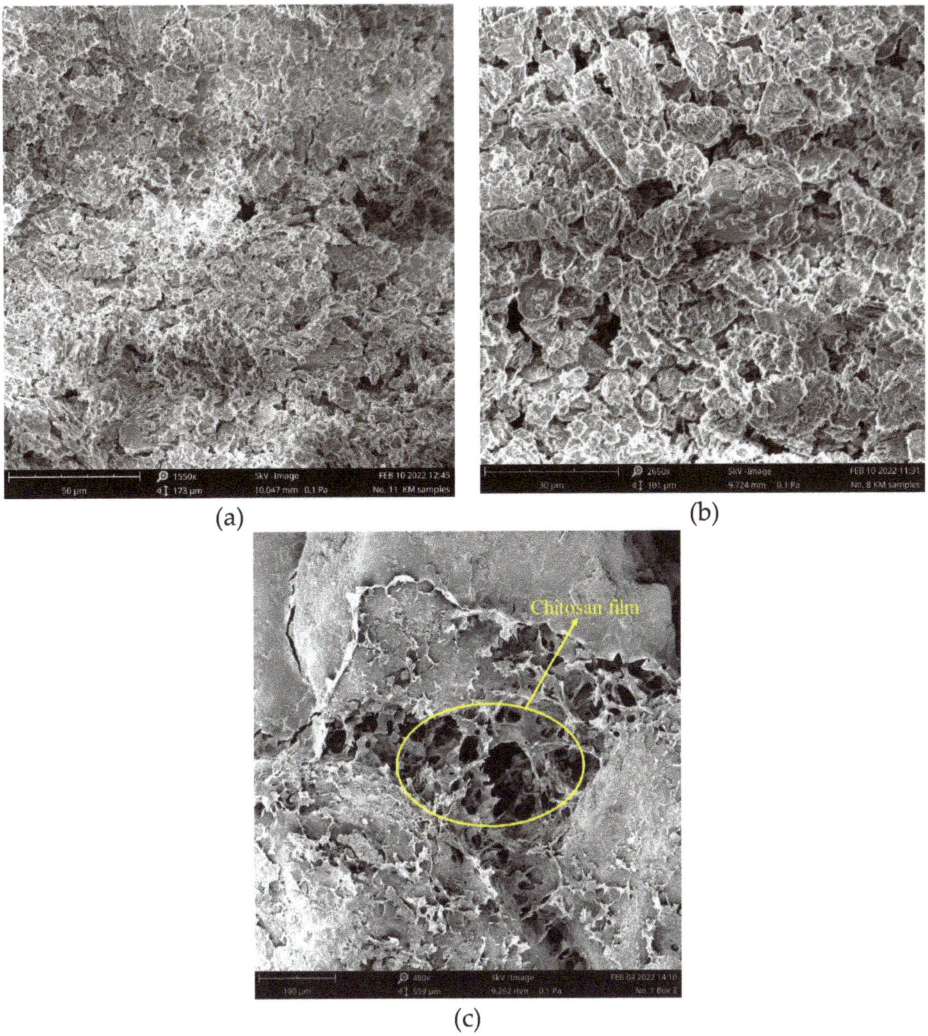

Figure 15. SEM images from untreated and treated samples. (**a**) Untreated kaolinite, (**b**) chitosan-treated kaolinite, (**c**) chitosan-treated sand.

The molecular structure of chitosan is also shown in Figure 16a. Chitosan, the deacetylated form of chitin, is one of the most common polymers found in nature and is composed of N-acetylated glucosamine and glucosamine units [39]. The degree of deacetylation (DD) determines the content of free amino groups (NH2) in the polysaccharides, which can be used to differentiate between chitin and chitosan [40]. Chitosan is a weak base that exhibits low solubility in organic solvents and is completely insoluble in neutral and alkaline environments. However, it dissolves in dilute aqueous acidic solutions (pH < 6.5), which can convert glucosamine units into the soluble form R-NH_3^+ according to the following chemical reaction:

$$R\text{-}NH_2 + AcOH \leftrightarrow R\text{-}NH_3^+ + AcO^-$$

where R-NH_2 represents the free amine form of chitosan, while AcOH denotes the undissociated form of acetic acid. R-NH_3^+ represents the protonated form of chitosan, which is soluble in water, and AcO^- is the acetate counter ion [41].

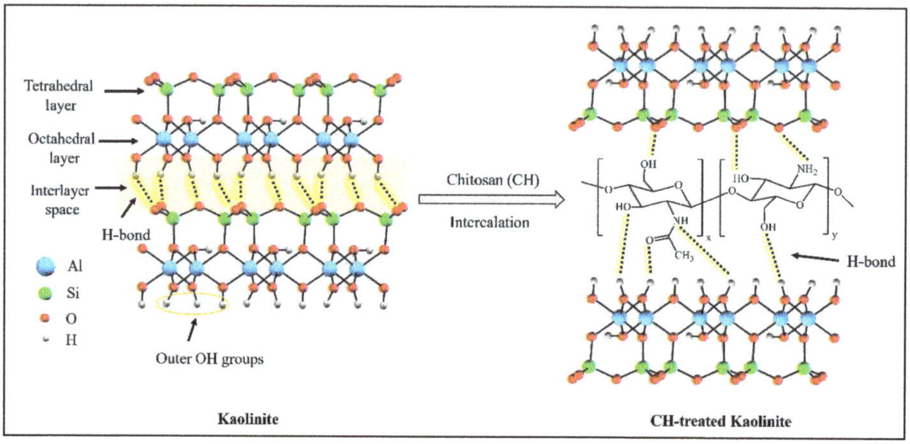

(a)

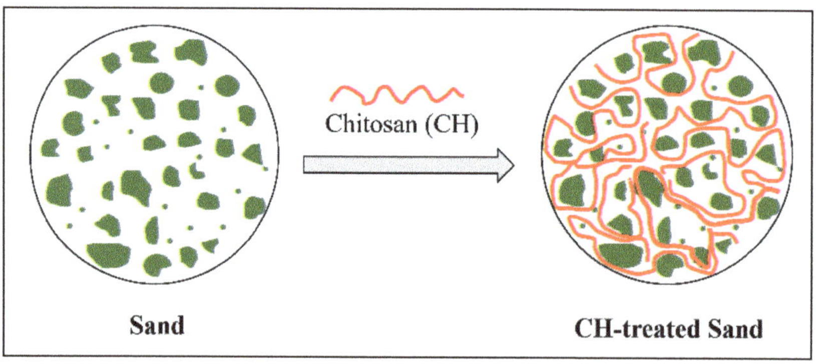

(b)

Figure 16. Interaction of chitosan biopolymer with soil. (**a**) Chitosan-treated kaolinite, (**b**) chitosan-treated sand.

The functional groups in both structures play an important role in the formation of the kaolinite–chitosan composite. Chitosan contains several active functional groups, including amine (-NH2), hydroxyl (-OH), and ketone (-C=O). These groups can form hydrogen bonds with the hydroxyl groups present on the surface of kaolinite (Figure 16a). This interaction improves the adhesion between chitosan and kaolinite in the composite. In acidic conditions (pH < 6.5), another potential interaction occurs between the NH3+

groups of chitosan and the negatively charged surface of kaolinite, involving electrostatic forces [42,43].

The major component of river sand is the SiO_2 compound. Unlike kaolinite, the sand does not have enough active and available functional groups to establish significant hydrogen bonding with chitosan functional groups. Two primary driving forces can be identified in studying the interactions between sand and chitosan in the composite: (1) interfacial mechanical interaction and (2) hydrophobic bonding. The interfacial mechanical interaction enhances the composite structure by creating polymer bonds around the sand grains, as illustrated in Figure 16b. In addition, a hydrophobic interaction occurs at the interface between the polymer fibers containing nonpolar carbon chains and the sand grains with uncharged surfaces [6].

The acid concentration and temperature are important factors that may affect the interactions in the composites. Heat treatment is important for modifying the microstructure and properties of chitosan composites. The temperature most probably modifies hydrogen bond distribution and improves chitosan–clay interactions. As shown in Figure 17a, increasing the temperature breaks down the hydrogen bonds that connect polymer chains. This encourages a more open structure, which enables cross-linking interactions. In addition, heat treatment reduces the moisture content and leads to the compaction of the composite structure. It should be noted that there is a temperature threshold known as the ceiling temperature. This temperature is critical because it marks the point above which the efficiency of the reaction or the strength of cross-linking interactions begins to decrease. Additionally, higher temperatures can enhance the mobility of chitosan chains, facilitating better interaction and bonding with soil particles. This can lead to increased tensile strength and durability of the composite. However, if the temperature exceeds the optimal range, it can cause thermal degradation of chitosan, weakening the composite structure. Proper temperature control is essential to maintaining the balance between enhanced interaction and preventing thermal damage. Furthermore, temperature variations can influence the rate of moisture evaporation, which plays a crucial role in the final stabilization of the soil composite [44–46].

Figure 17b illustrates the mechanism demonstrating the necessity of sufficient acid for dissolving chitosan. The adsorption of chitosan onto kaolinite is mainly driven by electrostatic attraction between positively charged chitosan molecules at low pH and negatively charged kaolinite surfaces. The pKa value of chitosan has been previously reported to be around 6.5–7.0. Most glucosamine units in chitosan are protonated at pH 3.5. Approximately half of the glucosamine units are protonated at pH 6.5, while most glucosamine units are not protonated at pH 8.5. These results show that increasing the acid concentration can increase the number of protonated glucosamine units, thereby enhancing the electrostatic interaction between kaolinite and chitosan. Moreover, under acidic conditions, the solubility of chitosan in water increases, facilitating the homogeneous adhesion of chitosan onto sand particles. Additionally, an optimal acid concentration is crucial for maximizing the dissolution of chitosan, ensuring that it evenly coats soil particles and improves composite integrity. However, excessively high acid concentrations can lead to over-protonation of chitosan, resulting in excessive positive charge that may repel soil particles rather than bind them effectively. This can decrease the overall stability and strength of the composite. Furthermore, very high acid concentrations can cause chitosan to precipitate out of the solution, reducing its effectiveness as a stabilizing agent. Therefore, maintaining an optimal acid concentration is key to achieving the best possible interaction and stabilization effect in chitosan-treated soils [47–49].

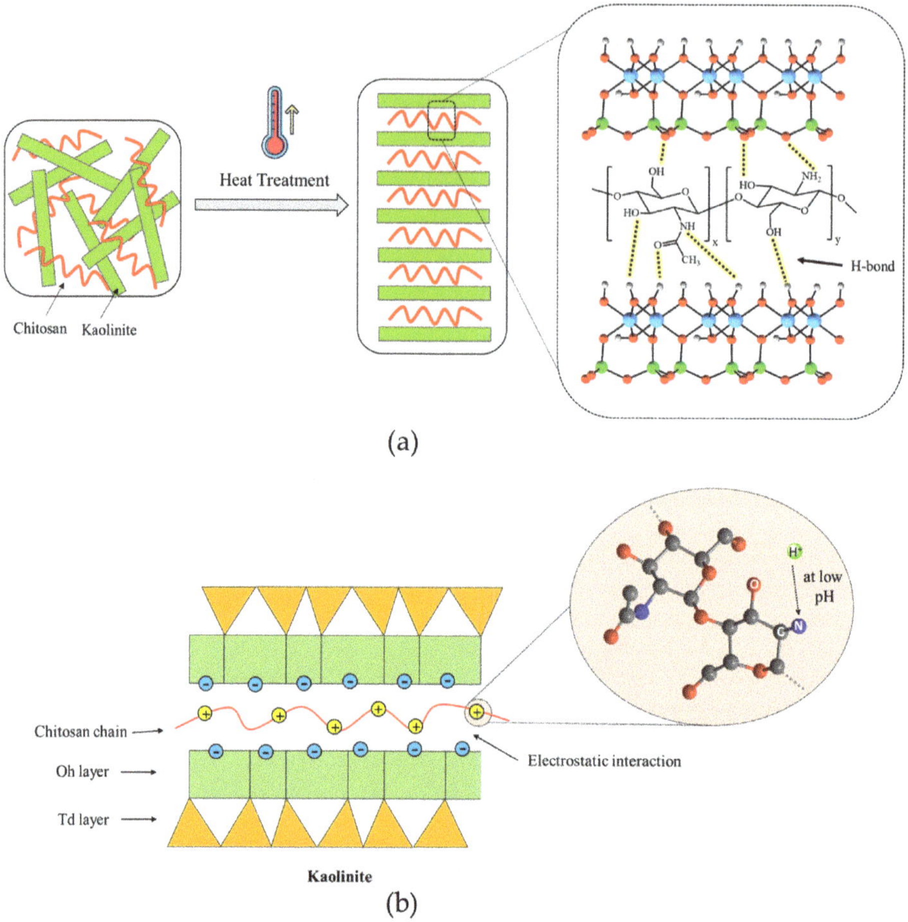

Figure 17. Effects of various parameters on the properties of chitosan-treated soil. (**a**) Temperature, (**b**) Acid concentration.

5. Comparison of Chitosan-Treated Soil Behavior with Other Biopolymer-Treated Soils

5.1. Biopolymer Content

This section presents chitosan's performance in improving soil compression strength compared to some marine and conventional biopolymers, including sodium alginate (SA), carrageenan (CA), and xanthan (XA). As seen in Figure 18a, chitosan significantly outperforms other biopolymers in UCS enhancement of kaolinite. At 0.5% biopolymer content, chitosan outperformed xanthan, although sodium alginate and carrageenan exhibited better results. However, at biopolymer contents higher than 0.5%, chitosan significantly increased the strength, demonstrating the best performance compared to the other biopolymers. Particularly in terms of the comparison with xanthan, the most commonly used biopolymer in soil treatment, chitosan presented an exceptional performance, highlighting its potential as a great alternative to other common additives.

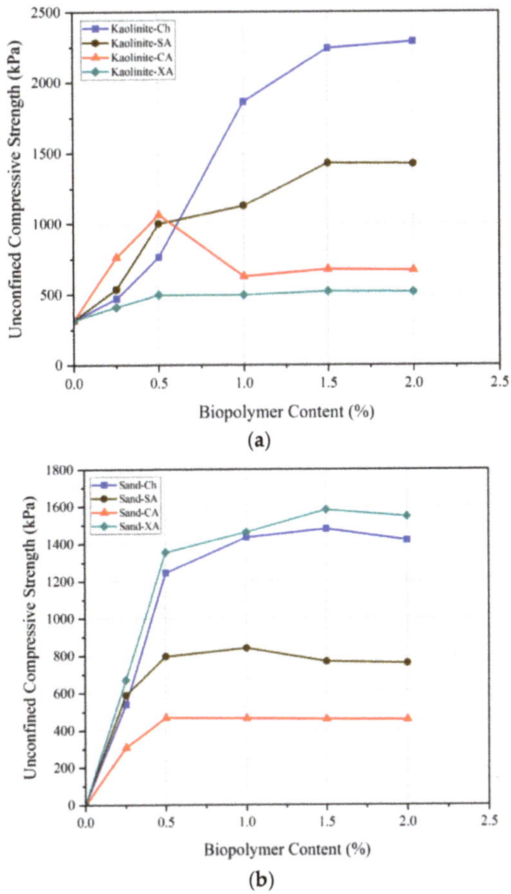

Figure 18. Different biopolymers' effect on UCS improvement of (**a**) kaolinite and (**b**) sand.

Chitosan also demonstrated remarkable effectiveness in stabilizing sandy soils. Its application led to a significant increase in UCS, from 0 kPa in untreated sand to 1480 kPa with only 1.5% chitosan. Even though xanthan achieved a UCS of 1581 kPa at 1.5% content, these results were not significantly higher than those achieved with chitosan. Also, carrageenan-treated sand reached a maximum UCS of 468.73 kPa at 0.5%, with minimal change at higher concentrations, while sodium alginate-treated sand peaked at 840 kPa with 1% content, showing a slight decline at higher dosages. The comparative results clearly establish chitosan as a highly effective biopolymer, outperforming others, including the commonly used xanthan, in kaolinite stabilization.

5.2. Long-Term Strength

This section compares the long-term compressive strength of soils treated with chitosan (CH), carrageenan (CA), sodium alginate (SA), and xanthan gum (XA) over 1, 14, and 120 days of curing. Figure 19 illustrates the development of compressive strength for each biopolymer after 120 days. In kaolinite (Figure 19a), chitosan-treated samples provided a UCS increase from 508 kPa at 1 day to 763 kPa at 14 days, followed by a slight decrease to 721 kPa at 120 days. This suggests that while chitosan provides strong initial stabilization, it experiences minimal strength reduction over time. On the other hand, carrageenan-treated kaolinite peaked at a higher UCS of 1259 kPa at 14 days but decreased more significantly to 1191 kPa by 120 days. Sodium alginate- and xanthan-treated kaolinite showed smaller UCS, with sodium alginate decreasing from 1000 kPa to 750 kPa and xanthan stabilizing around 548 kPa. Chitosan-treated kaolinite retained nearly 95% of its strength after 120 days, indicating its strong potential for use in medium-term projects.

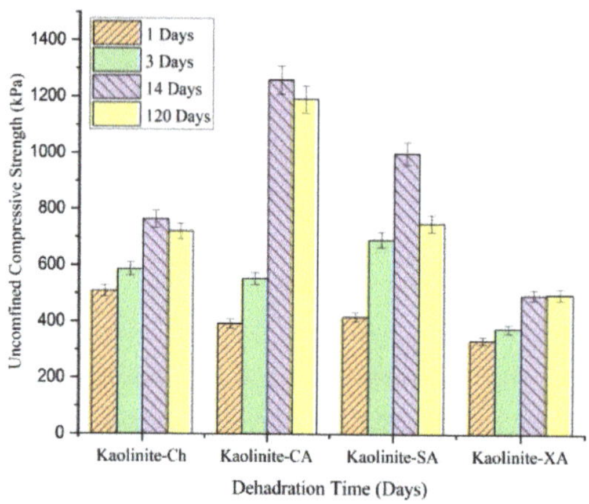

(a)

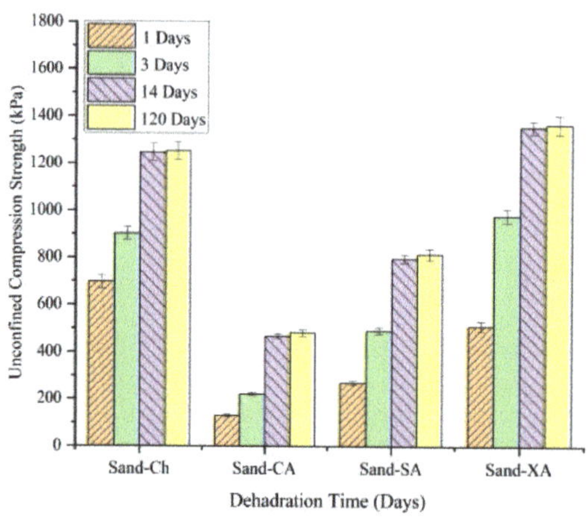

(b)

Figure 19. Long-term strength comparison of different biopolymers in stabilizing (**a**) kaolinite and (**b**) sand.

As shown in Figure 19b, all tested biopolymers maintained their strength effectively over a 120-day period. For sand, chitosan treatment improved the UCS from 697 kPa at 1 day to 1246 kPa at 14 days, with a slight increase to 1253 kPa at 120 days, indicating robust long-term strength. Although xanthan-treated sand outperformed chitosan with a UCS of 1366 kPa after 120 days, chitosan maintained its high strength. Sodium alginate and carrageenan gave relatively lower strengths. Overall, it can be concluded that chitosan provides a reliable performance and stability over time for both kaolinite and sand.

6. Conclusions

This study highlights the significant impact of chitosan biopolymer on the mechanical properties of sand and kaolinite soils. The key findings emphasize the effects of biopolymer content, long-term curing, temperature, and acid concentration on soil stabilization, particularly in enhancing unconfined compressive strength (UCS) and modulus of elasticity (EM).

The study demonstrates that chitosan significantly enhances the unconfined compressive strength (UCS) and elastic modulus (EM) of both sand and kaolinite soils. Sand achieved its maximum UCS at a chitosan content of 1.5%, reaching 1480 kPa, while kaolinite exhibited an optimal concentration of 2%, yielding a UCS of 2290 kPa. Although sand initially performed better due to its coarse-grained structure, facilitating effective mixing with chitosan, kaolinite surpassed it at higher biopolymer concentrations. This enhanced performance was attributed to stronger bonding mechanisms, including hydrogen bonding, electrostatic interactions, and hydrophobic bonding, which created a robust network that restricted soil particle movement and increased strength.

Long-term curing plays a critical role in the strength development of chitosan-treated soils, with both sand and kaolinite showing significant UCS increases over a 120-day period. The most notable strength gains occurred within the first seven days, particularly for sand, which exhibited a rapid increase in strength within one day due to its porous structure. UCS stabilization was observed after 14 days, highlighting the durability of chitosan-soil bonds over time. This strength enhancement was closely tied to the process of dehydration, where the reduction in moisture content reinforced the chitosan gel, allowing soils to achieve near-maximum strength within two weeks while maintaining structural integrity and low moisture levels over prolonged periods.

The efficacy of chitosan-stabilized soils is notably influenced by temperature. Experiments revealed that both sand and kaolinite experienced optimal performance at moderate curing temperatures around 45 °C, where optimal dehydration and stronger chemical bonding were observed. Beyond this range, excessive dehydration led to brittleness, with sand's UCS declining above 45 °C and kaolinite achieving its peak strength at 60 °C before decreasing at 80 °C. Additionally, high temperatures, particularly at 80 °C, diminished the modulus of elasticity, most notably in sand, due to the formation of brittle chitosan fibers and a reduced adhesion surface area. These findings underscore the critical importance of maintaining moderate curing temperatures to enhance the strength and elasticity of chitosan-treated soils.

The dissolution of chitosan in acidic solutions plays a critical role in soil stabilization, as evidenced by UCS tests conducted with varying acetic acid concentrations (0% to 5%). The results indicate that a 0.5% acid concentration was most effective in enhancing chitosan solubility and improving soil strength. At lower concentrations (0–0.1%), chitosan dissolution was inadequate, leading to only slight strength improvements. In contrast, higher acid concentrations (5%) did not significantly affect UCS, highlighting an optimal range for chitosan solubility. Furthermore, the study underscores that the typical levels of acid rain encountered in the environment are far lower than those required to dissolve chitosan, suggesting that environmental acidification poses minimal risk to the stability of chitosan-treated soils.

The use of chitosan greatly enhances the shear strength characteristics of kaolinite. The cohesiveness of the material increased from 70 kPa to 192 kPa, while the internal friction angle increased from 23.76° to 34.58°. These enhancements may be attributed to the improved bonding between particles and the increased cohesion of the soil. The compaction process led to the creation of chitosan–soil conglomerates, which enhanced interlocking and frictional resistance. This highlights the efficacy of chitosan as a soil stabilizer.

7. Future Recommendations

In order to enhance the use of chitosan in soil stabilization, more investigation is required to optimize the chitosan concentration, duration of curing, and temperature for different soil compositions. Conducting extensive field trials is crucial to confirm the results obtained in the laboratory, while conducting thorough investigations on the environmental effect will guarantee the sustainable use of resources. In addition, conducting research on the use of different acids to dissolve chitosan and evaluating long-term durability would improve its practical use. Cost–benefit assessments are used to assess the economic viability of large-scale undertakings.

Author Contributions: Conceptualization, R.W., D.E.L.O., H.F., P.X. and T.Y.; methodology, R.W., D.E.L.O., H.F., P.X. and T.Y.; validation, R.W., D.E.L.O., H.S., H.F., M.G. and T.Y.; resources, R.W., H.S., H.F. and M.G.; writing—original draft preparation, R.W., H.F. and H.S.; writing—review and editing, R.W., D.E.L.O., H.F. and P.X.; supervision, D.E.L.O., H.F. and P.X. All authors have read and agreed to the published version of the manuscript.

Funding: This research was supported by "the Fundamental Research Funds for the Central Universities" (B240201026).

Institutional Review Board Statement: Not applicable.

Data Availability Statement: The original contributions presented in the study are included in the article; further inquiries can be directed to the corresponding authors.

Conflicts of Interest: The funders had no role in the design of the study; in the collection, analyses, or interpretation of data; in the writing of the manuscript; or in the decision to publish the results.

References

1. Chang, I.; Lee, M.; Tran, A.T.P.; Lee, S.; Kwon, Y.M.; Im, J.; Cho, G.C. Review on biopolymer-based soil treatment (BPST) technology in geotechnical engineering practices. *Transp. Geotech.* **2020**, *24*, 100385. [CrossRef]
2. Izadi, R.; Mahinroosta, M.; Allahverdi, A.; Ghadir, P. Stabilizers based on nanoclay and blast furnace slag to reduce wind erosion of sandy soil green stabilization of sandy soil. *Int. J. Environ. Sci. Technol.* **2024**, *21*, 9075–9095. [CrossRef]
3. Saberian, M.; Perera, S.T.A.M.; Zhu, J.; Roychand, R.; Li, J.; Wang, G. Enhancing the Mechanical and Microstructural Properties of Low Expansive Clay Subgrade with Vinyl Acetate-Ethylene Polymer Incorporation. *J. Mater. Civil Eng.* **2024**, *36*, 04024161. [CrossRef]
4. Liu, C.; Lu, K.; Wu, Z.; Liu, X.; Garg, A.; Qin, Y.; Guoxiong, M.; Chao, L. Expansive soil improvement using industrial bagasse and low-alkali ecological cement. *Constr. Build. Mater.* **2024**, *423*, 135806. [CrossRef]
5. Jahandari, S.; Tao, Z.; Alim, M.A.; Li, W. Integral waterproof concrete: A comprehensive review. *J. Build. Eng.* **2023**, *78*, 107718. [CrossRef]
6. Fatehi, H.; Ong, D.E.; Yu, J.; Chang, I. Sustainable soil treatment: Investigating the efficacy of carrageenan biopolymer on the geotechnical properties of soil. *Constr. Build. Mater.* **2024**, *411*, 134627. [CrossRef]
7. Sharma, P.; Goyal, S.K. Review of the Environmental Impacts of Cement Production and a Sustainable Remedy. *J. Inst. Eng. (India) Ser. D* **2024**, 1–12. [CrossRef]
8. Ollikainen, M.; Lötjönen, S.; Tikkanen, T.; Ala-Harja, V.; Uusitalo, R.; Ekholm, P. Gypsum and structure lime amendments in boreal agricultural clay soils: Do climate emissions compromise water quality benefits? *Agric. Food Sci.* **2024**, *33*, 90–115. [CrossRef]
9. Miraki, H.; Shariatmadari, N.; Ghadir, P.; Jahandari, S.; Tao, Z.; Siddique, R. Clayey soil stabilization using alkali-activated volcanic ash and slag. *J. Rock Mech. Geotech. Eng.* **2022**, *14*, 576–591. [CrossRef]
10. Vafaie, S.; Salajegheh, E. A Comparative Study of Shearlet, Wavelet, Laplacian Pyramid, Curvelet, and Contourlet Transform to Defect Detection. *J. Soft Comput. Civil Eng.* **2023**, *7*, 1–42.
11. Li, J.; Zhu, F.; Wu, F.; Chen, Y.; Richards, J.; Li, T.; Li, P.; Shang, D.; Yu, J.; Viles, H.; et al. Impact of soil density on biomineralization using EICP and MICP techniques for earthen sites consolidation. *J. Environ. Manag.* **2024**, *363*, 121410. [CrossRef] [PubMed]
12. Zhang, K.; Tang, C.S.; Jiang, N.J.; Pan, X.H.; Liu, B.; Wang, Y.J.; Shi, B. Microbial-induced carbonate precipitation (MICP) technology: A review on the fundamentals and engineering applications. *Environ. Earth Sci.* **2023**, *82*, 229. [CrossRef] [PubMed]
13. Fatehi, H.; Bahmani, M.; Noorzad, A. Strengthening of Dune Sand with Sodium Alginate Biopolymer. In *Geo-Congress 2019: Soil Improvement*; American Society of Civil Engineers: Reston, VA, USA, 2019.

14. Shabani, K.; Bahmani, M.; Fatehi, H.; Chang, I. Improvement of the geotechnical engineering properties of dune sand using a plant-based biopolymer named serish. *Geomech. Eng.* **2022**, *29*, 535–548.
15. Fatehi, H.; Abtahi, S.M.; Hashemolhosseini, H.; Hejazi, S.M. A novel study on using protein based biopolymers in soil strengthening. *Constr. Build. Mater.* **2018**, *167*, 813–821. [CrossRef]
16. Fatehi, H.; Ong, D.E.; Yu, J.; Chang, I. Biopolymers as Green Binders for Soil Improvement in Geotechnical Applications: A Review. *Geosciences* **2021**, *11*, 291. [CrossRef]
17. Fatehi, H.; Ong, D.E.; Yu, J.; Chang, I. The effects of particle size distribution and moisture variation on mechanical strength of biopolymer-treated soil. *Polymers* **2023**, *15*, 1549. [CrossRef] [PubMed]
18. Zhang, J.; Liu, J. A Review on Soils Treated with Biopolymers Based on Unsaturated Soil Theory. *Polymers* **2023**, *15*, 4431. [CrossRef] [PubMed]
19. Tasuji, M.A.; Ghadir, P.; Hosseini, A.; Javadi, A.A.; Korayem, A.H.; Ranjbar, N. Experimental investigation of sandy soil stabilization using chitosan biopolymer. *Transp. Geotech.* **2024**, *46*, 101266. [CrossRef]
20. Ren, F.; Ding, H.; Dong, B.; Qian, X.; Liu, J.; Tan, J. Study on the improvement of soil properties using hydrophilic-hydrophobic biopolymer crosslinking. *Constr. Build. Mater.* **2024**, *415*, 135101. [CrossRef]
21. Vafaie, S.; Salajegheh, E. Comparisons of wavelets and contourlets for vibration-based damage identification in the plate structures. *Adv. Struct. Eng.* **2019**, *22*, 1672–1684. [CrossRef]
22. Mahdy, O.S.; Ali, A.B.; Mahdi, M.S.; Jasim, D.J.; Kazemi-Varnamkhasti, H.; Goli, M.; Salahshour, S.; Baghaei, S. Thermal performance of nanofluid natural convection magneto-hydrodynamics within a chamber equipped with a hot block. *Int. J. Thermofluids* **2024**, *24*, 100873. [CrossRef]
23. Firmansyah, D.A.; Somantri, A.K.; Sihombing, A.V.R.; Mase, L.Z.; Sundara, A. Mechanical Properties of Soft Clay Soil Improved with Nanomaterials and Chitosan Biopolymer. *Geotech. Eng. J. SEAGS AGSSEA* **2024**, *55*, 31–37.
24. Hamza, M.; Nie, Z.; Aziz, M.; Ijaz, N.; Ameer, M.F.; Ijaz, Z. Geotechnical properties of problematic expansive subgrade stabilized with xanthan gum biopolymer. *Road Mater. Pavement Des.* **2023**, *24*, 1869–1883. [CrossRef]
25. Aranaz, I.; Alcántara, A.R.; Civera, M.C.; Arias, C.; Elorza, B.; Heras Caballero, A.; Acosta, N. Chitosan: An overview of its properties and applications. *Polymers* **2021**, *13*, 3256. [CrossRef]
26. Madni, A.; Kousar, R.; Naeem, N.; Wahid, F. Recent advancements in applications of chitosan-based biomaterials for skin tissue engineering. *J. Bioresour. Bioprod.* **2021**, *6*, 11–25. [CrossRef]
27. Jamshidi, M.; Mokhberi, M.; Vakili, A.H.; Nasehi, A. Effect of chitosan bio-polymer stabilization on the mechanical and dynamic characteristics of marl soils. *Transp. Geotech.* **2023**, *42*, 101110. [CrossRef]
28. Ilman, B.; Balkis, A.P. Sustainable biopolymer stabilized earthen: Utilization of chitosan biopolymer on mechanical, durability, and microstructural properties. *J. Build. Eng.* **2023**, *76*, 107220. [CrossRef]
29. Adamczuk, A.; Kercheva, M.; Hristova, M.; Jozefaciuk, G. Impact of chitosan on water stability and wettability of soils. *Materials* **2021**, *14*, 7724. [CrossRef]
30. Shariatmadari, N.; Reza, M.; Tasuji, A.; Ghadir, P.; Javadi, A.A. Experimental study on the effect of chitosan biopolymer on sandy soil stabilization. In *E3S Web of Conferences*; EDP Sciences: Lisboa, Portugal, 2020.
31. ASTM International. *D698 Standard Test Methods for Laboratory Compaction Characteristics of Soil Using Standard Effort (12,400 ft-lbf/ft3 (600 kN-m/m3))*; ASTM International: West Conshohocken, PA, USA, 2021.
32. ASTM International. *D2850 Standard Test Method for Unconsolidated-Undrained Triaxial Compression Test on Cohesive Soils*; ASTM International: West Conshohocken, PA, USA, 2023.
33. ASTM International. *D4609 Standard Guide for Evaluating Effectiveness of Admixtures for Soil Stabilization*; ASTM International: West Conshohocken, PA, USA, 2008.
34. ASTM International. *D2166/D2166M Standard Test Method for Unconfined Compressive Strength of Cohesive Soil*; ASTM International: West Conshohocken, PA, USA, 2016.
35. ASTM International. *D7181 Standard Test Method for Consolidated Drained Triaxial Compression Test for Soils*; ASTM International: West Conshohocken, PA, USA, 2020.
36. Awad, M.E.; López-Galindo, A.; Setti, M.; El-Rahmany, M.M.; Iborra, C.V. Kaolinite in pharmaceutics and biomedicine. *Int. J. Pharm.* **2017**, *533*, 34–48. [CrossRef]
37. Hu, P.; Yang, H. Insight into the physicochemical aspects of kaolins with different morphologies. *Appl. Clay Sci.* **2013**, *74*, 58–65. [CrossRef]
38. Tombácz, E.; Szekeres, M. Surface charge heterogeneity of kaolinite in aqueous suspension in comparison with montmorillonite. *Appl. Clay Sci.* **2006**, *34*, 105–124. [CrossRef]
39. Hamed, A.A.; Abdelhamid, I.A.; Saad, G.R.; Elkady, N.A.; Elsabee, M.Z. Synthesis, characterization and antimicrobial activity of a novel chitosan Schiff bases based on heterocyclic moieties. *Int. J. Biol. Macromol.* **2020**, *153*, 492–501. [CrossRef] [PubMed]
40. Yeul, V.S.; Rayalu, S.S. Unprecedented chitin and chitosan: A chemical overview. *J. Polym. Environ.* **2013**, *21*, 606–614. [CrossRef]

41. Rekik, S.B.; Gassara, S.; Bouaziz, J.; Deratani, A.; Baklouti, S. Development and characterization of porous membranes based on kaolin/chitosan composite. *Appl. Clay Sci.* **2017**, *143*, 1–9. [CrossRef]
42. Bezerril, L.M.; De Vasconcelos, C.L.; Dantas, T.N.C.; Pereira, M.R.; Fonseca, J.L.C. Rheology of chitosan-kaolin dispersions. *Colloids Surf. A Physicochem. and Eng. Asp.* **2006**, *287*, 24–28. [CrossRef]
43. Monvisade, P.; Siriphannon, P. Chitosan intercalated montmorillonite: Preparation, characterization and cationic dye adsorption. *Appl. Clay Sci.* **2009**, *42*, 427–431. [CrossRef]
44. Sriupayo, J.; Supaphol, P.; Blackwell, J.; Rujiravanit, R. Preparation and characterization of α-chitin whisker-reinforced chitosan nanocomposite films with or without heat treatment. *Carbohydr. Polym.* **2005**, *62*, 130–136. [CrossRef]
45. Rubentheren, V.; Ward, T.A.; Chee, C.Y.; Nair, P.; Salami, E.; Fearday, C. Effects of heat treatment on chitosan nanocomposite film reinforced with nanocrystalline cellulose and tannic acid. *Carbohydr. Polym.* **2016**, *140*, 202–208. [CrossRef] [PubMed]
46. Rekik, S.B.; Gassara, S.; Bouaziz, J.; Baklouti, S.; Deratani, A. Fabrication, characterization and permeation studies of ionically cross-linked chitosan/kaolin composite membranes. *Period. Polytech. Chem. Eng.* **2023**. [CrossRef]
47. Lee, D.W.; Lim, C.; Israelachvili, J.N.; Hwang, D.S. Strong adhesion and cohesion of chitosan in aqueous solutions. *Langmuir* **2013**, *29*, 14222–14229. [CrossRef] [PubMed]
48. Sadighi, H.; Rowshanzamir, M.; Banitalebi-Dehkordi, M. A multi-aspect application of microwave radiation on rehabilitating and improving the geotechnical properties of polluted-sand-clay mixture. *J. Contam. Hydrol.* **2022**, *249*, 104040. [CrossRef] [PubMed]
49. Ouhadi, V.R.; Goli, M. Pore Fluid Dielectric Constant Effect on Geotechnical and Geo-Environmental Properties of Smectite and Kaolinite. *Soil Sediment Contam. Int. J.* **2024**, *33*, 1526–1548. [CrossRef]

polymers

MDPI

Article

The Effect of Alginate/Hyaluronic Acid Proportion on Semi-Interpenetrating Hydrogel Properties for Articular Cartilage Tissue Engineering †

Izar Gorroñogoitia [1,2], Sheila Olza [2,3,4], Ana Alonso-Varona [2] and Ane Miren Zaldua [1,*]

1 Leartiker S. Coop., 48270 Makina-Xemein, Spain; igorronogoitia@leartiker.com
2 Faculty of Medicine and Nursing, University of the Basque Country (UPV/EHU), 48940 Leioa, Spain; sheila.olza@ehu.eus (S.O.); ana.alonsovarona@ehu.eus (A.A.-V.)
3 E2S UPPA, CNRS, IPREM, Universite de Pau et des Pays de l'Adour, 64600 Anglet, France
4 MANTA-Marine Materials Research Group, E2S UPPA, Universit'e de Pau et des Pays de l'Adour, 64600 Anglet, France
* Correspondence: amzaldua@leartiker.com; Tel.: +34-946-169-089

† This Article is a Revised and Expanded Version of a Conference Poster Entitled "The Effect of Alginate/Hyaluronic Acid Proportion on Semi-Ipn Hydrogel Properties for Articular Cartilage Tissue Engineering", which was presented at European Society for Biomaterials 2023 (ESB) at Davos (Switzerland) in 4–8 September 2023.

Abstract: One of the emergent regenerative treatments for the restoration of the articular cartilage is tissue engineering (TE), in which hydrogels can functionally imitate the extracellular matrix (ECM) of the native tissue and create an optimal microenvironment for the restoration of the defective tissue. Hyaluronic acid (HA) is known for its potential in the field of TE as a regenerative material for many tissues. It is one of the major components of the articular cartilage ECM contributing to cell proliferation and migration. HA is the only non-sulphated glycosaminoglycan (GAG). However, herein, we use a HA presenting a high amount of sulphated glycosaminoglycans (sGAGs), altering the intrinsic properties of the material particularly in terms of biological response. Alginate (Alg) is another polysaccharide widely used in TE that allows stiff and stable hydrogels to be obtained when crosslinked with $CaCl_2$. Taking the benefit of the favourable characteristics of each biomaterial, semi-interpenetrating (semi-IPN) hydrogels had been developed by the combination of both materials, in which alginate is gelled, and HA remains uncrosslinked within the hydrogel. Varying the concentration of alginate and HA, the final rheological, viscoelastic, and mechanical properties of the hydrogel can be tailored, always seeking a trade-off between biological and physico-mechanical properties. All developed semi-IPN hydrogels have great potential for biomedical applications.

Keywords: hydrogel; alginate; hyaluronic acid; bioprinting; scaffold

1. Introduction

Nowadays there are several clinical treatment methods for articular cartilage damage, which include palliative, reparative, and regenerative treatments. The first one reduces the clinical symptoms (local pain) but healing of the tissue does not happen. In reparative treatments, a fibrocartilage is formed rather than the hyaline cartilage, leading to misfunction and a lack of adequate mechanical properties of the tissue. Among the regenerative treatments, autologous chondrocyte implantation (ACI) and matrix-assisted autologous chondrocyte implantation (MACI) are the most known and effective ones so far, in which isolated autologous chondrocytes are cultured in vitro and injected into the damaged site [1,2]. A new approach that can further achieve articular cartilage restoration satisfactorily is tissue engineering (TE), where scaffolds, cells, and signalling factors are involved [2–5]. Scaffolds can imitate the extracellular matrix (ECM) of native tissue, both structurally and functionally, and provide the suitable and supportive 3D microenvironment for the regeneration of defective tissue. The scaffold should be biocompatible and biodegradable and should present a porous structure that permits cell attachment, proliferation, and the maintenance of a differentiated phenotype. In addition, the degradation rate of the scaffold should match that of the new ECM synthesis, and they must have the ability to present similarities to biomechanical properties of native tissue. Scaffolds can be fabricated by 3D bioprinting that allows the fabrication of high precision complex constructs with customized geometries from 3D digital models. The most common technique for tissue fabrication is extrusion bioprinting (EBB), in which living constructs can be manufactured layer by layer with the precise positioning of bioinks [6–10].

Among all of the biomaterials available for cartilage regeneration, natural polysaccharide-based ones such as alginate (Alg) and hyaluronic acid (HA) have great potential. Due to their natural origin and abundance, they both are inexhaustible. However, the fact that they come from a natural source makes them have batch-dependent properties [3,11]. Thus, knowing in detail the intrinsic properties of each batch at every moment is essential since they will govern the hydrogel's final properties. One of the most abundant glycosaminoglycans (GAGs) in the native ECM of cartilage is HA, which plays a crucial role in the structural and functional properties of cartilage. Its major disadvantages, however, are its poor mechanical properties and degradability. HA is involved in the water adsorption and retention, lubrication, and compression bearing of native tissue, and it is known to interact with several cell surface receptors such as CD44 and more [3,12–18]. It consists of repeating disaccharide units of β-N-acetyl-D glucosamine (Glc-NAc) and D-Glucuronic acid (GlcA) and can be extracted from different sources: bovine vitreous humors, rooster combs, the skin of shark, umbilical cords, and bacteria [12,15]. It is the only non-sulphated GAG and can exhibit very high molecular weight (M_w) [1,10] (M_w affects its viscoelastic and biological properties [12,14,15]), unlike the sulphated GAGs (sGAGs). These sGAGs are characterized by having very low M_w and have key biological properties for cartilage tissue engineering (CTE) like cell recognition and signalling [13,17,19]. Another biopolymer widely used and explored in CTE is alginate [20] since it shows similar structure to GAGs. It is extracted from brown seaweed and is composed of β-D-mannuronic acid (M) and α-L-guluronic acid (G) repeating monomer units. Its chemical structure (M_w and M/G ratio) varies depending on its origin [21–25] and will determine the final properties of the scaffold, as demonstrated in several studies [24,26–28] and in our previous work [11]. The gelation of alginate is mediated by divalent ions such as calcium that reacts with carboxyl groups (COO^-) of G-blocks of alginate, forming a so-called "egg-box" structure [20–22,24]. Because of its ease of use, low cost, biocompatibility, versatility, and rapid gelation, robust hydrogels can be obtained. Nevertheless, one of the main drawbacks of alginate is its low bioactivity that could be addressed by chemical modification or combining with other polymers.

The combination of alginate and hyaluronic acid has demonstrated efficient application in CTE as reported by several authors. Janarthan et al. [29] used a mix of alginate and hyaluronic acid hydrogel to print different structures of various layers by 3D bioprinting, and the in vitro live/dead assay proved the excellent biocompatibility of the constructs using chondrocytes. Nedunchezian et al. [30] showed the chondrogenic capability of adipose-derived stem cells (ADSC) combined with alginate/hyaluronic acid (Alg/HA) hydrogel constructs printed by 3D bioprinting technology, and Antich et al. [31] developed a semi-interpenetrating (semi-IPN) Alg/HA hydrogel that successfully promoted chondrogenesis and maintained the chondrocyte phenotype using chondrocytes. So, these hybrid hydrogels are highly desirable 3D supportive microenvironments for cartilage repair.

The aim of this study is to develop semi-IPN hydrogels of alginate and hyaluronic acid and validate them as potential candidates for use in biomedical applications such as the restoration of articular cartilage injuries. Indeed, we used a HA that came from natural origin (Wharton's jelly extracted from the umbilical cord) and contained a high amount of sGAGs. This helped us to develop bioactive hydrogels as sGAGs acted as bioactive molecules for providing cell adhesion and proliferation. At the same time, this helped us to mimic the native cartilage even better in terms of composition, thus further approaching the targeting of functional properties. In addition, the alginate used in this work enabled us to develop stiff hydrogels when crosslinked with $CaCl_2$. So, exploiting the favourable properties and strengths of each biomaterial, novel Alg/HA hybrid semi-IPN hydrogels were developed for CTE, where alginate was crosslinked by ionic crosslinking and HA remained uncrosslinked within the hydrogel. Finally, to verify these hydrogels as possible candidates for CTE and asses their suitability in 3D bioprinting, rheological, mechanical, and biological characterization were performed, and the printability was evaluated.

2. Materials and Methods

Sodium alginate (Alg), calcium chloride ($CaCl_2$, anhydrous), phosphate-buffered saline (PBS), glutaraldehyde, and dimethyl sulfoxide (DMSO) were purchased from Sigma-Aldrich, St. Louis, MO, USA. Minimum essential medium (MEM), Trypsin-EDTA, foetal bovine serum (FBS), penicillin, non-essential amino acids (NEAA), sodium pyruvate (NaPyr), trypan blue stain, Presto Blue stain and propidium iodide (PI) were provided by ThermoFischer Scientific, Waltham, MA, USA. Murine fibroblast cells (L929), which are a commercial cell line, were obtained from American Type Culture Collection (ATCC, Manassas, VA, USA). Hexamethydisilazane was obtained from Electron Microscopy Sciences, Hatfield, PA, USA. Ethanol was obtained from ENMA S.L, Spain, calcein-AM was obtained from Abcam, Cambridge, UK and gentamicin was obtained from Gibco®, Cambridge, UK. Native whartonide hyaluronic acid (HA) was provided by Histocell S.L Regenerative Medicine, Derio, Spain. The properties of HA were indicated in the certificate of analysis (Table 1). The molecular weight (M_w) of sodium alginate was determined by POLYMAT, Donostia-San Sebastián, Spain (Table 2). Detailed information about the experimental process of obtaining the M_w of sodium alginate is included in the supplementary information.

Table 1. Native whartonide hyaluronic acid properties.

Designation	Source	Batch	M_w (kDa)	[a] sGAGs
HA	Wharton's jelly	210,041	<2000	Yes
		230,002 + 230,005		
		230,029 + 230,056		

[a] Sulphated glycosaminoglycans.

Table 2. Sodium alginate properties. M_w obtained from gel permeation chromatography (GPC) (polyethylene glycol (PEO) standards).

Designation	Source	Batch	M_w (kDa)
Alg	Brown algae	BCCD8789	392

2.1. Hydrogel Preparation and Crosslinking

Alginate and hyaluronic acid solutions as single components were prepared at different concentrations (2–4% w/w) in PBS. Alginate solutions were mixed vigorously under a propeller-type stirrer and using a magnetic stirrer in the case of hyaluronic acid solutions, both at room temperature until complete dissolution. Alginate/hyaluronic acid (Alg/HA) hybrid solutions were prepared by adding alginate powder first to PBS until complete dissolution under a propeller-type stirrer at room temperature and hyaluronic acid powder later at different concentrations (Table 3). The concentration of the formulations was selected based on their viscosity. Higher or lower concentrations of both polymers caused a too high or too low viscosity for bioprinting and processability. As shown in Table 3, the molar ratio of both polymers varied considerably, obtaining formulations with high alginate content in contrast to hyaluronic acid content. All solutions were stored at 2–8 °C to prevent polymer degradation.

Table 3. Alg/HA acid formulations obtained from the mixtures at different proportions in PBS.

Formulation	Alginate % (w/w)	HA % (w/w)	Molar Ratio Alg/HA
2Alg1HA	2	1	10:1
2Alg2HA	2	2	5:1
3Alg1HA	3	1	15:1
3Alg2HA	3	2	7.6:1
4Alg1HA	4	1	20:1
4Alg2HA	4	2	10:1

Alginate bulk gels and Alg/HA semi-interpenetrating (semi-IPN) bulk gels were prepared in a Petri dish by adding 100 mM $CaCl_2$ solution and were left to crosslink for 24 h. In semi-IPN hydrogels, only alginate was crosslinked while HA remained uncrosslinked. Hydrogels for compression testing and biological characterization were punched with a hole puncher of 6 mm diameter and the ones for viscoelastic measurements using a hole puncher with a diameter of 20 mm.

2.2. Rheological Characterization

The shear viscosity of all hydrogel precursors was measured as a function of shear rate at 23 and 37 °C using a HAAKE MARS III (ThermoFischer Scientific, Waltham, MA, USA) rheometer equipped with a peltier element for temperature control. A plate–plate geometry (20 mm, aluminium, gap = 1 mm) was used for the tests. Shear rate varied from 0.5 to 1000 s^{-1}. Flow data were adjusted to the Ostwald–de Waele and Carreau–Yasuda models. Figure S1 of the supplementary information shows the viscosity-curves of different HA batches. All of the flow parameters are summarized in Tables S1 and S2 of the supplementary information.

2.3. Viscoelastic and Mechanical Characterization

Dynamic shear experiments were performed for the determination of the storage modulus (G') of each hydrogel (h: 2–4 mm, D $\approx$ 20 mm) at 23 and 37 °C (conditioning time = 5 min) using the Peltier module integrated with parallel plates (20 mm, aluminium serrated, F_n = 0.5 N autotension). Time varied from 0 to 300 s, applying a frequency of 1 Hz and a constant deformation in the linear viscoelastic regime (0.01%).

The mechanical properties of all hydrogels (h: 2–4 mm, D: 5–6 mm) were determined by applying unconfined compression (UC) using Univert (CellScale, Waterloo, ON, Canada) equipped with a 10 N load cell. A pre-load of 0.1 N was first applied to set the zero point and then samples were compressed to 40% of deformation at a rate of 25%/min in air at room temperature and in a PBS bath at 37 °C (conditioning time = 5 min). Tangent modulus at a target strain of 10% was determined for each hydrogel.

Supplementary information provides an example of a time sweep experiment (Figure S2) and a stress–strain curve (Figure S3), as well as all of the experimental data (Table S3).

2.4. Scaffold Fabrication Process

Printed structures (0.84 × 30 × 30 mm, pore size: 5 mm^2) were fabricated layer by layer (4 layers) using a BIO V1 3D bioprinter (REGEMAT 3D, Granada, Spain) equipped with three syringes and one fused deposition modelling (FDM) extruder, consisting of hardware and Designer software (REGEMAT 3D, Granada, Spain) that are connected by an electronic control unit (ECU). Scaffolds were printed in glass slides (76 × 52 mm) at room temperature using a 5 mL syringe and a 27G conical nozzle inner diameter. Nozzle speed was set to 5 mm/s, the offset to 0.5 mm, and the flow rate to 0.6 µL/s.

2.5. Printability Evaluation

The printability of fabricated structures was evaluated qualitatively and quantitatively. The latter was performed by measuring the experimental filament diameter (D_{exp}), area of pores, and perimeter of pores of printed structures. Please find further details about the quantitative evaluation in Table S4 of the supplementary information. The qualitative evaluation was carried out by visual 3D studies to check the quality and printability of the fabricated scaffolds (4 layers) and accomplished before and after immersing the scaffolds in 100 mM $CaCl_2$ solution. The structures were compared with their homologues to analyse the influence of hyaluronic acid addition in the printability and quality of printed constructs.

2.6. Preparation of Cell-Laden Solution and Bioprinting of Constructs

To prepare the cell-laden Alg/HA solution for bioprinting, a 3Alg2HA hydrogel was selected. Briefly, L929 cells were resuspended in 200 µL MEM cell culture medium and were mixed with a 3Alg2HA hybrid solution to a concentration of 1 × 10^6 cells/mL using a syringe and carefully stirring until obtaining a homogeneous distribution of the cells. The printing process was performed using sterile material. The cell-laden bioink was printed using a BIO V1 3D bioprinter (REGEMAT 3D, Granada, Spain) with a printing conical nozzle of 22G inner diameter and a flow rate of 2 µL/s. The syringe was set to 37 °C and glass slides (76 × 52 mm) at room temperature were used as a printing bed. The offset was set to 1 mm. After printing the constructs, they were crosslinked with 100 mM of $CaCl_2$ for 5 h and were cultured in MEM supplemented with 1 mM of $CaCl_2$ for 1, 3, and 7 days. Cell-laden 3Alg2HA hydrogels without printing in the bioprinter were used as control samples.

2.7. In Vitro Cytotoxicity Assay

In vitro cytotoxicity was determined following the International Standard Organization (ISO) 10993-5:2009 normative [32], using L929 mouse fibroblast cells. As complete medium MEM was used, it was supplemented with 1% *v*/*v* of non-essential amino acids, sodium pyruvate 1 mM, 1% *v*/*v* of penicillin and streptomycin (100 U/mL), and 0.1% *v*/*v* of gentamicin plus a 10% *v*/*v* of bovine foetal serum. The extracted culture medium was prepared under sterile conditions following the ISO 10993-12:2021 [33], immersing each sample of alginate and Alg/HA hydrogel (≈0.2 g, diameter: 6 mm, height: 5 mm) previously sterilized by ultraviolet (UV) light for 30 min, in 1 mL of complete medium for

24 h at 37 °C and 5% CO_2 in a cell culture cabinet. L929 cells were seeded on a Sarstedt 96-well cell culture plate at a density of 4×10^3 cells/well in complete medium (100 µL) and incubated at 37 °C for 24 h. For the blank, complete medium without cells was used. As a positive control, 10% *v*/*v* of DMSO in complete medium was used and as a negative control, only complete medium with cells was used, which were also incubated at the same conditions as the samples. After the incubation time, the complete medium was removed from the wells and extractive medium and controls (100 µL) were added to the corresponding wells and incubated at 37 °C and 5% CO_2. The cell viability was determined after 24, 48, and 72 h using the Presto Blue™ (ThermoFischer Scientific, Waltham, MA, USA) cell viability reagent, which was added to the wells and incubated for 3 h before the absorbance measurement at 570 and 600 nm using the microplate reader Synergy HT (Biotek, Shoreline, WA, USA). Cell viability (%) was calculated as follows:

$$\text{Viability } (\%) = \frac{A_{test}}{A_{control}} \times 100 \quad (1)$$

where A_{test} is the absorbance of the sample cells and $A_{control}$ is the absorbance of the negative control cells. *A* value above 70% was considered without any cytotoxic potential according to ISO 10993-5:2009 [32].

2.8. Cell Morphology and Adhesion

The morphology and the adhesion of the L929 cells on the surface of the materials was assessed by a Hitachi S-3400 (Hitachi, Chiyoda, Japan) scanning electron microscopy (SEM). Prior to cell seeding, samples were sterilized with UV for 30 min, pre-wetted in complete medium and incubated at 37 °C for 24 h. Later, hydrogel samples were placed in 24-well ultralow attachment plates (Corning, NY, USA) and cells with a density of 5×10^4 cells/well were seeded on the top surface of each hydrogel. Before acquiring SEM images at the time of 72 h, the samples were washed three times with PBS, fixed with 2% *v*/*v* glutaraldehyde in 0.01 M PBS buffer for 30 min at 4 °C and finally rinsed with PBS. The dehydration of samples was carried out through a rising series of graded aqueous ethanol solutions and hexamethyldisilazane was employed to desiccate them for 10 min. Finally, the samples were coated with a thin layer of gold by sputtering.

2.9. Cell Viability of Seeded Cells on Hydrogels and Bioprinted Cells

In order to evaluate the viability of cells adhered to the materials (seeded cells) a live/dead assay was performed. Cells cultured for 7 days on the surface of hydrogels were washed in 1X PBS and incubated with a mix of 4 µM of calcein-AM and 5 µM propidium iodide PBS solution for 20 min at room temperature in dark conditions. The samples were then observed under a fluorescence microscope (ZEISS Apotome 3) to visualize live cells (stained green; $\lambda_{ex\text{-}em}$ = 495–515 nm) and dead cells (stained red; $\lambda_{ex\text{-}em}$ = 535–615 nm).

The viability of cells encapsulated in the bioprinted constructs and control samples (not printed constructs) was analysed following the same procedure, but constructs were cultured for 1, 3, and 7 days. A confocal fluorescence microscope (ZEISS LSM 800, ZEISS, Oberkochen, Germany) was used for sample observation.

All images were analysed by ImageJ software (1.54k, Wayne Rasband, Bethesda, MD, USA). The quantitative evaluation of seeded cells was carried out by splitting the images into four sections and performing the statistical analysis. All images were quantified using the colour threshold technique and particle counting. The cell viability (%) was calculated as (number of viable cells/total number of cells) × 100.

2.10. Statistical Analysis

All data are presented as mean ± standard deviation (SD) (error bars) of at least three sample replicates. Statistical comparisons between the different groups were made by one-way analysis of variance (ANOVA) followed by Tukey's multiple post-hoc comparisons test and repeated measures ANOVA followed by Dunnett's multiple comparisons test using Prism® statistical software (9.5.1, GraphPad, San Diego, CA, USA). * $p < 0.05$; ** $p < 0.01$; *** $p < 0.001$; and **** $p < 0.0001$. Values of $p < 0.05$ were statistically significant.

3. Results

The intrinsic properties of alginate (Alg) and hyaluronic acid (HA) polymers are shown in Tables 1 and 2. As mentioned previously, the HA provided by Histocell presented a molecular weight (M_w) of ≤2000 kDa and was composed of sulphated glycosaminoglycans (sGAGs) that will act as bioactive molecules, allowing the interaction with cells and favouring the biological properties of hydrogels. Three different HA batches were used, the first one for the non-biological characterization and the remaining two for the biological characterization. With the aim of minimizing differences in properties between the batches as much as possible, the M_w of 230,002 + 230,005 and 230,029 + 230,056 batches was adjusted to the M_w of the 210,041 batch by combining two different extracts. Please find further details in Figure S1 of the Supplementary Information. The sodium alginate exhibited a M_w of 392 kDa allowing us to develop printable and robust hydrogels when gelled with the crosslinking agent calcium chloride ($CaCl_2$).

3.1. Rheological Characterization

Rheological properties such as shear-thinning ability and fast recovery are essential characteristics for a hydrogel precursor to obtain structures with high shape fidelity in 3D bioprinting applications [34].

The shear viscosity of all hydrogel precursors was determined by rotational shear-test experiments where flow curves were obtained at 23 and 37 °C; in Figure 1, data at 37 °C are presented. From these viscosity curves, Newtonian viscosity (η_0) was determined, whereas k and n parameters were determined from stress curves). All parameters (η_0, k, and n) and their obtention are summarized in the supplementary information.

In relation to single-component Alg and HA solutions, Table S1 of the supplementary information shows the Ostwald–de Waele parameters obtained from the power-law model (Equation (S1) of the Supplementary Information) and Newtonian viscosity obtained from Carreau–Yasuda model (Equation (S2) of the Supplementary Information) for all solutions at different concentrations. It can be observed that in both cases, the higher the concentration of Alg or HA, the higher the zero-shear viscosity (η_0) due to a higher amount of polymer chains dissolved in the medium at high concentrations, resulting in more entanglements among the polymer chains. This effect was much more pronounced on Alg, where a small change in concentration led to a high change in viscosity (10 times higher when concentration was doubled), whereas in the case of HA, viscosity could almost not be modulated with concentration. The zero-shear viscosity of Alg was greater than the HA at all concentrations. Considering that the viscosity is determined by the molecular weight (M_w), HA should present higher viscosity than Alg since it displayed a M_w of ≤2000 kDa (Table 1) and Alg, however, was 392 kDa (Table 2). Nevertheless, HA contained a high amount of sGAGs, triggering a reduction in viscosity due to the very low M_w of these sGAGs compared to HA (non-sulphated). The M_w of the sGAGs can vary from 10 to 50 kDa [13], contributing to the low viscosity obtained. Moreover, the n values for HA were close to one which indicated almost a Newtonian behaviour, while Alg exhibited pseudoplastic behaviour ($n < 1$) [35,36], as it can also be observed in Figure 1.

This is favourable for bioprinting since it can reduce cell death during the extrusion process where shear forces are produced. The greater the concentration, the more the pseudoplastic behaviour is pronounced. Regarding temperature, zero-shear viscosity decreased slightly at 37 °C (Table S1), mainly in Alg samples where the temperature effect was greater at higher Alg concentrations due to a possible breaking of polymer entanglements [37–39]. The influence of temperature for HA could be considered negligible.

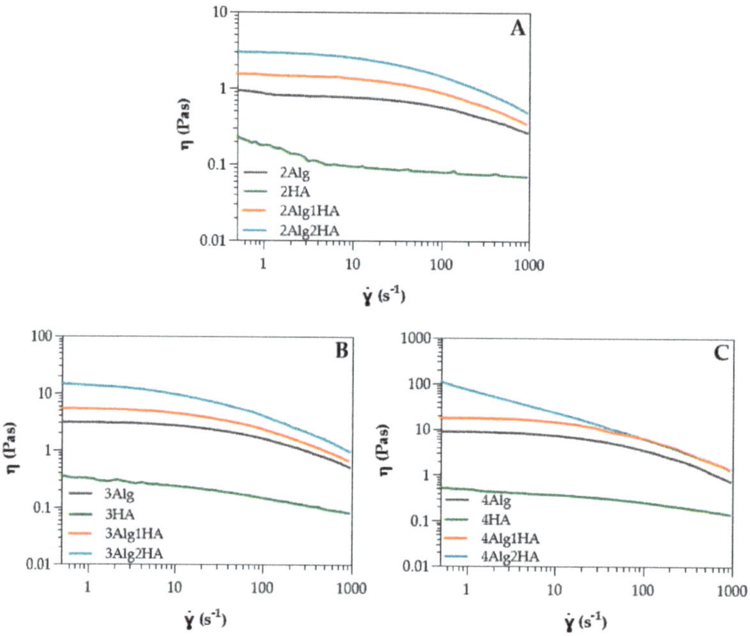

Figure 1. Shear viscosity of HA and Alg solutions at (**A**) 2% (*w*/*w*) and its homologues, (**B**) 3% (*w*/*w*) and its homologues, and (**C**) 4% (*w*/*w*) and its homologues at 37 °C (n = 3 per group).

With respect to alginate/hyaluronic acid (Alg/HA) hybrid solutions, the Ostwald–de Waele parameters and Newtonian viscosity are presented in Table S2 of the supplementary information. Initially, intermediate zero-shear viscosity values of the hybrid solutions could be expected, however, zero-shear viscosity increased when both materials were mixed, obtaining higher values than Alg (Table S2 and Figure 1). For example, the viscosity of 4Alg2HA went up enough (>100 Pas) that the behaviour of the curve changed and therefore, it did not have a Newtonian zone (see Figure 1C, blue data). This synergetic effect is attributed to interactions between Alg chains and HA chains, which are forming physical entanglements and hydrogen bonds (H bonds) (see Figure 2), causing an increase in viscosity [40–42]. Apart from this, the viscosity value increased with increasing both Alg and HA concentrations where the interactions between Alg polymer chains and HA polymer chains were higher. All formulations presented a shear-thinning behaviour ($n < 1$) [35,36], closer to the alginate value than the HA value. This means that the interaction between both polymers is destroyed upon the application of shear and the reduction in viscosity is high. Finally, all mixtures exhibited lower viscosity at higher temperatures (Table S2), concluding that the temperature energy might break the H bonds [43–46] and polymer chain entanglements [37–39] formed within the biopolymer hybrid solutions, thus increasing the mobility between polymer chains.

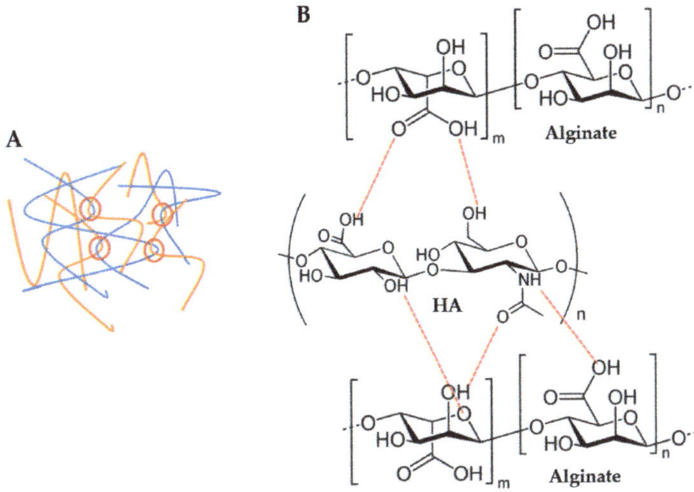

Figure 2. Possible physical interactions between alginate and hyaluronic acid (HA) polymer chains: (**A**) polymer chain entanglements (red circles) and (**B**) H bonds (red dashed lines).

3.2. Viscoelastic and Mechanical Characterization

The viscoelastic and mechanical properties of all hydrogels crosslinked with 100 mM of $CaCl_2$ at 23 and 37 °C are displayed in Table S3 of the Supplementary Information. It should be noted that dynamic shear tests at 37 °C were performed by heating the lower plate of the rheometer (thermal conduction) whereas the compression tests were carried out using a saline bath (PBS) at 37 °C (thermal convection) with the aim of simulating physiological conditions. Concerning viscoelastic measurements, at 37 °C, a lower storage modulus G' was obtained because the thermal energy could be breaking the physical or intermolecular interactions such as chain entanglements and H bonds between Alg and HA polymer chains [46]. In respect to mechanical tests, the steeper decrease in the tangent modulus at 37 °C was related not only to thermal energy, but also to ion exchange, since the PBS bath contained different inorganic salts, where Na^+ ions participate in the ion exchange with the Ca^{2+} of the hydrogel (Ca^{2+}/Na^+ exchange) [47–50], which caused its partial decrosslinking. An increase in bath temperature will induce an acceleration in the exchange of these ions and thus, in the degradation of the hydrogel [49]. In addition, there could be a possible release of HA into the medium as it was uncrosslinked. If shear modulus and tangent modulus are compared, the decrease in the modulus was higher for the tangent modulus, especially at higher alginate concentrations where the modulus dropped by $\approx$40%; however, the shear modulus only reduced its value by $\approx$10%. The fact that both thermal energy and ion exchange were present in mechanical tests may have contributed to the steeper decrease in the modulus obtained.

For the study of the influence of Alg and HA concentrations in viscoelastic and mechanical properties, the storage modulus G' (A) and tangent modulus at 10% of strain (B) of all hydrogels at 37 °C are displayed in Figure 3. Adding HA to the formulation revealed that the storage modulus G' and tangent modulus decreased, especially when 2% (w/w) HA was added (white data, Figure 3). This is the result of the interaction between carboxyl groups of Alg (G-blocks) and amide groups of HA, as demonstrated by FTIR and rheological analysis by several authors [40–42,51], which hinders the gel formation by Ca^{2+} chelation since there is a reduction in the available crosslinking points [52–54] (see Figure 2B). Consequently, a minor crosslinking density was obtained in the Alg chains and thus, a lower modulus. Regarding the effect of Alg concentration, results show that a higher storage modulus G' and tangent modulus were obtained when Alg concentration increased in the formulation due to a higher crosslinking density where more crosslinking points could be found in Alg polymer chains.

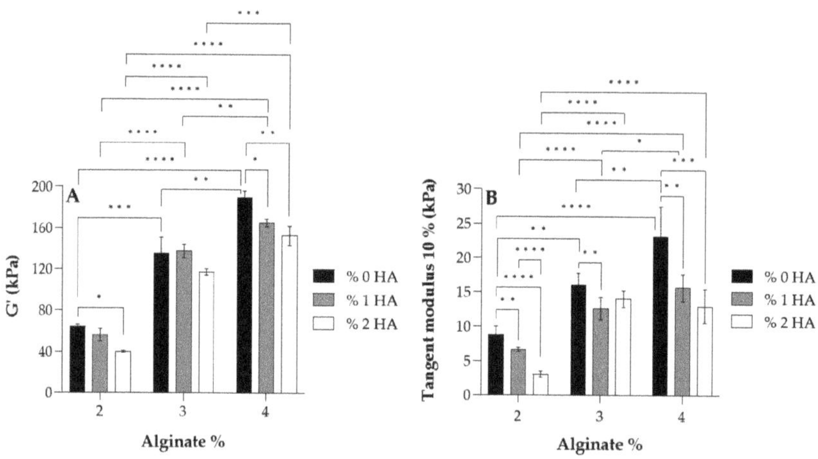

Figure 3. Values of the storage modulus G' (**A**) and tangent modulus at 10% of strain (**B**) for each alginate and its homologues crosslinked with 100 mM of $CaCl_2$ at 37 °C. Symbols denote statistically significant differences (* $p < 0.05$; ** $p < 0.01$; *** $p < 0.001$; and **** $p < 0.0001$).

According to the literature, articular cartilage presents a dynamic modulus of 100–3000 kPa [55], matching that of most hydrogels developed in this study, with the group with 2% Alg being the exception since they showed lower values of the modulus (<100 kPa). Further, almost all hydrogels possessed a tangent modulus > 10 kPa, which is in accordance with Naranda et al. [55] who mentioned that natural scaffolds used in cartilage tissue engineering (CTE) exhibit compressive Young's moduli values in the range of 10–250 kPa, corresponding to very low mechanical properties compared to native cartilage tissue (240–1000 kPa [56,57]).

3.3. Printability Evaluation

The qualitative printability evaluation was performed with all developed hydrogels and was divided into three groups by Alg concentration. So, the influence of HA addition in the printability of hydrogels was studied. All hydrogel solutions showed suitable viscosity (η = 1–23 Pas) for 3D bioprinting according to He et al. [58], who remarked that the appropriate viscosity range for a good processability is 0.3–100 Pas. The 4Alg2HA hydrogel was the exception since it presented a very high viscosity (>100 Pas) as mentioned before in Section 3.1, showing a viscosity value out of range, and therefore, making it unsuitable for 3D bioprinting. Consequently, this hybrid hydrogel was rejected for further studies. It should be noted that printability is not only determined by the viscosity of hydrogels because the surface tension of the printing bed plays a crucial role in the quality of printed constructs as well [9,59,60].

Figure 4 shows the different imprint structures fabricated using a needle inner diameter of 27G. A general overview revealed that the addition of HA had a great effect on printability due to the higher ability of the hydrogel precursor to retain the shape of the structure. When HA was added to the formulation, the viscosity of the hydrogel increased, preventing the filament from collapsing. Accordingly, the formulations containing pure Alg in all groups displayed a lower printability than their homologues, causing a slight collapse of the filament and obtaining more irregular and less defined round-like pores, mainly with 2Alg and 3Alg samples. Formulations containing HA, however, presented better printability as a result of their higher viscosity that helped to maintain the shape of the structures. Consequently, good-shaped scaffolds were fabricated with square-like uniform pores. The printability of the hydrogels was enhanced at higher HA concentration as can be observed in the 2Alg2HA and 3Alg2HA samples.

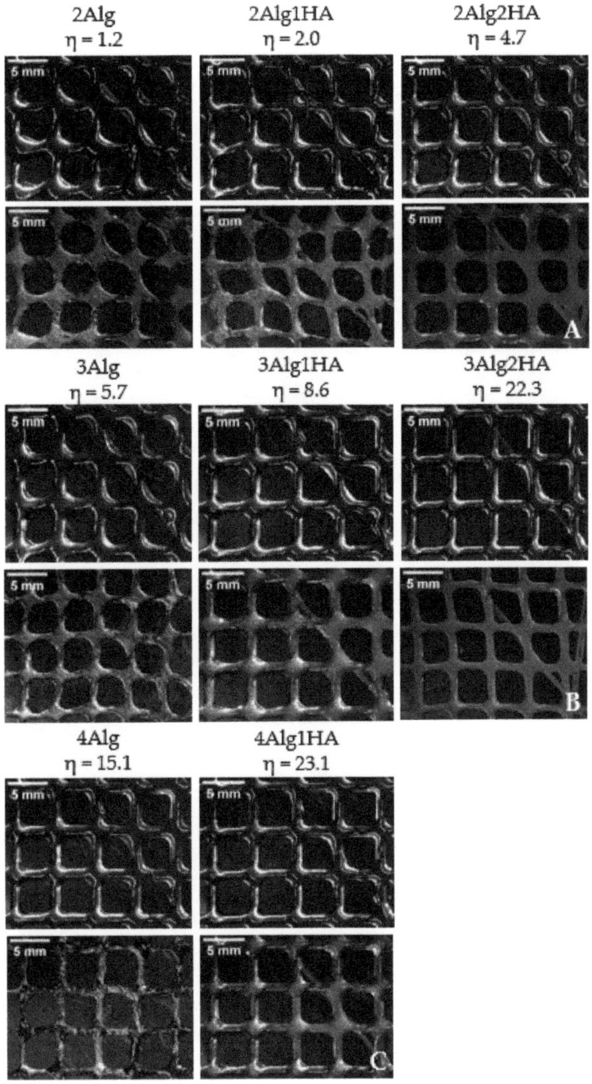

Figure 4. Printed scaffolds of alginate samples and their homologues. Upline: uncrosslinked samples and downline: crosslinked samples. (**A**) Group of 2% (*w*/*w*) alginate (**B**) 3% (*w*/*w*) alginate (**C**) 4% (*w*/*w*) alginate. (Magnification: 0.61×. Needle inner diameter: 27G).

The suitable viscosity for 3D bioprinting proposed in our previous work by Gorroñogoitia et al. [11] was set to the range of 10–80 Pas, but after the results obtained with the new hybrid formulations developed, the viscosity range could be further adjusted to 1–80 Pas for a good processability or even to 5–80 Pas for a more accurate adjustment, thanks to the interaction between the two biopolymers.

3.4. In Vitro Citotoxicity

The in vitro biocompatibility of hydrogels was evaluated by short–term cytotoxicity assays (Figure 5). As mentioned, the cell viability was measured by incubating L929 murine fibroblast cells for 24, 48, and 72 h in hydrogel extract medium. The results show that none of the developed hydrogels demonstrated signs of cytotoxicity since they all presented cell viability values above the 70% of acceptance limit set down by ISO 10993-5:2009 standard [32] throughout the entire experiment. Furthermore, after 24 h the cell viability increased above 100%, reaching values up to 120% and values of 90% or higher were obtained after 72 h of cell cultivation (Figure 5).

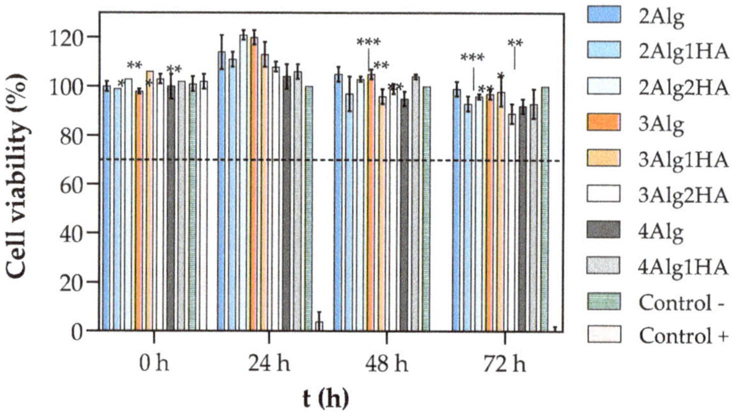

Figure 5. Viability of L929 cells treated with extract liquid of hydrogels crosslinked with 100 mM of $CaCl_2$: group of 2% (w/w) alginate (blue), 3% (w/w) alginate (orange), 4% (w/w) alginate (grey), negative control (complete medium with cells, green), and positive control (10% v/v of DMSO in complete medium, red). The dashed line represents the acceptance limit established by ISO 10993-5:2009 [32] (70% of the negative control value). Symbols denote statistically significant differences in comparison to 24 h (* $p < 0.05$; ** $p < 0.01$; and *** $p < 0.001$).

It is well known that cells are adhered by calcium bonds; therefore, calcium promotes the intercellular interaction. At the same time, calcium is involved in many cellular activities such as cell proliferation and provides a more favourable environment for cell survival [61–64]. Consequently, the calcium ions that could have been released from the Alg/HA hydrogel to the surrounding medium during the preparation of the liquid extract could be inducing the proliferation of cells and thus, higher cell viability was obtained in the first 24 h. In addition, Herrero-Mendez et al. [19,65] demonstrated that the sulphated glycosaminoglycans (sGAGs), which are known for having specific binding sites for growth factors and cell molecules that are involved in cell adhesion and proliferation [13,17,19], can promote an acceleration in the proliferative rate of cells. Just like calcium, these sGAG molecules could have been liberated to the medium since they remained uncrosslinked within the hydrogels, and consequently, they might be contributing to the higher cell viability obtained at 24 h.

3.5. Cell Morphology and Adhesion

The study of the cells' morphology and adhesion was performed by the scanning electron microscopy (SEM) analysis that is depicted in Figure 6. As can be observed, in samples containing 2, 3, and 4% alginate (Figure 6A–C), cells were not able to acquire flattened morphology, nor establish cytoplasmic projections adhering to the surface of samples that contained Alg. Accordingly, cells adopted a spherical shape, forming cell aggregates, which was the result of a strong cell–cell interaction rather than cell–matrix interaction, as reported previously by other authors. It is already known that Alg, despite being biocompatible, lacks cell-interactive domains, making it a non-adhesive biomaterial since cells do not have any attachment point with the material itself. To overcome this intrinsic characteristic, Alg can be chemically modified with cell-specific binding motifs such as the RGD peptide that is used as a ligand to promote cell adhesion [24,25,66,67]. Eventually it can also be blended with other materials to induce cell adhesion. For example, Sarker et al. [68] used a hybrid hydrogel composed of alginate and gelatine that was capable of adhering to NHDF cells after 7 days of incubation, unlike a pure alginate hydrogel.

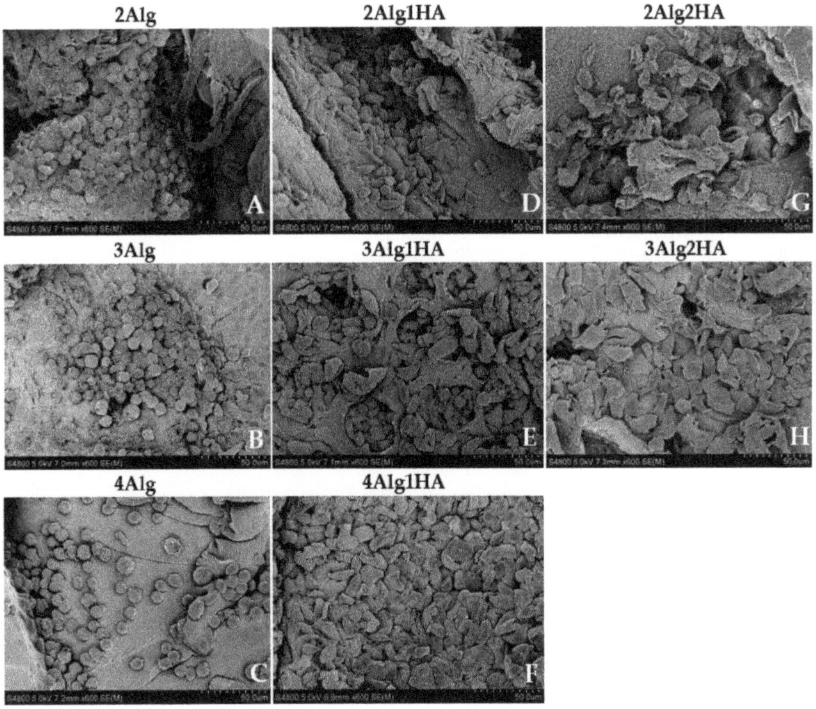

Figure 6. Morphological assessment of L929 cells on alginate (Alg) hydrogels and their homologues crosslinked with 100 mM of $CaCl_2$ after 3 days of incubation. Top line: 2% (*w*/*w*) alginate and homologues (**A**,**D**,**G**), middle line: 3% (*w*/*w*) alginate and homologues (**B**,**E**,**H**), and bottom line: 4% (*w*/*w*) alginate and homologues (**C**,**F**) (magnification: 600×).

Regarding hydrogels containing HA, cells acquired more flattened morphology than single-alginate hydrogels due to cell–matrix interaction that corroborated the adhesion of cells into the material surface (Figure 6D–H). For example, this was not achieved by several authors [69–71] who used a thiol-modified HA or a non-sulphated HA in combination with other biopolymers and the cell adhesion experiments revealed that cells were not able to adhere to the hydrogel surface, acquiring a spherical morphology rather than spindle-like one. The results obtained in this study can be attributed to the ability of sGAGs to bind (non-covalently) through sulphate groups (negatively charged) to growth factors and cell molecules that are involved in many biological activities and functions such as cell proliferation, differentiation, and cell–cell and cell–matrix communication and signalling [13,17,19]. This is reflected in the work of Herrero-Mendez et al. [65] who revealed that adipose-derived stem cells (ASC) and fibroblasts were able to adhere to the hydrogel (HA-enriched + sGAG enriched) and were even viable and capable of repairing chondral and dermal defects by the synthesis of Col-II and Col-I/Fibronectin in vitro, respectively.

3.6. Cell Viability of Seeded Cells on Hydrogels and Bioprinted Cells

The viability of murine fibroblasts (L929) after seeding on the surface of hydrogels was examined by live/dead fluorescence staining, which is represented in Figure 7. Most of the cells seeded in all hydrogels were seen to remain viable (stained green) after 7 days of incubation with intact cell membranes and only a few of them died (stained red). The following are the cell viability values for all samples % 91 $\pm$ 5 (2Alg), % 91 $\pm$ 7 (2Alg1HA), % 94 $\pm$ 3 (2Alg2HA), % 86 $\pm$ 10 (3Alg), % 90 $\pm$ 3 (3Alg1HA), % 88 $\pm$ 7 (3Alg2HA), and % 90 $\pm$ 10 (4Alg1HA). As also observed by SEM images, cells were found to be agglomerated and forming clusters on Alg single hydrogels (Figure 7A–C) confirming that cell–cell interactions were stronger than cell–material interactions. Nevertheless, this was not the case for Alg/HA hybrid hydrogels, where cells were able to adhere to the material and even proliferate after 7 days of incubation. Furthermore, 7 days after cultivation, cells covered most of the Alg/HA hybrid hydrogel surface (Figure 7D–H); in contrast to Alg single hydrogels where very few cells remain attached, especially in the 3 and 4% alginate hydrogels (Figure 7B,C). Accordingly, it could be concluded that the addition of HA in the formulation of hydrogels, in terms of bioactivity, was favourable for all single Alg hydrogels but mainly for the 3 and 4% alginate ones. Finally, the higher the HA concentration, the greater the cell density on the surface of the hydrogel due to a higher availability of cell-binding sites, thus, increasing the number of adherent cells, as can be observed in Figure 7G,H.

For studying the suitability of the developed hydrogels as precursors for bioprinting, Alg2HA hydrogel was selected because its rheological behaviour and its printability had demonstrated that it could be a good candidate for this purpose. In addition, it showed good cell viability and adhesion. As aforementioned, the bioink was printed using a BIO V1 3D bioprinter with a printing conical nozzle of 22G inner diameter and a flow rate of 2 µL/s, at 37 °C. The cell viability of encapsulated cells in bioprinted constructs was assessed by staining live (green) and dead (red) cells with calcein-AM and propidium iodide. Figure 8 shows the fluorescent images of all samples, both the ones that were not printed (control samples, Figure 8A) and the printed samples (Figure 8B), 1, 3 and 7 days after culture. It can be observed that cells were homogeneously distributed along the constructs resulting in a satisfactory mixing of the cells in the 3Alg2HA hydrogel, respectively. The results obtained show that cells remained viable (stained green) after printing, concluding that the extrusion process did not negatively affect cell viability and thus, the suitability of the 3Alg2HA hydrogel precursor as a good candidate for bioprinting was confirmed due to its appropriate viscosity with shear-thinning properties (Figure 1B, blue data) that avoided cell death. The cell viability remained high during the experiment, showing values for control samples and bioprinted samples of % 77 and % 72 at day 1, % 89 and % 94 at day 3, and % 79 and % 92 at day 7. Apparently, the L929 cells encapsulated in the 3Alg2HA hydrogel did not proliferate over time, showing a constant density of live cells throughout the incubation time.

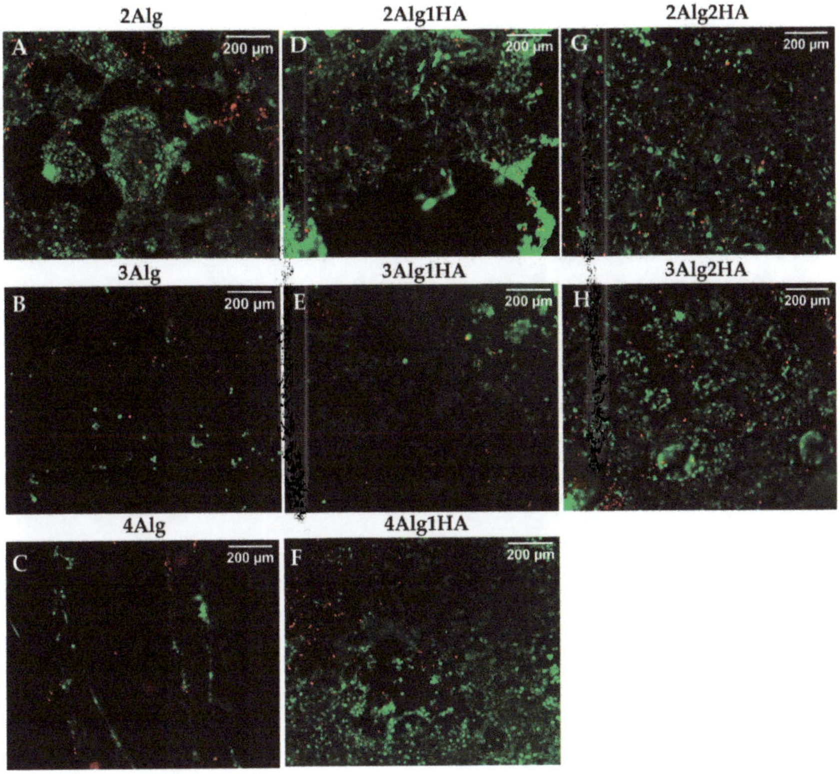

Figure 7. Live/dead cell viability staining of L929 cells on the surface of alginate (Alg) hydrogels and their homologues crosslinked with 100 mM of $CaCl_2$ after 7 days of incubation. Top line: 2% (*w*/*w*) alginate and homologues (**A**,**D**,**G**), middle line: 3% (*w*/*w*) alginate and homologues (**B**,**E**,**H**), and bottom line: 4% (*w*/*w*) alginate and homologues (**C**,**F**). Dead cells are stained red and live cells are stained green (magnification: 10×).

Although the cell viability was good after printing, it is important to note that during the culturing period of the constructs in MEM, the uncrosslinked HA could have been released to the medium, causing the partial destruction of the scaffold. However, to fully verify that HA is liberated from the hydrogel network, degradation experiments and the quantification of HA should be performed. Moreover, if cells create their own extracellular matrix over time and its synthesis is synchronized to the degradation rate of the hydrogel, there should not be a problem. Anyway, one of the solutions to avoid HA leaching out of the hydrogel network is to crosslink it and obtain IPN hydrogels, instead. However, this was not the aim of this work.

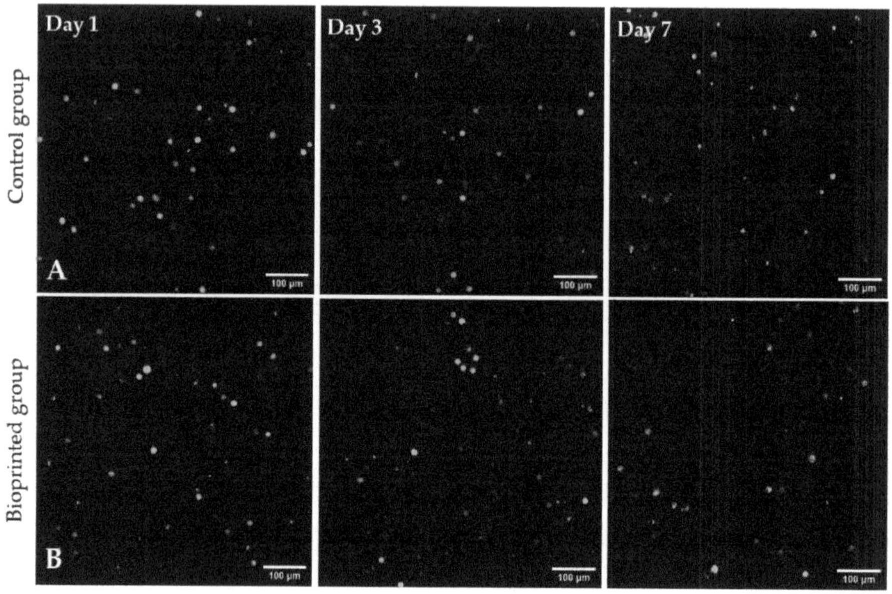

Figure 8. Live/dead images of L929 cells encapsulated in the 3Alg2HA hydrogel after 1, 3, and 7 days of incubation: (**A**) control samples and (**B**) bioprinted constructs. Dead cells are stained red and live cells are stained green (magnification: 10×).

4. Discussion and Conclusions

Some research studies have proved the suitability of alginate (Alg) and hyaluronic acid (HA) as mixed biomaterials for cartilage tissue application, as mentioned in the introduction. However, with the aim of adding value and novelty to this investigation, it is important to note that those studies used a different HA from the one used in this work. The HA used in this study differs in composition from the rest of the investigations since it contains a high percentage of sulphated glycosaminoglycans (sGAGs), enhancing the bioactivity properties of the final hydrogels, and at the same time, it allows us to obtain a higher resemblance of the extracellular matrix (ECM) of the native tissue because they are components of the cartilage itself. In addition, the final properties of alginate hydrogels are determined by its microstructure as reported by several authors [11,24,26–28]; at the same time, the mechanical properties affect the biological behaviour of cells, so it is very important to have the control of the microstructure of alginate. The alginate used in this work was selected based on the research of our previous work [11], enabling us to develop robust hydrogels.

The semi-interpenetrating (semi-IPN) hydrogels composed of Alg and HA that were developed in this study demonstrated that the final properties of hydrogels could be tailored by varying the concentration of both biopolymers, always seeking a trade-off between mechanical and biological properties. HA was responsible for promoting key biological properties in hydrogels while the presence of Alg was favourable for mechanical properties and hydrogel stability. Except for the 4Alg2HA hybrid hydrogel, which showed too high of a viscosity, the rest of the hydrogels exhibited a suitable viscosity for 3D bioprinting, making them good candidates for bioink applications. In addition, the development of hybrid hydrogels by the combination of Alg and HA, demonstrated a synergetic effect in viscosity but not in mechanical and viscoelastic properties. According to printability results, the stability and quality of printed scaffolds was enhanced by the interaction of the two biopolymers compared to single-alginate scaffolds. The in vitro cytotoxicity assay showed that hydrogels were biocompatible and cell viability experiments by live/dead staining revealed that seeded cells on the surface of hydrogels and encapsulated cells in the printed constructs after bioprinting were viable after 7 days of incubation. Finally, cell adhesion was improved by the incorporation of HA in the formulation. Despite the great potential of all hydrogels in tissue engineering (TE) applications and regenerative medicine, values of the compressive modulus of the hydrogels were far from those of native articular cartilage; the hybrid hydrogels showed a tangent modulus between 3 and 16 kPa, which is still considerably lower than that of native articular cartilage (240–1000 kPa). However, this problem could be addressed either by developing interpenetrating (IPN) hydrogels [72–75] via the crosslinking of both biomaterials or by adding nanoparticles such as cellulose nanocrystals (CNC) to the formulation [76–80], in order to reinforce mechanically the aforementioned hydrogels. Eventually, some other studies suggest using a high M_w alginate along with a low M_w alginate for the improvement of the mechanical properties [27,81]. Finally, further degradation experiments to analyse the degradation rate of the hydrogels and experiments using chondrocytes or mesenchymal stem cells (MSC) should be carried out so that a 3D cartilage model could be developed by 3D bioprinting. Once this is accomplished, the suitability of these hydrogels in the field of cartilage tissue engineering (CTE) could be confirmed.

Supplementary Materials: The following supporting information can be downloaded at https://www.mdpi.com/article/10.3390/polym17040528/s1, Figure S1: Viscosity curves for the hyaluronic acid batches at a concentration of % 2 (w/w) in PBS at 23 °C, Figure S2: Storage modulus G' (squares) and loss modulus G'' (triangles) as a function of time for hydrogels: 2Alg (blue), 2Alg1HA (orange), and 2Alg2HA (grey) crosslinked with 100 mM of $CaCl_2$ at 37 °C (n = 3 per group), Figure S3: Stress–strain curves for alginate hydrogel at 2% (w/w) and its homologues crosslinked with 100 mM of $CaCl_2$ at 37 °C (n = 5 per group), Table S1: Ostwald–de Waele parameters and zero-shear viscosity of alginate and hyaluronic acid single-component solutions at different concentrations and two different temperatures, Table S2: Ostwald–de Waele parameters and zero-shear viscosity of alginate/hyaluronic acid hybrid solutions at different concentrations and two different temperatures, Table S3: Values of the storage modulus G' and tangent modulus at 10% strain for each alginate and its homologues crosslinked with 100 mM of $CaCl_2$, Table S4: Printability parameters for all printed hydrogels. References [35,82] are cited in the supplementary materials.

Author Contributions: Conceptualization, I.G., A.M.Z. and A.A.-V.; methodology, I.G. and S.O.; investigation, I.G.; writing—original draft preparation, I.G.; writing—review and editing, I.G., S.O., A.M.Z. and A.A.-V.; visualization, I.G.; supervision, S.O., A.M.Z. and A.A.-V. All authors have read and agreed to the published version of the manuscript.

Funding: This research received partial funding from the University of The Basque Country UPV/EHU (grant references GIU21/054 and PES50/21).

Data Availability Statement: The original contributions presented in this study are included in the article/Supplementary Material. Further inquiries can be directed to the corresponding author.

Acknowledgments: We would like to thank POLYMAT (Donostia-San Sebastián, Spain) for characterizing the sodium alginate and Histocell S.L Regenerative Medicine (Derio, Spain) for providing us with the native whartonide hyaluronic acid. We would like also to acknowledge General Research Services from the University of the Basque Country (SGIker, UPV/EHU).

Conflicts of Interest: Authors Izar Gorroñogoitia and Ane Miren Zaldua were employed by the company Leartiker S. Coop. The remaining authors declare that the research was conducted in the absence of any commercial or financial relationships that could be construed as a potential conflict of interest.

References

1. Chartrain, N.A.; Gilchrist, K.H.; Ho, V.B.; Klarmann, G.J. 3D Bioprinting for the Repair of Articular Cartilage and Osteochondral Tissue. *Bioprinting* **2022**, *28*, e00239. [CrossRef]
2. Kwon, H.; Brown, W.E.; Lee, C.A.; Wang, D.; Paschos, N.; Hu, J.C.; Athanasiou, K.A. Surgical and Tissue Engineering Strategies for Articular Cartilage and Meniscus Repair. *Nat. Rev. Rheumatol.* **2019**, *15*, 550–570. [CrossRef] [PubMed]
3. Wei, W.; Dai, H. Articular Cartilage and Osteochondral Tissue Engineering Techniques: Recent Advances and Challenges. *Bioact. Mater.* **2021**, *6*, 4830–4855. [CrossRef] [PubMed]
4. Wasyłeczko, M.; Sikorska, W.; Chwojnowski, A. Review of Synthetic and Hybrid Scaffolds in Cartilage Tissue Engineering. *Membranes* **2020**, *10*, 348. [CrossRef]
5. Rahmani Del Bakhshayesh, A.; Babaie, S.; Tayefi Nasrabadi, H.; Asadi, N.; Akbarzadeh, A.; Abedelahi, A. An Overview of Various Treatment Strategies, Especially Tissue Engineering for Damaged Articular Cartilage. *Artif. Cells Nanomed. Biotechnol.* **2020**, *48*, 1089–1104. [CrossRef]
6. Murphy, S.V.; Atala, A. 3D Bioprinting of Tissues and Organs. *Nat. Biotechnol.* **2014**, *32*, 773–785. [CrossRef] [PubMed]
7. Vijayavenkataraman, S.; Yan, W.-C.; Lu, W.F.; Wang, C.-H.; Fuh, J.Y.H. 3D Bioprinting of Tissues and Organs for Regenerative Medicine. *Adv. Drug Deliv. Rev.* **2018**, *132*, 296–332. [CrossRef] [PubMed]
8. Sun, W.; Starly, B.; Daly, A.C.; Burdick, J.A.; Groll, J.; Skeldon, G.; Shu, W.; Sakai, Y.; Shinohara, M.; Nishikawa, M.; et al. The Bioprinting Roadmap. *Biofabrication* **2020**, *12*, 022002. [CrossRef] [PubMed]
9. Tian, S.; Zhao, H.; Lewinski, N. Key Parameters and Applications of Extrusion-Based Bioprinting. *Bioprinting* **2021**, *23*, e00156. [CrossRef]
10. Mobaraki, M.; Ghaffari, M.; Yazdanpanah, A.; Luo, Y.; Mills, D.K. Bioinks and Bioprinting: A Focused Review. *Bioprinting* **2020**, *18*, e00080. [CrossRef]
11. Gorroñogoitia, I.; Urtaza, U.; Zubiarrain-Laserna, A.; Alonso-Varona, A.; Zaldua, A.M. A Study of the Printability of Alginate-Based Bioinks by 3D Bioprinting for Articular Cartilage Tissue Engineering. *Polymers* **2022**, *14*, 354. [CrossRef] [PubMed]
12. López-Ruiz, E.; Jiménez, G.; de Cienfuegos, L.Á.; Antich, C.; Sabata, R.; Marchal, J.A.; Gálvez-Martín, P. Advances of Hyaluronic Acid in Stem Cell Therapy and Tissue Engineering, Including Current Clinical Trials. *Eur. Cell Mater.* **2019**, *37*, 186–213. [CrossRef] [PubMed]
13. Sodhi, H.; Panitch, A. Glycosaminoglycans in Tissue Engineering: A Review. *Biomolecules* **2020**, *11*, 29. [CrossRef] [PubMed]
14. Dovedytis, M.; Liu, Z.J.; Bartlett, S. Hyaluronic Acid and Its Biomedical Applications: A Review. *Eng. Regen.* **2020**, *1*, 102–113. [CrossRef]
15. Snetkov, P.; Zakharova, K.; Morozkina, S.; Olekhnovich, R.; Uspenskaya, M. Hyaluronic Acid: The Influence of Molecular Weight on Structural, Physical, Physico-Chemical, and Degradable Properties of Biopolymer. *Polymers* **2020**, *12*, 1800. [CrossRef] [PubMed]
16. Wang, M.; Deng, Z.; Guo, Y.; Xu, P. Designing Functional Hyaluronic Acid-Based Hydrogels for Cartilage Tissue Engineering. *Mater. Today Bio* **2022**, *17*, 100495. [CrossRef] [PubMed]
17. Menezes, R.; Vincent, R.; Osorno, L.; Hu, P.; Arinzeh, T.L. Biomaterials and Tissue Engineering Approaches Using Glycosaminoglycans for Tissue Repair: Lessons Learned from the Native Extracellular Matrix. *Acta Biomater.* **2023**, *163*, 210–227. [CrossRef]
18. Luo, Y.; Tan, J.; Zhou, Y.; Guo, Y.; Liao, X.; He, L.; Li, D.; Li, X.; Liu, Y. From Crosslinking Strategies to Biomedical Applications of Hyaluronic Acid-Based Hydrogels: A Review. *Int. J. Biol. Macromol.* **2023**, *231*, 123308. [CrossRef] [PubMed]
19. Herrero-Mendez, A.; Palomares, T.; Castro, B.; Herrero, J.; Alonso-Varona, A. Generation of Tunable Glycosaminoglycan Hydrogels to Mimic Extracellular Matrices: Generation of Tunable GAG Hydrogels to Mimic ECM. *J. Tissue Eng. Regen. Med.* **2016**, *10*, 1000–1011. [CrossRef]
20. Farokhi, M.; Jonidi Shariatzadeh, F.; Solouk, A.; Mirzadeh, H. Alginate Based Scaffolds for Cartilage Tissue Engineering: A Review. *Int. J. Polym. Mater. Polym. Biomater.* **2020**, *69*, 230–247. [CrossRef]
21. Abka-khajouei, R.; Tounsi, L.; Shahabi, N.; Patel, A.K.; Abdelkafi, S.; Michaud, P. Structures, Properties and Applications of Alginates. *Mar. Drugs* **2022**, *20*, 364. [CrossRef] [PubMed]
22. Hurtado, A.; Aljabali, A.A.A.; Mishra, V.; Tambuwala, M.M.; Serrano-Aroca, Á. Alginate: Enhancement Strategies for Advanced Applications. *Int. J. Mol. Sci.* **2022**, *23*, 4486. [CrossRef] [PubMed]
23. Pahlevanzadeh, F.; Mokhtari, H.; Bakhsheshi-Rad, H.R.; Emadi, R.; Kharaziha, M.; Valiani, A.; Poursamar, S.A.; Ismail, A.F.; RamaKrishna, S.; Berto, F. Recent Trends in Three-Dimensional Bioinks Based on Alginate for Biomedical Applications. *Materials* **2020**, *13*, 3980. [CrossRef]
24. Neves, M.I.; Moroni, L.; Barrias, C.C. Modulating Alginate Hydrogels for Improved Biological Performance as Cellular 3D Microenvironments. *Front. Bioeng. Biotechnol.* **2020**, *8*, 665. [CrossRef] [PubMed]
25. Kang, S.-M.; Lee, J.-H.; Huh, Y.S.; Takayama, S. Alginate Microencapsulation for Three-Dimensional In Vitro Cell Culture. *ACS Biomater. Sci. Eng.* **2021**, *7*, 2864–2879. [CrossRef]
26. Drury, J.L.; Dennis, R.G.; Mooney, D.J. The Tensile Properties of Alginate Hydrogels. *Biomaterials* **2004**, *25*, 3187–3199. [CrossRef]

27. Lee, K.Y.; Mooney, D.J. Alginate: Properties and Biomedical Applications. *Progress. Polym. Sci.* **2012**, *37*, 106–126. [CrossRef] [PubMed]
28. Enobakhare, B.; Bader, D.L.; Lee, D.A. Concentration and M/G Ratio Influence the Physiochemical and Mechanical Properties of Alginate Constructs for Tissue Engineering. *J. Appl. Biomater. Biomech.* **2006**, *4*, 87–96. [PubMed]
29. Janarthanan, G.; Kim, J.H.; Kim, I.; Lee, C.; Chung, E.-J.; Noh, I. Manufacturing of Self-Standing Multi-Layered 3D-Bioprinted Alginate-Hyaluronate Constructs by Controlling the Cross-Linking Mechanisms for Tissue Engineering Applications. *Biofabrication* **2022**, *14*, 035013. [CrossRef]
30. Nedunchezian, S.; Banerjee, P.; Lee, C.-Y.; Lee, S.-S.; Lin, C.-W.; Wu, C.-W.; Wu, S.-C.; Chang, J.-K.; Wang, C.-K. Generating Adipose Stem Cell-Laden Hyaluronic Acid-Based Scaffolds Using 3D Bioprinting via the Double Crosslinked Strategy for Chondrogenesis. *Mater. Sci. Eng. C* **2021**, *124*, 112072. [CrossRef]
31. Antich, C.; de Vicente, J.; Jiménez, G.; Chocarro, C.; Carrillo, E.; Montañez, E.; Gálvez-Martín, P.; Marchal, J.A. Bio-Inspired Hydrogel Composed of Hyaluronic Acid and Alginate as a Potential Bioink for 3D Bioprinting of Articular Cartilage Engineering Constructs. *Acta Biomater.* **2020**, *106*, 114–123. [CrossRef] [PubMed]
32. *ISO 10993-5:2009*; Biological Evaluation of Medical Devices—Part 5: Tests for In Vitro Cytotoxicity. International Organization for Standardization: Geneve, Switzerland, 2009. Available online: https://www.iso.org/standard/36406.html (accessed on 15 April 2024).
33. ISO 10993-12:2021; Biological Evaluation of Medical Devices—Part 12: Sample Preparation and Reference Materials. Available online: https://www.iso.org/standard/75769.html (accessed on 15 April 2024).
34. Serafin, A.; Culebras, M.; Collins, M.N. Synthesis and Evaluation of Alginate, Gelatin, and Hyaluronic Acid Hybrid Hydrogels for Tissue Engineering Applications. *Int. J. Biol. Macromol.* **2023**, *233*, 123438. [CrossRef]
35. Yeung, R.A.; Kennedy, R.A. A Comparison of Selected Physico-Chemical Properties of Calcium Alginate Fibers Produced Using Two Different Types of Sodium Alginate. *J. Mech. Behav. Biomed. Mater.* **2019**, *90*, 155–164. [CrossRef]
36. *Understanding Rheology*; Oxford University Press: New York, NY, USA, 2001; ISBN 978-0-19-514166-5.
37. Kong, D.; Yang, M.; Zhang, X.; Du, Z.; Fu, Q.; Gao, X.; Gong, J. Control of Polymer Properties by Entanglement: A Review. *Macro Mater. Eng.* **2021**, *306*, 2100536. [CrossRef]
38. Jiang, M.; Liu, B.-W.; He, F.-M.; Zhang, Q.; Wang, A.; Guo, D.-M.; Zhao, H.-B.; Chen, L.; Wang, Y.-Z. High-Performance Flame-Retardant Aliphatic Polyamide via Enhanced Chain Entanglement. *Chem. Eng. J.* **2023**, *455*, 140637. [CrossRef]
39. Li, J.; Cao, X.; Liu, Y.; Chen, Q. Thermorheological Complexity of Poly(Vinyl Alcohol)/Borax Aqueous Solutions. *J. Rheol.* **2020**, *64*, 991–1002. [CrossRef]
40. Oerther, S.; Maurin, A.-C.; Payan, E.; Hubert, P.; Lapicque, F.; Presle, N.; Dexheimer, J.; Netter, P.; Lapicque, F. High Interaction Alginate-Hyaluronate Associations by Hyaluronate Deacetylation for the Preparation of Efficient Biomaterials. *Biopolymers* **2000**, *54*, 273–281. [CrossRef]
41. Khlibsuwan, R.; Khunkitti, W.; Pongjanyakul, T. Alginate-Caseinate Composites: Molecular Interactions and Characterization of Cross-Linked Beads for the Delivery of Anticandidals. *Int. J. Biol. Macromol.* **2018**, *115*, 483–493. [CrossRef] [PubMed]
42. Athamneh, T.; Amin, A.; Benke, E.; Ambrus, R.; Leopold, C.S.; Gurikov, P.; Smirnova, I. Alginate and Hybrid Alginate-Hyaluronic Acid Aerogel Microspheres as Potential Carrier for Pulmonary Drug Delivery. *J. Supercrit. Fluids* **2019**, *150*, 49–55. [CrossRef]
43. Li, T.; Zhang, C.; Xie, Z.; Xu, J.; Guo, B.-H. A Multi-Scale Investigation on Effects of Hydrogen Bonding on Micro-Structure and Macro-Properties in a Polyurea. *Polymer* **2018**, *145*, 261–271. [CrossRef]
44. He, Y.; Zhu, B.; Inoue, Y. Hydrogen Bonds in Polymer Blends. *Progress. Polym. Sci.* **2004**, *29*, 1021–1051. [CrossRef]
45. Ti, Y.; Chen, D. Temperature Dependence of Hydrogen Bond in Fe-OCAP/Polyurethane Blends. *J. Appl. Polym. Sci.* **2013**, *130*, 2265–2271. [CrossRef]
46. Lamm, M.E.; Song, L.; Wang, Z.; Rahman, M.A.; Lamm, B.; Fu, L.; Tang, C. Tuning Mechanical Properties of Biobased Polymers by Supramolecular Chain Entanglement. *Macromolecules* **2019**, *52*, 8967–8975. [CrossRef]
47. Teixeira, M.C.; Lameirinhas, N.S.; Carvalho, J.P.F.; Valente, B.F.A.; Luís, J.; Pires, L.; Oliveira, H.; Oliveira, M.; Silvestre, A.J.D.; Vilela, C.; et al. Alginate-Lysozyme Nanofibers Hydrogels with Improved Rheological Behavior, Printability and Biological Properties for 3D Bioprinting Applications. *Nanomaterials* **2022**, *12*, 2190. [CrossRef] [PubMed]
48. Hunt, N.C.; Smith, A.M.; Gbureck, U.; Shelton, R.M.; Grover, L.M. Encapsulation of Fibroblasts Causes Accelerated Alginate Hydrogel Degradation. *Acta Biomater.* **2010**, *6*, 3649–3656. [CrossRef] [PubMed]
49. Giuseppe, M.D.; Law, N.; Webb, B.; Macrae, R.A.; Liew, L.J.; Sercombe, T.B.; Dilley, R.J.; Doyle, B.J. Mechanical Behaviour of Alginate-Gelatin Hydrogels for 3D Bioprinting. *J. Mech. Behav. Biomed. Mater.* **2018**, *79*, 150–157. [CrossRef]
50. Bociaga, D.; Bartniak, M.; Grabarczyk, J.; Przybyszewska, K. Sodium Alginate/Gelatine Hydrogels for Direct Bioprinting—The Effect of Composition Selection and Applied Solvents on the Bioink Properties. *Materials* **2019**, *12*, 2669. [CrossRef] [PubMed]
51. Yang, W.; Xu, H.; Lan, Y.; Zhu, Q.; Liu, Y.; Huang, S.; Shi, S.; Hancharou, A.; Tang, B.; Guo, R. Preparation and Characterisation of a Novel Silk Fibroin/Hyaluronic Acid/Sodium Alginate Scaffold for Skin Repair. *Int. J. Biol. Macromol.* **2019**, *130*, 58–67. [CrossRef] [PubMed]

52. Kolar, S.; Jurić, S.; Marijan, M.; Vlahoviček-Kahlina, K.; Vinceković, M. Applicability of Alginate-Based Composite Microspheres Loaded with Aqueous Extract of Stevia Rebaudiana Bertoni Leaves in Food and Pharmaceutical Products. *Food Biosci.* **2022**, *50*, 101970. [CrossRef]
53. Wang, Y.-Q.; Zhang, Q.; Liu, J.-C.; Yan, J.-N.; Wang, C.; Lai, B.; Zhang, L.-C.; Wu, H.-T. Construction and Characterization of Alginate/Calcium β-Hydroxy-β-Methylbutyrate Hydrogels: Effect of M/G Ratios and Calcium Ion Concentration. *Int. J. Biol. Macromol.* **2024**, *273*, 133162. [CrossRef] [PubMed]
54. Díez-García, I.; Lemma, M.R.D.C.; Barud, H.S.; Eceiza, A.; Tercjak, A. Hydrogels Based on Waterborne Poly(Urethane-Urea)s by Physically Cross-Linking with Sodium Alginate and Calcium Chloride. *Carbohydr. Polym.* **2020**, *250*, 116940. [CrossRef] [PubMed]
55. Naranda, J.; Bračič, M.; Vogrin, M.; Maver, U. Recent Advancements in 3D Printing of Polysaccharide Hydrogels in Cartilage Tissue Engineering. *Materials* **2021**, *14*, 3977. [CrossRef]
56. Messaoudi, O.; Henrionnet, C.; Bourge, K.; Loeuille, D.; Gillet, P.; Pinzano, A. Stem Cells and Extrusion 3D Printing for Hyaline Cartilage Engineering. *Cells* **2020**, *10*, 2. [CrossRef] [PubMed]
57. Santos-Beato, P.; Midha, S.; Pitsillides, A.A.; Miller, A.; Torii, R.; Kalaskar, D.M. Biofabrication of the Osteochondral Unit and Its Applications: Current and Future Directions for 3D Bioprinting. *J. Tissue Eng.* **2022**, *13*, 204173142211334. [CrossRef] [PubMed]
58. He, Y.; Yang, F.; Zhao, H.; Gao, Q.; Xia, B.; Fu, J. Research on the Printability of Hydrogels in 3D Bioprinting. *Sci. Rep.* **2016**, *6*, 29977. [CrossRef] [PubMed]
59. Gillispie, G.; Prim, P.; Copus, J.; Fisher, J.; Mikos, A.G.; Yoo, J.J.; Atala, A.; Lee, S.J. Assessment Methodologies for Extrusion-Based Bioink Printability. *Biofabrication* **2020**, *12*, 022003. [CrossRef]
60. Miri, A.K.; Mirzaee, I.; Hassan, S.; Mesbah Oskui, S.; Nieto, D.; Khademhosseini, A.; Zhang, Y.S. Effective Bioprinting Resolution in Tissue Model Fabrication. *Lab. Chip* **2019**, *19*, 2019–2037. [CrossRef] [PubMed]
61. Rouzaire-Dubois, B.; Dubois, J.M. Calcium-Dependent Proliferation of NG108-15 Neuroblastoma Cells. *Gen. Physiol. Biophys.* **2004**, *23*, 231–239. [PubMed]
62. Cao, N.; Chen, X.B.; Schreyer, D.J. Influence of Calcium Ions on Cell Survival and Proliferation in the Context of an Alginate Hydrogel. *ISRN Chem. Eng.* **2012**, *2012*, 1–9. [CrossRef]
63. Rosińska, K.; Bartniak, M.; Wierzbicka, A.; Sobczyk-Guzenda, A.; Bociaga, D. Solvent Types Used for the Preparation of Hydrogels Determine Their Mechanical Properties and Influence Cell Viability through Gelatine and Calcium Ions Release. *J. Biomed. Mater. Res.* **2023**, *111*, 314–330. [CrossRef]
64. Berridge, M.J.; Lipp, P.; Bootman, M.D. The Versatility and Universality of Calcium Signalling. *Nat. Rev. Mol. Cell Biol.* **2000**, *1*, 11–21. [CrossRef] [PubMed]
65. Herrero-Mendez, A.; Palomares, T.; Castro, B.; Herrero, J.; Granado, M.H.; Bejar, J.M.; Alonso-Varona, A. HR007: A Family of Biomaterials Based on Glycosaminoglycans for Tissue Repair: HR007 Biomaterials for Clinical Repair of Chondral or Dermal Defects. *J. Tissue Eng. Regen. Med.* **2017**, *11*, 989–1001. [CrossRef]
66. Bhat, A.; Hoch, A.I.; Decaris, M.L.; Leach, J.K. Alginate Hydrogels Containing Cell-interactive Beads for Bone Formation. *FASEB J.* **2013**, *27*, 4844–4852. [CrossRef] [PubMed]
67. Chun, H.J.; Reis, R.L.; Motta, A.; Khang, G. (Eds.) *Biomimicked Biomaterials: Advances in Tissue Engineering and Regenerative Medicine*; Advances in Experimental Medicine and Biology; Springer: Singapore, 2020; Volume 1250, ISBN 978-981-15-3261-0.
68. Sarker, B.; Singh, R.; Silva, R.; Roether, J.A.; Kaschta, J.; Detsch, R.; Schubert, D.W.; Cicha, I.; Boccaccini, A.R. Evaluation of Fibroblasts Adhesion and Proliferation on Alginate-Gelatin Crosslinked Hydrogel. *PLoS ONE* **2014**, *9*, e107952. [CrossRef]
69. Shu, X.Z.; Ghosh, K.; Liu, Y.; Palumbo, F.S.; Luo, Y.; Clark, R.A.; Prestwich, G.D. Attachment and Spreading of Fibroblasts on an RGD Peptide-Modified Injectable Hyaluronan Hydrogel. *J. Biomed. Mater. Res.* **2004**, *68A*, 365–375. [CrossRef] [PubMed]
70. Safikhani, M.M.; Asefnejad, A.; Aghdam, R.M.; Rahmati, S. Fabrication, and Characterization of Crosslinked Sodium Alginate/Hyaluronic Acid/Gelatin 3Dprinted Heparin-Loaded Scaffold. *J. Polym. Res.* **2024**, *31*, 121. [CrossRef]
71. Godesky, M.D.; Shreiber, D.I. Hyaluronic Acid-Based Hydrogels with Independently Tunable Mechanical and Bioactive Signaling Features. *Biointerphases* **2019**, *14*, 061005. [CrossRef]
72. Baei, P.; Daemi, H.; Mostafaei, F.; Azam Sayahpour, F.; Baharvand, H.; Baghaban Eslaminejad, M. A Tough Polysaccharide-Based Cell-Laden Double-Network Hydrogel Promotes Articular Cartilage Tissue Regeneration in Rabbits. *Chem. Eng. J.* **2021**, *418*, 129277. [CrossRef]
73. Aldana, A.A.; Houben, S.; Moroni, L.; Baker, M.B.; Pitet, L.M. Trends in Double Networks as Bioprintable and Injectable Hydrogel Scaffolds for Tissue Regeneration. *ACS Biomater. Sci. Eng.* **2021**, *7*, 4077–4101. [CrossRef]
74. Matricardi, P.; Di Meo, C.; Coviello, T.; Hennink, W.E.; Alhaique, F. Interpenetrating Polymer Networks Polysaccharide Hydrogels for Drug Delivery and Tissue Engineering. *Adv. Drug Deliv. Rev.* **2013**, *65*, 1172–1187. [CrossRef]
75. Chen, P.; Xia, C.; Mo, J.; Mei, S.; Lin, X.; Fan, S. Interpenetrating Polymer Network Scaffold of Sodium Hyaluronate and Sodium Alginate Combined with Berberine for Osteochondral Defect Regeneration. *Mater. Sci. Eng. C* **2018**, *91*, 190–200. [CrossRef]

76. Li, Z.; Liao, Y.; Li, D.; Wang, H.; Sun, X.; Chen, X.; Yan, H.; Lin, Q. Design and Properties of Alginate/Gelatin/Cellulose Nanocrystals Interpenetrating Polymer Network Composite Hydrogels Based on in Situ Cross-Linking. *Eur. Polym. J.* **2023**, *201*, 112556. [CrossRef]
77. Han, C.; Wang, X.; Ni, Z.; Ni, Y.; Huan, W.; Lv, Y.; Bai, S. Effects of Nanocellulose on Alginate/Gelatin Bio-Inks for Extrusion-Based 3D Printing. *BioRes* **2020**, *15*, 7357–7373. [CrossRef]
78. Dutta, S.D.; Hexiu, J.; Patel, D.K.; Ganguly, K.; Lim, K.-T. 3D-Printed Bioactive and Biodegradable Hydrogel Scaffolds of Alginate/Gelatin/Cellulose Nanocrystals for Tissue Engineering. *Int. J. Biol. Macromol.* **2021**, *167*, 644–658. [CrossRef] [PubMed]
79. Lin, L.; Jiang, S.; Yang, J.; Qiu, J.; Jiao, X.; Yue, X.; Ke, X.; Yang, G.; Zhang, L. Application of 3D-Bioprinted Nanocellulose and Cellulose Derivative-Based Bio-Inks in Bone and Cartilage Tissue Engineering. *IJB* **2022**, *9*, 637. [CrossRef]
80. Domingues, R.M.A.; Gomes, M.E.; Reis, R.L. The Potential of Cellulose Nanocrystals in Tissue Engineering Strategies. *Biomacromolecules* **2014**, *15*, 2327–2346. [CrossRef]
81. Kong, H.-J.; Lee, K.Y.; Mooney, D.J. Decoupling the Dependence of Rheological/Mechanical Properties of Hydrogels from Solids Concentration. *Polymer* **2002**, *43*, 6239–6246. [CrossRef]
82. Dodero, A.; Vicini, S.; Alloisio, M.; Castellano, M. Sodium Alginate Solutions: Correlation between Rheological Properties and Spinnability. *J. Mater. Sci.* **2019**, *54*, 8034–8046. [CrossRef]

polymers

MDPI

Article

Impact of Molar Composition on the Functional Properties of Glutinous Rice Starch–Chitosan Blend: Natural-Based Active Coating for Extending Mango Shelf Life

Chawakwan Nitikornwarakul [1,2], Rodjanawan Wangpradid [1] and Natthida Rakkapao [3,4,*]

1 Faculty of Innovative Agriculture and Fishery Establishment Project, Prince of Songkla University, Surat Thani Campus, Surat Thani 84000, Thailand; chawakwan.n@ku.th (C.N.); rodjanawannn@gmail.com (R.W.)
2 Department of Food Science and Technology, Faculty of Agro-Industry, Kasetsart University, Bangkok 10900, Thailand
3 Department of Applied Chemistry, Faculty of Science and Industrial Technology, Prince of Songkla University, Surat Thani Campus, Surat Thani 84000, Thailand
4 Membrane Science and Technology Research Center, Faculty of Science, Prince of Songkla University, Hat Yai Campus, Songkhla 90110, Thailand
* Correspondence: natthida.r@psu.ac.th

Abstract: This study investigates natural-based blends of glutinous rice starch (GRS) and chitosan (CS), varying their molar composition (0:100, 30:70, 50:50, 70:30, and 100:0) to explore their interaction dynamics. Our findings illustrate the versatility of these blends in solution and film forms, offering applications across diverse fields. Our objective is to understand their impact on coatings designed to extend the post-harvest shelf life of mangoes. Results reveal that increasing chitosan content in GRS/CS blends enhances mechanical strength, hydrophobicity, and resistance to *Colletotrichum gloeosporioides* infection, a common cause of mango anthracnose. These properties overcome limitations of GRS films. Advanced techniques, including FTIR analysis and stereo imaging, confirmed robust interaction between GRS/CS blend films and mango cuticles, improving coverage with higher chitosan content. This comprehensive coverage reduces mango dehydration and respiration, thereby preserving quality and extending shelf life. Coating with a GRS/CS blend containing at least 50% chitosan effectively prevents disease progression and maintains quality over a 10-day storage period, while uncoated mangoes fail to meet quality standards within 2 days. Moreover, increasing the starch proportion in GRS/CS blends enhances film density, optical properties, and reduces reliance on acidic solvents, thereby minimizing undesirable changes in product aroma and taste.

Keywords: starch; chitosan; blend composition; mango; green active coating; functional biopolymer; post-harvest preservation; anthracnose inhibition; antifungal; shelf-life extension

1. Introduction

Fruits and vegetables serve as vital repositories of valuable nutrients, yet they continue to respire and transpire post-harvest, rendering them highly perishable. Horticultural crops worldwide experience significant post-harvest losses, estimated at around 1.3 billion tons annually, with developing countries facing more challenges [1]. In response to the imperative need for sustainable solutions in the food industry, natural-based active films have emerged as a promising alternative for extending the shelf life of fresh produce.

Comprised of biodegradable polymers and food-grade additives, these films are applied as thin layers in the form of wraps or coatings on food surfaces. The expanding environmental consciousness, coupled with regulatory restrictions on the use of agrochemicals and an escalating consumer demand for safe foods, has spurred a robust exploration of biodegradable films in post-harvest applications. Leveraging various biomaterials, including polysaccharides (such as starch, cellulose, chitosan, pectin, and alginate), proteins (including whey protein, casein, gelatin, collagen, soy protein, wheat gluten, and corn zein), and lipids (comprising natural waxes, acetylated monoglycerides, and resins) [2], these innovative films exhibit barrier properties that regulate water and gas exchange, thereby mitigating post-harvest losses.

Mango (*Mangifera indica* L.) stands as the second most economically significant tropical fruit commodity worldwide, with a global annual production reaching nearly 59 million tons in 2022 [3]. The fruit is renowned for its exceptional taste, aroma, and rich nutritional content. Classified as climacteric fruits, mangoes undergo a dramatic increase in respiration and ethylene production during ripening [4]. Upon harvesting from the tree, mangoes ripen quickly within 3–9 days at ambient temperatures [5]. During the onset of ripening, certain physiochemical changes take place, leading to the softening of the fruit tissue, thus increasing its vulnerability to microbial invasion [6,7]. Furthermore, mangoes are susceptible to various challenges, including water loss; microbial decay such as anthracnose and stem end rot; chilling injury; and mechanical damage, all of which hinder their handling and transportation [5]. Considering these challenges, preserving mango quality presents a significant challenge. Various approaches have been employed to maintain the quality of exported mangoes, including heat treatment, cold storage management, controlled atmosphere storage, 1-methylcyclopropene, ethylene, methyl jasmonate, and edible coatings [8]. Recent research has focused on utilizing natural materials as fruit coatings to reduce post-harvest diseases, preserve quality, and extend the shelf life of mangoes.

Starch stands out for its versatility, abundance, cost-effectiveness, edibility, and renewability, making it a prime choice for biodegradable film. Its unique characteristic and gelatinization properties have led to extensive utilization in this field [9]. Starch-based films have found success in delaying ripening and reducing weight/firmness loss in various fruits, such as avocados [10], pears [11], grapes [12], and cherry tomatoes [13]. Improving the barrier properties of starch-based films has been suggested by elevating starch crystallinity or amylopectin content. Furthermore, starch films with high amylopectin exhibit enhanced tensile strength, reduced moisture sensitivity, and increased crystallinity [14]. Glutinous rice starch (GRS), with its high amylopectin content (98–99%), exhibits exceptional gas barrier properties, rendering it an optimal candidate for investigation in this study. However, GRS alone may lack essential physicochemical, mechanical, and antimicrobial properties required for industrial applications. Thus, blending it with other biopolymers becomes imperative to enhance its properties, like water vapor and gas barrier abilities, along with antimicrobial activity.

Chitosan (poly-ß-(1,4)-D-glucosamine) has emerged as a prominent candidate in post-harvest applications, owing to its film-forming properties, biocompatibility, biodegradability, oxygen barrier ability, mechanical strength, and remarkable antimicrobial properties, particularly its anti-fungal ability [15]. The presence of reactive groups in chitosan, such as -OH and -NH_2, plays a crucial role in its antimicrobial properties. Yan et al. [16] elucidated that the -NH_3^+ group of deacetylated chitosan interacts electrostatically with the phosphoryl groups of microbial cell membranes, resulting in cell leakage. Chitosan also exhibits anti-fungal properties by directly impeding fungal growth and activating specific biological processes in plant tissue [17]. Its ability to inhibit gas exchange between fruits and their surroundings, thereby reducing respiration rates, positions chitosan as a compelling option for extending the post-harvest longevity of diverse produce, such as berry fruit [18], fresh fig [19], and mango [20].

Combining starch and chitosan presents a compelling strategy to enhance the inherent properties of both polymers. Starch, though versatile, often falls short in mechanical strength, water resistance, and antimicrobial efficacy. Conversely, chitosan, prized for its antimicrobial ability, can introduce undesirable traits such as a yellowish hue and subtle alterations in taste and odor due to the use of acidic solvents. These factors directly influence consumer safety and acceptance. By synergizing these polymers, we aim to overcome these limitations and develop environmentally friendly, edible coatings or packaging solutions.

Our research explores this innovative approach to extend the shelf life of produce and food, ensuring both safety and sustainability.

The utilization of composite starch/chitosan films has proven effective in enhancing the shelf life of fresh produce. Numerous studies have illustrated the beneficial impact of these films on extending shelf life. For instance, a cassava starch–chitosan coating mitigated weight loss and microbial spoilage in black mulberries stored at 5 °C for 16 days [21]. Starch films incorporating chitosan nanoparticles efficiently inhibited microbial growth in cherry tomatoes [22]. Furthermore, the application of a purple yam starch/chitosan film on apples for 4 weeks preserved fruit quality by reducing weight loss [23]. This study introduces a novel approach by proposing a coating composed of GRS and chitosan, with the aim of extending the shelf life of mangoes by mitigating anthracnose infection. Our primary objective is to explore the mechanical and functional characteristics of the GRS-CS film across various molar ratios, focusing on its potential as a biodegradable coating tailored specifically for mango preservation. Given the significant economic importance of Nam Dok Mai mangoes in Thailand and their widespread global distribution, this research offers a promising technology to enhance their shelf life in export markets. By addressing critical challenges related to mango preservation and transportation, our findings contribute to the advancement of post-harvest preservation strategies for this valuable commodity.

2. Materials and Methods

2.1. Materials

Glutinous rice starch (GRS) was sourced from Thai Flour Industry Co., Ltd. (Bangkok, Thailand). GRS has a repeat unit molecular weight of 162 Da. Chitosan flakes prepared from squid pens were purchased from Bio21 Co., Ltd. (Chonburi, Thailand). The chitosan used in this study possesses a high molecular weight exceeding 1000 kDa, along with a 95% degree of deacetylation, and the molecular weight of a repeat unit of chitosan is 161 Da. Mature mango fruit, which were harvested at 95–110 days old (*Mangifera indica* L. cv. Nam Dok Mai), were purchased from a commercial orchard in Ratchaburi, Thailand. The selected mangoes were uniform in size, shape, color, ripeness and lack of blemishes, injuries, and disease symptoms.

2.2. Methods

The study was structured into two main experimental phases: (I) exploring the impact of diverse molar proportions of starch to chitosan on GRS/CS solution and film properties; and (II) implementing GRS/CS active coatings with varying molar ratios on mango surfaces. This investigation aims to assess the feasibility of employing GRS/CS blends as active coatings to preserve the quality of post-harvest agricultural products, particularly fruits and vegetables.

2.2.1. Experiment Part I: Assessing the Effects of Varying Molar Proportions of GRS to Chitosan on GRS/CS Solution and Film Properties

The GRS (2% w/v) and chitosan (2% w/v) solutions were prepared separately. GRS was dissolved in deionized water and stirred at 75–80 °C for 1 h, while chitosan solution was prepared in 1 M acetic acid and stirred at room temperature (~27 °C) for 48 h. The GRS and chitosan solutions were then mixed in different molar ratios (100:0, 70:30, 50:50, 30:70, 0:100) to yield a total volume of 100 mL, as outlined in Table 1, resulting in the formulation of coating solutions named GRS100/CS0, GRS70/CS30, GRS50/CS50, GRS30/CS70, and GRS0/CS100. The mixed polymer solutions were stirred at room temperature (~27 °C) for 30 min until clear and homogeneous solutions were obtained, and the pH of the solutions was recorded. To form films, the coating solution (30 mL) was carefully poured onto a 9 cm diameter plastic *Petri* dish and dried in a hot air oven at 50–55 °C for 24 h. The dried films were peeled off, placed in plastic Ziplock bag, and stored at 53 ± 1% RH and 27 ± 1 °C in desiccators prior to any testing.

Table 1. Quantities of polymers and solvents for GRS/CS blends with varied molar ratios (total volume: 100 mL).

Sample Named	Ratio of GRS and Chitosan (%mol)		Weight of GRS and Chitosan (g)		Volume of Solvent (mL)	
	GRS	Chitosan	GRS	Chitosan	H_2O	CH_3COOH (0.1 M)
GRS100/CS0	100	0	1.98	0.00	100	0
GRS70/CS30	70	30	1.39	0.60	70	30
GRS50/CS50	50	50	0.99	1.00	50	50
GRS30/CS70	30	70	0.60	1.40	30	70
GRS0/CS100	0	100	0.00	2.00	0	100

Properties of GRS/CS Blended Solution

Viscosity

The viscosity of the coating solutions was measured using a Brookfield viscometer, Model LVD V3T (Ametex Brookfield, Middleborough, MA, USA). A 200 mL sample of each coating solution was placed in the viscometer and measured using spindle LV-2 at a speed of 40 rpm for 3 min at 27 °C. The viscosity of the coating solutions was obtained as the average of 3 replications.

Effect of coating solution on Colletotrichum gloeosporioides mycelial growth

The fungal strain *C. gloeosporioides* was obtained from the Plant Protection Research and Development office, Department of Agriculture, Bangkok, Thailand. A pure culture of *C. gloeosporioides* was sub-cultured on potato dextrose agar (PDA, Difco^{TM}, Franklin Lakes, NJ, USA) in *Petri* dishes (9 cm diameter) and incubated at room temperature for 14 days until mycelial growth reached the dish's margin.

The in vitro antifungal activity of the coating solution was evaluated using PDA amended with different concentrations of the coating solution. A 7 mm diameter agar plug from the margin of a 14-day-old pure culture of *C. gloeosporioides* was transferred to the center of each PDA plate supplemented with the desired concentration of GRS/CS coating solution (100:0, 70:30, 50:50, 30:70, 0:100 (% mol)). PDA plates containing 0.1% *v/v* aqueous acetic acid solution (pH 5.6) served as the negative control. Three replicates of each coating solution were prepared. All plates were then incubated at room temperature for 14 days, and the radial mycelial growth was measured daily in two perpendicular directions. The results are expressed as the percentage of mycelial growth inhibition (MGI), calculated using Equation (1), where D_c represents the colony diameter in the control plate and D_s represents the colony diameter in the PDA supplemented with a coating solution.

$$\text{MGI } (\%) = [(D_c - D_s)/D_c] \times 100 \quad (1)$$

Morphology, Physical, and Mechanical Properties of GRS/CS Blend Films

Morphology

Scanning electron microscopy (SEM; Quanta400, FEI, Brno-Černovice, Czech Republic) was used to study variations in film morphology resulting from different GRS and chitosan blend compositions. Sample films were mounted with conductive adhesive tape, sputter-coated with gold, and observed at an accelerating voltage of 15 kV. The analysis encompassed both top-surface and cross-sectional morphology. Images of the film's top surface were captured at 500× magnification, while images of the cross-section were captured at 1500×, 12,000×, and 30,000× magnifications.

A confocal microscope with white light lasers (STELLARIS 5, Leica, Germany) was used to examine the surface morphology and fluorescence properties of pure GRS, chitosan, and the GRS50/CS50 blend film. These samples, placed on glass slides, were observed at 63× magnification under a Transmitted Light Detector (TLD) and using a fluorescent contrasting method in their overlay channels. The resulting images present an overlay of fluorescence using excitation and emission wavelengths around 405 nm and 470 nm, respectively, consistent with the reported characteristics of the chitosan oligomer [24].

Density

The 1-inch diameter GRS/CS polymer films, punched into discs by a cork borer, were placed in a hot air oven at 105 °C for 24 h. The thickness of each film sample was measured using a Mitutoyo Thickness Gauge (Kawasaki, Kanagawa, Japan), while the weight was also recorded. The polymer film density (ρ) was calculated by dividing the weight (m) by the volume (v) using the following equation: $\rho = m/v$. Reported data represent the average of 3 samples per replicate across 3 replicates.

Water solubility (WS)

WS measurements were conducted on 1-inch diameter discs of GRS/CS polymer films, obtained by punching with a cork borer. The discs were subjected to 24 h treatment in a hot air oven at 105 °C, and the initial dry weight ($W_{initial}$) of each disc was recorded. Subsequently, the dried films were immersed in 50 mL of distilled water in sealed beakers at room temperature for 24 h. After removing the film residues from the beakers, they were dried at 105 °C for 24 h and reweighed (W_{final}) to determine the dry matter. WS of each film was calculated using Equation (2). The reported data represent the average of 3 samples per replicate across 3 replicates.

$$WS = ((W_{initial} - W_{final})/W_{initial}) \times 100 \quad (2)$$

Water contact angle

The water contact angle, as an indicator of the film's hydrophilic properties, was measured using a Data Physics Instruments OCA15 (GmbH, Filderstadt, Germany). Each film applied to the mango surface was tested in triplicate using the sessile drop observation. In this method, 1 μL droplets of distilled water were analyzed at various locations on the film's surface.

Mechanical properties

The mechanical properties, including tensile strength and elongation, were assessed according to the standard test method of thin plastic sheeting [25]. Strips of GRS/CS films measuring 10 mm in width and 60 mm in length were prepared for testing. A Universal Testing Machine (Tinius Olsen, Horizon program, Salfords, Surrey, England) was utilized with an initial grip separation of 50 mm, an initial gauge length of 25 mm, and a probe speed of 50 mm/min. The reported data of each GRS/CS sample represent the average of 3 samples per replicate, with 3 replicates.

2.2.2. Experiment Part II: Implementing GRS/CS Active Coatings with Varying Molar Ratios on Mango Surfaces

The maturity of mangoes was determined based on specific gravity [26]. Selected mature mangoes were washed with calcium hypochlorite solution (200 ppm), followed by rinsing with distilled water, and were then air-dried at room temperature for 45 min. Subsequently, the mangoes were immersed in different GRS/CS coating solutions (GRS100/CS0, GRS70/CS30, GRS50/CS50, GRS30/CS70, and GRS0/CS100) for 1 min and air-dried for 30 min at room temperature. The uncoated mangoes served as the control. The control and GRS/CS-coated mangoes were stored at room temperature for 10 days. The qualities of the mangoes, including appearance, total soluble solids (TSS), titratable acidity (TA), and the TSS/TA ratio, were evaluated at day 0 and at 2-day intervals over the 10-day storage period. Each treatment was replicated 3 times.

Characterization of GRS/CS film coated on mango surface

The adherence of the coating on the mango surface was assessed through stereo microscopic observation. Cross-sectional slices were obtained from the middle of the mango cheek using a stainless-steel blade for each treatment (GRS100/CS0, GRS70/CS30, GRS50/CS50, GRS30/CS70, and GRS0/CS100), as well as for the uncoated mangoes. The samples surfaces were examined at 45× magnification using a Leica Stereo Microscope (Model S Apo Stereozoom 1.0×–8.0×, Singapore).

The efficiency of polymer coating on mango surface and the interaction between chitosan and GRS were examined by analyzing the FTIR spectra of the samples. A Spectrum Two FTIR spectrometer ((PerkinElmer, Shelton, CT, USA)) equipped with an ATR sampling accessory was employed for this purpose. The infrared spectra of each treatment were recorded in 4000–400 cm^{-1}, using 32 scans and a resolution of 4 cm^{-1}.

2.2.3. Statistical Analysis

The experiments were performed based on a completely randomized experimental design (CRD). Three replicates per treatment were applied in each experimental stage and data are expressed as mean ± standard deviation. The statistical comparisons of qualities of film, percent of mycelial inhibition, and qualities of coated mango were performed by one-way analysis of variance (ANOVA) using IBM SPSS Statistics 22 for Windows User (Chicago, IL, USA). The significance of differences among treatment means was compared by Duncan's Multiple Range Test with 95% confidence level.

3. Result and Discussion

3.1. Part I: Effects of Varying Molar Compositions of GRS to Chitosan on GRS/CS Solution and Film Properties

3.1.1. Physical and Chemical Properties of GRS/CS Blend Solutions and Films

From Table 1, it is evident that the pH decreases with the chitosan content. Chitosan, being insoluble in water, dissolves effectively in weak acid solvents such as formic, acetic, propionic, and lactic acids [27]. In this study, acetic acid was chosen as the solvent due to its safety for consumption and its ability to produce chitosan films with desirable properties. Compared to other weak acids, acetic acid yields films with low oxygen and water vapor permeability, hydrophobicity, mechanical strength, and excellent clarity [27], making it ideal for coating or packaging fruits and vegetables intended for consumption.

However, a notable limitation of using chitosan films prepared with acetic acid as a coating or packaging material for food products is the persistent pungent odor and sour taste of acetic acid, which remains in the film and cannot be entirely eliminated. This odor and taste can transfer to the food or produce coated with the film, potentially affecting consumer acceptance [28].

To mitigate these limitations, diluting chitosan with GRS can be employed to reduce the concentration of acetic acid in the film. This approach results in a film with modified properties that are better suited for use as a coating or packaging material to extend the shelf life of food, vegetables, and fruits while minimizing undesirable sensory effects.

The viscosity of the coating solution plays a critical role in determining film formation and performance, thereby influencing its efficacy for active coating applications on produce. It directly affects film thickness and uniformity on surfaces, while also playing a significant role in determining the film's barrier properties against gases and water vapor. Therefore, understanding the viscosity of the coating solution is important for optimizing the functional properties of resulting films and ensuring their efficacy in packaging.

From Table 2, the viscosity analysis revealed that the pure GRS solution exhibited a viscosity of 27.3 cp, markedly lower than the pure chitosan solution at 667.5 cp, consistent with prior research [29]. This pronounced disparity in viscosities indicates that using GRS alone may not be optimal for its intended application as a coating to extend the post-harvest shelf life of fruits and vegetables. The low viscosity of the GRS solution could hinder its ability to form a cohesive film on produce surfaces, potentially compromising its effectiveness in preserving quality. This lower viscosity of GRS can be attributed to its composition, characterized by a low amylose content (1–2%) and a high amylopectin content (98–99%) [30]. Amylopectin, being highly branched with (1→4)-linked *α-D*-glucosyl units in chains connected by (1→6) linkages (4–5%), exhibits a semi-crystalline nature [31]. In the case of GRS, amylopectin consists of shorter chains (19–20 glucose units), limiting its capacity to form entanglements and consequently reducing viscosity. Conversely, chitosan, with its linear chains composed of long segments exceeding 6200 units of *N*-acetylglucosamine and glucosamine, can form numerous entanglements, resulting in higher viscosity. Blending GRS or another waxy starch with chitosan increases viscosity of the starch due to several factors. Chitosan can form hydrogen bonds with starch molecules, leading to a network structure and increased viscosity [32].

Table 2. Physical and chemical properties of GRS/CS blend solutions and films with varying molar compositions.

	Properties	GRS100/CS0	GRS70/CS30	GRS50/CS50	GRS30/CS70	GRS0/CS100
Solution	pH	6.81 ± 0.03	5.63 ± 0.07	5.22 ± 0.11	4.55 ± 0.08	3.68 ± 0.04
	Viscosity (cP) at 29 ± 1 °C	27.3 ± 0.8	34.5 ± 0.5	66.5 ± 0.7	136.5 ± 0.4	667.5 ± 0.6
Film	Appearance of film					
	Density (g/cm^3)	1.88 ± 0.12	1.61 ± 0.08	1.53 ± 0.06	1.41 ± 0.03	1.09 ± 0.10
	Water solubility (%)	N/A	N/A	26.75 ± 0.50	19.94 ± 1.12	12.31 ± 0.59

Additionally, the introduction of chitosan into the GRS solution creates more entanglements between polymer chains, further increasing viscosity. The enhanced viscosity promotes better film formation on produce surfaces, leading to improved adhesion and coverage, thereby enhancing the effectiveness of the coating in extending the shelf life of produce. An exponential increase in viscosity observed in GRS/CS blend solutions with higher chitosan concentrations suggests a significant role of intermolecular hydrogen bonding between starch and chitosan. This phenomenon likely contributes to the enhanced interactions within the polymer matrix, leading to greater viscosity. Consequently, the viscosity of the polymer blend solution can be adjusted as needed to suit the intended use by modifying the proportions of GRS and chitosan. Notably, the viscosity of the polymer solutions significantly influences the adhesion and thickness of the polymer coating on fruit and vegetable surfaces. This observation aligns closely with the experimental results obtained from employing GRS/CS blend solution as a coating to extend the shelf life of Nam Dok Mai mangoes in this research, in that the film thickness increased notably with higher chitosan proportion.

The optical properties of a polymer play a vital role in determining its suitability and performance for preserving the quality and marketability of produce. A coating with desirable optical characteristics enhances the visual appearance, facilitates quality control, and boosts consumer confidence, ultimately contributing to the success of the product in the marketplace. Table 2 provides valuable insights into the optical properties of the films, particularly the mixed GRS/CS solution with high chitosan content, which exhibits a viscosity suitable for coating fruits and vegetables. However, there are color limitations to consider, as pure chitosan film tends to have a slight yellowish tint attributed to carotenoid pigments, primarily astaxanthin, which forms strong bonds with the chitin molecule and interact with proteins in the exoskeleton's epithelial layer [33]. Rigorous chemical treatment is required to produce colorless chitosan products, albeit compromising some properties [33]. This yellow hue may pose limitations when used as a coating or wrapping, distorting product color and potentially influencing consumer decisions. Conversely, pure

GRS film exhibits clarity without coloration. Blending with colorless GRS reduces the yellow tint of the chitosan film, with the GRS/CS film showing a moderate yellow hue due to dilution effect, while optical clarity remains consistent across varying ratios. This underscores the remarkable amorphicity of both pure and blended polymer films, making GRS/CS blend film more suitable for coating or wrapping. Despite the yellow tint in films with high chitosan content, they maintain aesthetic appeal enhancing overall appearance as observed in the mango coating study.

The density of the polymer coating affects its ability to act as a barrier against external factors, such as moisture, gases, and contaminants. This property is crucial for extending the shelf life of produce by minimizing moisture loss and preventing the ingress of oxygen, which can accelerate deterioration. The density of the chitosan film, measured at 1.09 g/cm^3, aligns with the existing literature [34] and falls below that of the GRS film at 1.88 g/cm^3, similar to high amylopectin corn starch film at 1.74 g/cm^3 [35]. As the starch content increases in GRS/CS blended films, the density shows a linear rise attributed to amylopectin molecules integrating into the interstitial spaces of chitosan chains. This incorporation of glutinous rice starch in the GRS/CS blend fosters a more compact arrangement of polymer chains within the film matrix, resulting in enhanced packing density and, consequently, increased film density. Moreover, interactions, particularly hydrogen bonding between chitosan and GRS molecules, facilitate the formation of a dense and cohesive film structure.

An increase in chitosan content leads to decreased water solubility, reflecting its intrinsic hydrophobic nature. This is evidenced by the water contact angle of the films, which is further discussed. Consequently, blending chitosan with GRS enhances various physical properties of the film, including optical characteristics, density, and hydrophobicity, rendering it more suitable for specific applications. By adjusting the proportions of GRS and chitosan, the properties of GRS/CS blend films can be finely adjusted to meet distinct application requirements. This blending strategy not only improves the film's physical attributes but also enables the optimization of properties tailored to specific applications. Thus, the integration of chitosan with GRS offers a versatile approach for enhancing film properties and broadening its potential utility across diverse applications.

3.1.2. Effect of GRS/CS Coating Solution on *C. gloeosporioides* Mycelial Growth

The *C. gloeosporioides* is a prevalent plant pathogen found worldwide, notably thriving in tropical and subtropical climates. It causes anthracnose disease across a diverse array of crops, including almond, avocado, apple, coffee, guava, mango, strawberry, papaya, banana, passion fruit, citrus, grapes, and cashews [36]. Anthracnose represents a significant threat to mango production worldwide, affecting both the economic viability of mango farming and the availability of high-quality mangoes for consumers [37].

This study investigated the impact of the molar ratio of GRS/CS on its inhibitory efficacy against *C. gloeosporioides* strains isolated from mangoes. The mycelial growth inhibition (MGI) of coating solution by radial growth on PDA after incubation for 14 days is shown in Figure 1. The results reveal that the pure GRS solution facilitated extensive fungal growth, surpassing even the control group, as starch serves as a nutrient source for fungi and other microorganisms. Consequently, pure starch (GRS100/CS0) is unsuitable for food or fruit coatings or packaging due to its susceptibility to microbial proliferation. However, antibacterial properties were contributed by the chitosan admixture. MGI values for GRS/CS blends notably increased with rising chitosan content ($p < 0.05$), with complete inhibition observed for the pure chitosan (GRS0/CS100). This underscores chitosan's efficacy in anthracnose inhibition, a pivotal factor in mango deterioration globally. Previous studies have supported chitosan's efficacy, with concentrations of 1.5–2% yielding over 70% inhibition of *C. gloeosporioides* [38,39]. Furthermore, chitosan effectively inhibited *C. gloeosporioides* development in other crops, such as papaya [40] and manila mango [41]. Chitosan's antifungal effects depend on electrostatic interactions between its protonated amino groups and the negatively charged phospholipids of fungal membranes [42]. Chitosan induces membrane permeabilization, disruption, and release of cellular contents. Additionally, chitosan can penetrate fungal cells, inhibiting DNA/RNA synthesis, disrupting protein synthesis, and altering gene expression [43].

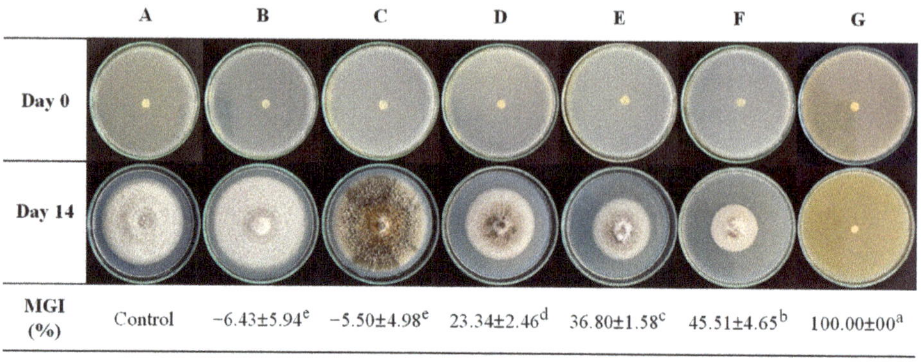

Figure 1. Effect of coating solution on *Colletotrichum gloeosporioides* mycelial growth and the percent of mycelial growth inhibition; control (**A**) (without coating solution in PDA), 0.1% acetic acid added (**B**), coating solution added in PDA: GRS100/CS0 (**C**), GRS70/CS30 (**D**), GRS50/CS50 (**E**), GRS30/CS70 (**F**), and GRS0/CS100 (**G**). Footnote: Values are presented as mean ± standard deviation. Values with the same superscript are not significantly different ($p \geq 0.05$).

3.1.3. Morphology of the GRS/CS Blend Films

The structure and morphology of a polymer blend film strongly influence its mechanical and barrier properties, which in turn are crucial aspects for its use as a coating or packaging material to extend the shelf life of agricultural and food products. Therefore, analyzing these aspects provides valuable insights into the film's performance and suitability for various applications.

SEM images were utilized to examine the surface characteristics of pure GRS, chitosan, and GRS/CS films at different blend ratios (Figure 2a–e). The images reveal smooth surfaces across all compositions, indicating excellent compatibility and no microphase separation. Cross-sectional analysis at 1500× (Figure 2f–j) unveiled thickness variations corresponding to film density reduction with increasing chitosan content, consistent with Table 2. At higher magnifications of 12,000× (Figure 2k–o) and 30,000× (Figure 2p–t), the pure starch film (GRS100/CS0) displayed numerous cracks, indicative of its inherent brittleness. These cracks can be attributed to various factors, including the formation of dense and rigid structures within the film matrix due to the waxy starch's high amylopectin content. Moreover, the drying process and non-uniform distribution of starch particles contribute to crack formation [44]. These cracks significantly impact the mechanical and barrier properties of the GRS film, rendering it inadequate for coating or packaging perishable produce. However, the incorporation of chitosan mitigates crack formation, with an increase in chitosan content corresponding to a reduction in crack prominence and improvement in mechanical properties. Furthermore, the presence of chitosan in the film coating applied to mango fruit may reduce the diffusion of water vapor and oxygen gas, thereby delaying mango ripening and minimizing the appearance of shriveling.

Pure chitosan films display a characteristic corrugated appearance under SEM analysis, characterized by uniform density devoid of cracks or air bubbles (Figure 2o,t), in contrast to GRS-mixed films. This uniform morphology correlates with the improved mechanical and barrier properties observed for chitosan films. The corrugated appearance may be attributed to chitosan's linear heteropolysaccharide structure and its interactions with neighboring molecules, resulting in irregular arrangements within the film matrix. Furthermore, structural modifications occurring during film formation contribute to this distinctive morphology [45].

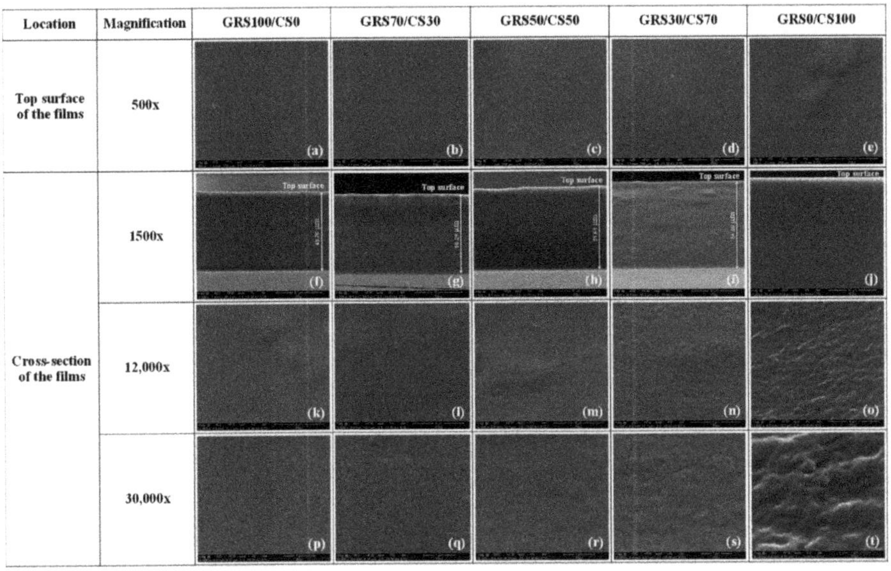

Figure 2. SEM micrographs of GRS/CS blend films with varying molar proportions. Top surfaces of the films at 500× magnification, scale bar: 100 μm (**a–e**). Cross-sections of the films at 1500× magnification, scale bar: 30 μm (**f–j**). Cross-sections of the films at 12,000× magnification, scale bar: 4 μm (**k–o**). Cross-sections of the films at 30,000× magnification, scale bar: 1 μm (**p–t**).

Fluorescence microscopy is a valuable tool for assessing the miscibility and morphology of polymer blends. By labeling specific components with fluorescent dyes or probes, it enables visualization of polymer distribution, enhancing our understanding of blend structure and performance. However, a limitation is the need for suitable fluorescent labels or probes, which may affect observed morphology and lead to misinterpretation [46]. In our study, we present a novel approach utilizing fluorescence microscopy to examine the miscibility and morphology of the GRS/CS composite film, offering an initial evaluation of its applicability for amine-containing polymers without the use of dyes or fluorescent probes.

Despite lacking typical fluorescence-associated structures, chitosan exhibited fluorescence, particularly in its oligomeric form [24]. Studies suggest that certain polymers with amine groups can display fluorescence due to the reaction between amine groups and CO_2, forming fluorescent carbamato anions ($NHCOO^-$) [24]. The amino groups of chitosan interact with airborne CO_2, generating fluorescent carbamato anions, with observed excitation and emission wavelengths at approximately 400 nm and 470 nm, respectively [24]. Fluorescence intensity is directly correlated with the concentration of chitosan oligomers [24]. Microscopic analysis of the pure chitosan film (Figure 3a,c) revealed a distinctive corrugated pattern accompanied by consistent fluorescence throughout (Figure 3b). In contrast, the pure GRS film (Figure 3g,i) displayed a uniform appearance without fluorescence (Figure 3h) due to the absence of fluorescence chromophores and amine groups in its structure. The minimal fluorescence observed in Figure 3h may be attributed to proteins on the microbial cell surface, where amino acids react with atmospheric CO_2, akin to the behavior observed in chitosan.

Microscopic images of the GRS50/CS50 blend film (Figure 3d) showed a uniform appearance similar to the pure GRS film. The fluorescence image (Figure 3e) and overlay image (Figure 3f) of the blend film confirmed excellent compatibility with no phase separation at the microscopic level. The decreased fluorescence intensity observed in the blend film (Figure 3e), relative to the pure chitosan, indicates a lower concentration of glucosamine units in the polymer composite. This is likely due to a dilution effect. This finding reinforces the idea that changes in chitosan content directly influence fluorescence properties, emphasizing the successful blending and compatibility of GRS and chitosan within the composite film structure.

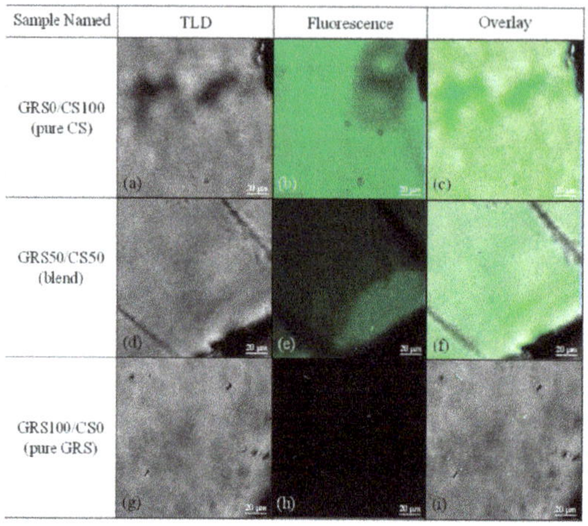

Figure 3. Microscope images of pure chitosan (**a–c**), pure GRS (**g–i**), and the GRS50/CS50 blend film (**d**,**e**,**f**). The sample films' top surfaces captured at 63x magnification, scale bar: 20 μm, using different modes: Transmitted Light Detector (TLD) (**a**,**d**,**g**), fluorescence (**b**,**e**,**h**), and an overlay of TLD and fluorescence (**c**,**f**,**i**).

3.1.4. Water Contact Angle of GRS/CS Film

Hydrophobicity plays a crucial role in determining the essential properties of active coatings aimed at extending the shelf life of produce, thereby preserving their freshness and quality. It exerts a profound influence on key characteristics including water vapor barrier properties [47,48], microbial growth inhibition [47], adhesion to produce surfaces [48], and gas permeability [47,48]. Striking the right balance among these properties is crucial for developing effective coatings that can maximize the shelf life and quality of fruits and vegetables [48]. Hence, this study investigates the water contact angle to elucidate the hydrophobic nature of the films. Additionally, the effect of the molar ratio between GRS and chitosan on the hydrophobicity of GRS/CS composite films is explored, aiming to identify compositions suitable for various applications.

The water contact angles provide insight into the influence of the polymer composition on the hydrophobicity of GRS/CS composite films, shown in Figure 4. The pure GRS film displays high hydrophilicity, characterized by a low contact angle of 68°, primarily attributed to the abundance of hydroxyl groups in starch. This hydrophilic nature of starch coatings predisposes them to water absorption, potentially enhancing moisture permeation, which could accelerate spoilage [49]. Furthermore, when using starch films under high-humidity atmospheres, such as coatings for fruits and vegetables, the heightened hydrophilicity may elevate gas permeability, thereby impacting gas-exchange dynamics and influencing the produce deterioration rate [50]. Consequently, the hydrophilic GRS film is considered unsuitable for applications as coatings to extend the shelf life of post-harvest produce.

In contrast, the pure chitosan film exhibited the highest hydrophobicity, with a contact angle of 102°, attributed to the presence of hydrophobic acetyl groups in incompletely deacetylated chitosan [51]. This hydrophobic characteristic contributes to excellent water vapor barrier properties, effectively preventing produce dehydration and maintaining optimal texture and appearance over extended periods [51]. Additionally, in high-humidity atmospheres, the hydrophobic coatings typically demonstrate low gas permeability, facilitating controlled gas exchange of oxygen and carbon dioxide [50]. This regulation of gas exchange helps manage respiration rates in produce, thereby slowing down the ripening process and extending shelf life. Moreover, hydrophobic coating acts as a barrier against moisture, inhibiting the growth of microorganisms like bacteria and fungi [47]. These findings are corroborated by a 10-day storage study, which underscores the efficacy of chitosan coatings in delaying ripening by mitigating fruit transpiration.

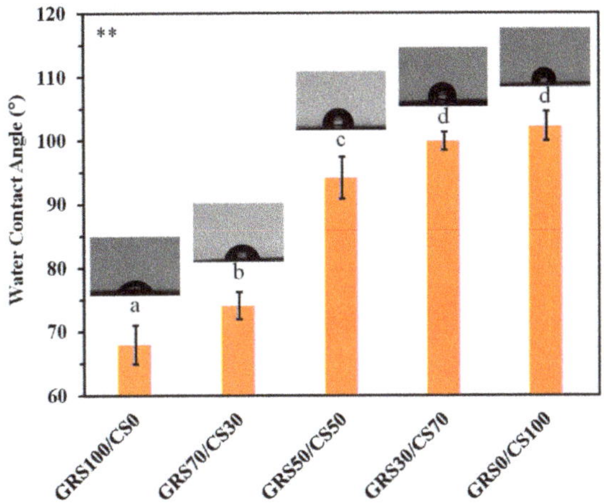

Figure 4. The water droplet profiles and the water contact angles for pure GRS, pure chitosan, and the GRS/CS blend films with different molar proportions coated on the mango surface. Footnote: ** $p \leq 0.01$ differences are highly significant, different letters above the bars indicate statistically significant difference.

The hydrophobicity of the GRS film can be enhanced by blending with chitosan, and the hydrophobic nature of GRS/CS composite films is improved with increased chitosan content. Consequently, the hydrophobicity of the GRS/CS coating can be readily tailored to accommodate the respiration rate and dehydration characteristics of each type of produce, thereby preserving produce quality for an extended duration. These findings are consistent with those of a 10-day storage study, wherein higher chitosan content resulted in prolonged ripening delay and reduced shriveling appearance of mango fruits.

3.1.5. Mechanical Properties of the GRS/CS Film

The mechanical properties of the active coating polymer significantly influence its ability to provide durable, flexible, and effective protection to produce against physical damage, microbial contamination, and environmental factors in general [52]. Optimizing these properties is crucial for developing coatings that meet the specific requirements of different types of produce and storage conditions, ultimately enhancing their shelf life and marketability.

Analysis of the stress–strain curves depicted in Figure 5a reveals distinct mechanical properties of the films. The GRS film exhibits characteristics of rigid and brittle fracture, attributed to the robust hydrogen bonding present in the polysaccharide chains, as well as its specific composition and molecular structure [53]. GRS, predominantly composed of amylopectin with minimal amylose content, demonstrates a semi-crystalline nature [54]. This molecular arrangement fosters tightly packed structures within the film matrix, resulting in reduced flexibility and limited mobility of the polymer chains, thereby displaying rigidity and brittleness during mechanical testing [53]. Conversely, the chitosan film shows a ductile response with a notable plastic range [55], making it suitable for applications requiring flexibility and strength, such as produce coatings.

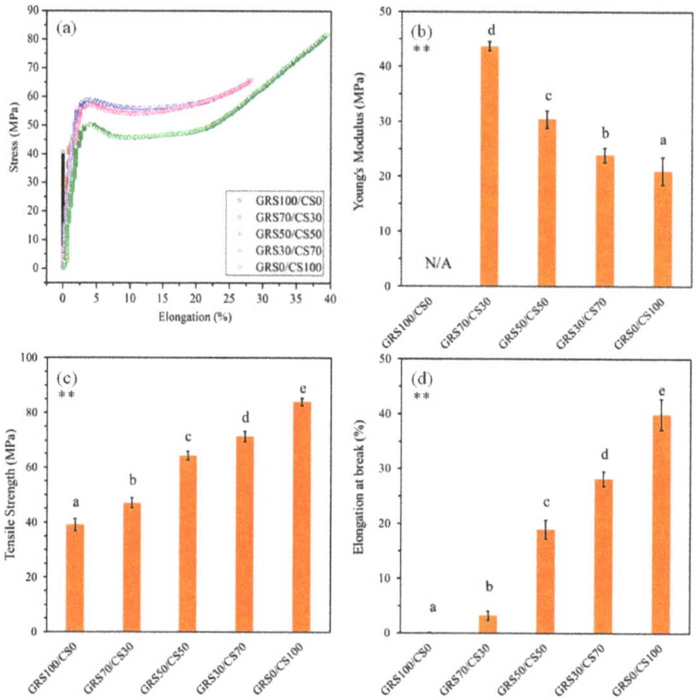

Figure 5. Stress–strain curves (**a**), Young's modulus (**b**), tensile strength (**c**), and elongation at break (**d**) of the GRS/CS blend films with different molar proportions. Footnote: ** $p \leq 0.01$ differences are highly significant, different letters above the bars indicate statistically significant difference.

Blending GRS with chitosan enhances the mechanical properties of starch films, rendering them suitable for use as active coatings for produce. The mechanical properties of GRS/CS blend films can be tailored for specific applications by adjusting the proportions of starch and chitosan. Increasing chitosan content improves flexibility and strength, resulting in films with enhanced adhesion to produce surfaces, resilience to temperature changes, resistance to physical damage, and maintenance of barrier integrity against moisture, gases, and microbes. Films with a higher starch content, such as the 70GRS/30CS blend, retain some brittleness but exhibit improved tensile strength and elongation compared to pure GRS. Conversely, films with more chitosan, such as the 50GRS/50CS and 30GRS/70CS blends, demonstrate ductile behavior similar to pure chitosan. Tensile strength and elongation increase, while modulus decreases with chitosan content, highlighting the tunability of mechanical properties in GRS/CS blend films.

3.2. Part II: Implementing GRS/CS Active Coatings with Alternative Molar Ratios on Mango Surfaces

3.2.1. Characterization of GRS/CS Coating on Mango Surface

This experimental section aims to evaluate the efficacy of GSR/CS as an active coating and investigate the impact of the starch to chitosan ratio on the suitability of polymer blend as a coating for maintaining post-harvest mango quality. This section places particular emphasis on critical parameters such as film thickness and adhesion to the mango surface. These factors play crucial roles in determining the efficacy of an active coating in preservation of post-harvest produce, contributing to the establishment of a robust barrier against external factors, ensuring uniform protection and enhanced durability during handling and storage. In Figure 6, the mango cross-section exhibits a uniformly coated surface throughout the polymer film, clearly indicated by the white arrow. Conversely, Figure 6b depicts a thinly coated mango surface with the GRS film, barely visible due to several contributing factors. The low viscosity of the GRS coating solution, coupled with its high film density and hydrophilic nature, leads to weak interaction with the hydrophobic cutin compounds present on the mango surface, resulting in poor adhesion. Thus, GRS alone proves inadequate as a coating for extending the fruit's shelf life.

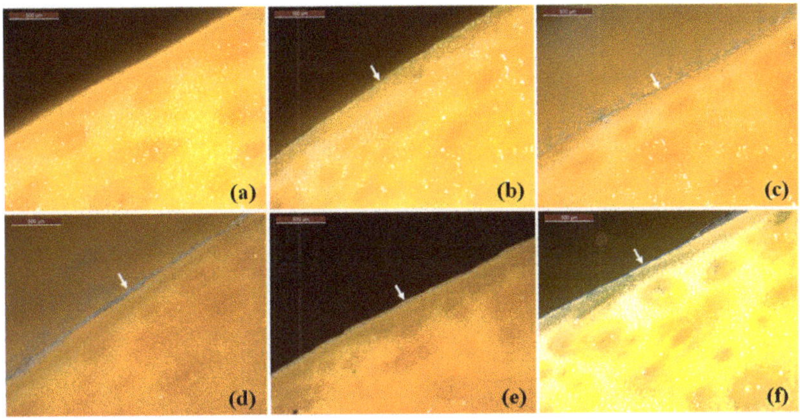

Figure 6. Optical stereo microscope images of cross-sections: film free mango surface (**a**), and the mango surfaces coated with GRS100/CS0 (**b**), GRS70/CS30 (**c**), GRS50/CS50 (**d**), GRS30/CS70 (**e**), and GRS0/CS100 (**f**). The scale bar is 500 μm. The white arrow indicates the GRS/CS film coated on the mango surface.

Blending GRS with chitosan significantly enhances film adhesion and thickness on the mango surface, as observed in our experimental findings. The thickness of the GRS/CS film coating increases significantly with chitosan content. This trend is particularly evident in the pure chitosan film, which emerges as the thickest among all coatings. Figure 6f illustrates the film with a thickness of approximately 10 μm on the mango surface, clearly visible under the microscope and shown in the image with a white arrow. This can be attributed to chitosan's hydrophobic nature, fostering strong interaction with the hydrophobic cutin compounds found in mango lenticels—macroscopic openings, approximately 0.2 mm in size, formed during fruit growth when stomata rupture [56]. Mango lenticels primarily comprise natural wax, facilitating gaseous exchange and transpiration. The robust adhesion of chitosan to mango lenticels ensures the coating's integrity, effectively protecting against physical damage and microbial contamination [57]. Conversely, hydrophilic coatings like starch may exhibit weaker adhesion to fresh produce surfaces, potentially compromising freshness preservation. Therefore, GRS/CS polymer composite films with higher chitosan content are more likely to be produce effective coatings. Adjusting the starch to chitosan ratio allows us to fine-tune film thickness and coating efficiency to suit different produce types.

The experimental findings align with previous studies on the efficacy of polymer film coatings on mango lenticels. Figure 7a illustrates the lenticels on the free surface of mangoes, revealing varied performances of polymer coatings in covering these structures. This divergence can be attributed to the hydrophobicity and viscosity of the polymer solutions. The GRS coating, characterized by low viscosity and high hydrophilicity, only partially covers the lenticels, as depicted in Figure 7b. Consequently, the pure GRS film proves inadequate in effectively controlling gas and water vapor diffusion, failing to delay produce ripening and to reduce shriveling. In contrast, the hydrophobic nature of chitosan enables effective coverage of mango lenticels, as depicted in Figure 7d.

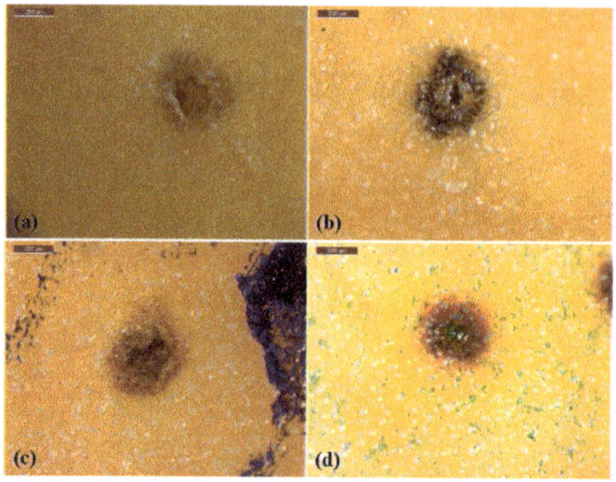

Figure 7. Optical stereo microscope images of lenticels on film free mango surface (**a**), and on mango surfaces coated with GRS100/CS0 (**b**), GRS50/CS50 (**c**), and GRS0/CS100 (**d**). The scale bar is 200 μm.

Blending GRS with the hydrophilic chitosan, which also exhibits high viscosity, enhances lenticel coverage, as demonstrated in Figure 7c. This renders the GRS/CS blend film more suitable for use as a produce coating. These findings are consistent with FTIR analysis and a 10-day storage study, indicating that chitosan coating delays mango ripening by reducing respiration and alleviates wilting by minimizing transpiration. In contrast, the GRS coating, which incompletely covers the lenticels, lacks efficiency in delaying ripening and reducing shriveling.

This study effectively utilized ATR-FTIR spectroscopy to assess the efficiency and effectiveness of polymer film coatings on produce or food surfaces, offering a valuable analytical tool within food science and technology. By employing FTIR spectroscopy, we gained insights into the chemical composition, molecular structure, and interaction of the polymer film with the surface of the produce or food. Our findings underscore the potential of FTIR spectroscopy as an effective method applicable to various food and product-related analyses. Figure 8 shows the FTIR spectra of the pure polymers and GRS/CS blended polymer films with different molar compositions coated on mango surface compared to the spectra of uncoated mango surface. The characterization of uncoated mango surface by FTIR spectroscopy has provided significant information on the nature of functional groups present in the cuticle matrix.

A broad band around 3352 cm^{-1} assigned to the stretching of hydroxyl (O-H) groups mainly contributed by the polysaccharide and the non-esterified hydroxyl groups of cutin [58]. The asymmetric stretching of CH_3 groups was found at 2956 cm^{-1}, and the asymmetric and symmetric stretching of CH_2 groups were found at 2916 and 2849 cm^{-1}, respectively. The CH_2 scissoring at 1473 and 1463 cm^{-1}, as well as the CH_2 rocking at 730 and 720 cm^{-1}, were attributed to aliphatic compounds, i.e., cutin, waxes, and cutan, in the plant cuticle [59]. The C=O stretching at 1729 cm^{-1} and the shoulder at 1714 cm^{-1}, as well as the asymmetric and symmetric C-O-C stretching at 1165 and 1105 cm^{-1}, were associated with ester and carboxylic acid groups of the cutin matrix [60].

The FTIR spectrum of mango surface coated with pure GRS film (GRS100/CS0) still exhibits characteristic peaks of the mango cutin layer at 2916, 2849, 1473, 1463, 730, and 720 cm^{-1}. This indicates that the GRS film covers the mango surface partially or in certain areas. Additionally, the spectrum displayed characteristic peaks of the GRS, including a broad band at 3327 cm^{-1} attributed to the stretching vibration of hydrogen-bonded hydroxyl groups (O-H) [61]. Another distinctive peak at 1640 cm^{-1} was assigned to the O-H stretching of hydroxyl groups [62]. The spectrum also exhibited noticeable absorbances at 1152, 1103, 1080, and 1021 cm^{-1}, which can be attributed to C-O and C-C stretching with some contribution from C-OH [63].

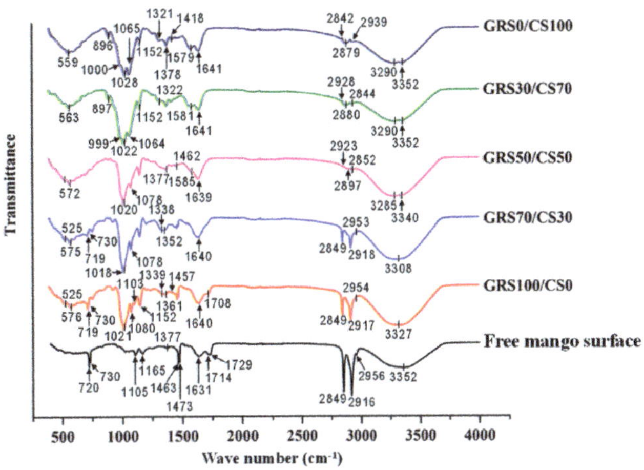

Figure 8. FTIR spectra of the film free mango surface compared with mango surfaces coated with pure GRS (GRS100/CS0), pure chitosan (GRS0/CS100), and the GRS/CS blend films with different molar proportions.

Coating mango surfaces with a GRS/CS polymer film with an increased proportion of chitosan showed an improved tendency to adhere and cover the mango surface effectively. This was evident from the intensity of the characteristic peaks of the cutin layer at 2916, 2849, 1473, 1463, 730, and 720 cm^{-1}, which decreased as the quantity of chitosan in the coating increased. The GRS/CS polymer film achieved complete coverage of the mango surface when the chitosan content in the coating reached 50% by moles. This is observed in the absence of the cutin peaks in the FTIR spectrum of mango coated with GRS50/CS50, GRS30/CS70, and pure chitosan (GRS0/CS100) films. In addition to the decrease in intensity of the cutin peaks as described above, a minor shift in the peak positions was also observed. The asymmetric stretching of CH_3 groups at 2956 cm^{-1} and CH_2 rocking at 720 cm^{-1} showed a slight decrease in wavenumber, while the asymmetric stretching of CH_2 groups at 2916 cm^{-1} exhibited a slight increase in wavenumber, corresponding to the increase in chitosan content in the coating. These FTIR results agree well with the microscope images shown in Figure 6.

The polymer coating adhesion to the mango surface is attributed to the interaction between the hydroxyl and carboxylic acid groups of the cutin layer and the active hydroxyl groups of GRS and chitosan. This is supported by the observed peak shifts of the hydroxyl and carboxylic acid groups of the cutin layer when coated with pure GRS. For instance, the stretching of hydroxyl groups of cutin shifted from 3352 to 3327 cm^{-1}, and the C=O stretching of carboxylic acid groups of cutin shifted from 1729 and 1714 cm^{-1} to 1708 cm^{-1}, indicating these interactions. These peak shifts towards lower wavenumbers reflect the stronger bonding between the hydroxyl and carboxylic acid groups of the cutin layer, which is further enhanced by increasing the amount of chitosan in the coating. Coating with pure chitosan was found to completely cover the mango surface, as confirmed by the absence of cutin peaks in the FTIR spectrum of GRS0/CS100 sample. Instead, characteristic peaks associated with chitosan were observed. These include the broad bands at 3352 and 3290 cm^{-1} corresponding to N-H and O-H stretching, respectively [64]. The absorption bands at 2939 and 2879 cm^{-1} correspond to C-H symmetric and asymmetric stretching, respectively, which are associated with the vibration of the characteristic pyranose ring of polysaccharides [65]. Residual *N*-acetyl groups were identified by the band at 1641 cm^{-1}

(C=O stretching of amide I), while the band at 1579 cm^{-1} corresponds to N-H bending of the primary amine [66]. CH_2 bending and CH_3 symmetrical deformations were confirmed by bands at 1418 and 1378 cm^{-1}, respectively [67]. The band at 1152 cm^{-1} is assigned to the asymmetric stretching of the C-O-C in glycosidic linkage, while C-O stretching for primary alcohol was observed at 1065 and 1028 cm^{-1} [62]. The band at 896 cm^{-1} belongs to CH out-of-plane bending vibration in the monosaccharide ring [67].

Comparing the FTIR spectra of mango surfaces coated with pure chitosan film (GS0/CS100) and the GRS/CS polymer blend films, it was observed that the presence of GRS in the coating led to increased freedom and disorder of the chitosan chains. This can be seen from the shift in peaks associated with the vibration of the pyranose ring towards higher wavenumbers as the GRS content increased. It is speculated that the interaction between chitosan, GRS, and mango cutin occurs primarily through the hydroxyl groups of chitosan. This is evident from the shifts in the O-H stretching at 3352 cm^{-1} and the C-O stretching at 1065 and 1028 cm^{-1} towards lower wavenumbers with increasing GRS content.

3.2.2. Quality Aspects of the Mango Coated with GRS/CS

Mango samples, treated with film coatings having various GRS/CS ratios, were stored at room temperature for 10 days, with monitoring of physical appearance and TSS/TA every 2 days (Figure 9). Initially, all the samples exhibited firm skin without disease symptoms. The control and GRS100/CS0 samples developed significant anthracnose symptoms, expressed as dark, sunken lesions on days 2 and 4, respectively, consistent with observations by Hadthamard et al. [68] of uncoated samples. The GRS70/CS30 showed symptoms of stem end rot on day 6. These post-harvest diseases typically manifest several days after fruit collection, significantly influencing consumer decisions. This study affirms that coating with chitosan mole fraction from 50% up effectively prevents disease throughout storage. These findings align with previous observations, indicating enhanced inhibition of *Colletotrichum* spp. with increasing chitosan proportions. All GRS-CS treatments satisfactorily reduced dehydration and delayed mango ripening. Coated samples demonstrated less shriveling and color change than the control samples, attributed to the hydrophobicity of chitosan (high contact angle), which effectively covered lenticels and limited water migration. The primary function of the coating material is to reduce moisture and gas permeation between fruit and its environment, resulting in lower respiration rates, reduced water migration, and delayed ripening [69].

Figure 9. The appearances of mangoes without and with the experimental GRS/CS coatings during storage at room temperature for 10 days.

During ripening, mangos undergo both physical changes, such as changes in peel and flesh color and loss of tissue firmness, as well as chemical changes, including heightened aroma, increased nutritional value, reduced acidity, and higher sugar content. Gas concentrations, particularly of oxygen (O_2) and carbon dioxide (CO_2), play a pivotal role in fruit ripening by influencing respiration rates. Effective gas control through the barrier properties of coating materials stands as a crucial solution for extending shelf life. Polysaccharide coating forms a tightly packed hydrogen-bonded network and serves as an effective oxygen barrier [70]. The application of chitosan coating successfully reduces the respiration rate and retards the ripening of mango and avocados [38,71]. Total soluble solids (TSS) relate to fruit sugar content, increasing as starch converts to sugar during ripening. On the other hand, titratable acidity (TA) indicates fruit acidity, decreasing with ripening. The ratio of TSS and TA is widely used to represent fruit taste, as a greater ratio means the fruit is sweeter and less sour. Initially, all treatments had TSS of 14–16°Brix and TA of 0.2–0.4%. The TSS/TA ratio (Figure 10) showed two distinct patterns: non-chitosan (control and GRS coating) and chitosan coating (various chitosan fractions). The non-chitosan group displayed a typical increasing TSS/TA pattern during storage, indicating ongoing ripening, with control and GRS100/CS0 reaching 896.5 and 689.6, respectively, after 10 days. Conversely, the chitosan-coated group exhibited a stable TSS/TA trend, signifying constant ripening. Chitosan-coated mangoes had TSS/TA in the range of 50–250, indicative of a turning ripened stage for Nam Dok Mai mango [72]. This finding demonstrates the efficacy of chitosan coating in delaying mango ripening. Similarly, other mango varieties (Tommy Atkins and Apple) exhibited delayed ripening when treated with 2% chitosan. Silva et al. [73] proposed that chitosan coating delays sugar accumulation and starch conversion. The findings of this study validate the efficacy of GRS-CS coating in extending the shelf life of mangoes. Uncoated mangoes met selling standards for 2 days, while those treated with at least 50% chitosan in coating maintained their quality throughout the entire 10-day storage period.

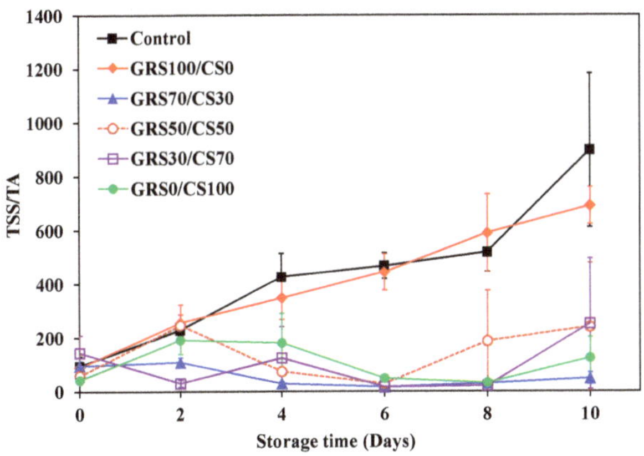

Figure 10. TSS/TA ratio for mangoes without and with the experimental GRS/CS coatings during storage at room temperature for 10 days. Vertical bars represent standard deviations.

4. Conclusions

This study highlights the potential of GRS/CS blends as natural active coatings for extending mango shelf life, with properties adjustable by varying the molar ratio of GRS to chitosan. This blending strategy presents a versatile approach for enhancing film properties across diverse food industry applications. The results show that pH decreased with chitosan content in GRS/CS blend solutions due to acetic acid solvent for chitosan dissolution. To address odor and taste limitations of acetic acid, blending chitosan with GRS was proposed to dilute acetic acid concentration in the film. Moreover, blending chitosan with pure GRS increased viscosity, facilitating film formation on produce surfaces and improving coating effectiveness. GRS/CS blend films with higher chitosan content exhibited higher density, increased hydrophobicity, and improved mechanical properties, contributing to better moisture, gas, and contaminant barrier properties and inhibiting microbial growth. Chitosan's hydrophobic nature fostered strong interaction with mango lenticels, ensuring coating integrity and protection against physical damage and microbial contamination. The GRS/CS coatings with alternative molar proportions on mango surfaces significantly improved film adhesion and thickness compared to GRS alone. FTIR spectroscopy confirmed polymer film interactions with mango surfaces, highlighting its potential in food science and technology. Coating with GRS/CS blends effectively reduced post-harvest diseases, delayed ripening, and maintained mango quality during storage. Overall, GRS/CS blends offer a promising solution for extending mango shelf life, with potential for similar applications to other produce types.

Author Contributions: Conceptualization, C.N. and N.R.; Data curation, C.N., R.W. and N.R.; Formal analysis, R.W.; Funding acquisition, C.N.; Investigation, C.N., R.W. and N.R.; Methodology, C.N. and N.R.; Project administration, R.W.; Resources, R.W.; Supervision, N.R.; Validation, C.N. and N.R.; Visualization, C.N., R.W. and N.R.; Writing—original draft, C.N., R.W. and N.R.; Writing—review and editing, C.N. and N.R. All authors have read and agreed to the published version of the manuscript.

Funding: This research received financial support from Prince of Songkla University, Surat Thani Campus, in 2024. Additionally, funding was provided by Prince of Songkla University, Suratthani Campus, Thailand, through grant number SIT561167S, and by the National Science and Technology Development Agency (NSTDA), Thailand, through grant number JRA-CO-2563-13578-TH.

Institutional Review Board Statement: This study did not involve human participants or animals; hence, ethical approval was not required.

Data Availability Statement: Data are contained within the article. The original contributions presented in the study are included in the article, further inquiries can be directed to the corresponding author.

Acknowledgments: We express our sincere gratitude for the financial support provided by Prince of Songkla University, Surat Thani Campus, Thailand, and the National Science and Technology Development Agency (NSTDA), Thailand. We also wish to express our sincere appreciation to Seppo Karrila for his invaluable copyediting service. Additionally, we extend our thanks to H. Boonyaputthikul of Leica Microsystems, HISTOCENTER (THAILAND) CO., LTD, for their assistance with fluorescence microscopy characterization.

Conflicts of Interest: The authors declare no conflicts of interest. The funders had no role in the design of the study; in the collection, analyses, or interpretation of data; in the writing of the manuscript; or in the decision to publish the results.

References

1. Kumari, C.; Sharma, M.; Kumar, V.; Sharma, R.; Kumar, V.; Sharma, P.; Irfan, M. Genome editing technology for genetic amelioration of fruits and vegetables for alleviating post-harvest loss. *Bioengineering* **2022**, *9*, 176. [CrossRef] [PubMed]
2. Barbosa, C.H.; Andrade, M.A.; Vilarinho, F.L.; Fernando, A.L.; Silva, A.S. Active edible packaging. *Encyclopedia* **2021**, *1*, 360–370. [CrossRef]
3. Shahbandeh, M. Global Mango Production 2000–2022. 2024. Available online: https://www.statista.com/statistics/577951/world-mango-production/ (accessed on 2 April 2024).
4. de Oliveira, K.Á.R.; da Conceição, M.L.; de Oliveira, S.P.A.; Lima, M.D.S.; de Sousa Galvão, M.; Madruga, M.S.; de Souza, E.L. Postharvest quality improvements in mango cultivar Tommy Atkins by chitosan coating with *Mentha piperita* L. essential oil. *J. Hortic. Sci. Biotechnol.* **2020**, *95*, 260–272. [CrossRef]
5. Hoque, M.I.; Chowhan, S.; Kamruzzaman, M. Physiological changes and shelf life of mango (*Mangifera indica* L.) Influenced by post harvest treatments. *SAARC J. Agric.* **2018**, *15*, 219–226. [CrossRef]
6. Busatto, N.; Vittani, L.; Farneti, B.; Khomenko, I.; Caffini, M.; Faccini, S.; Costa, F. Physiological and molecular characterization of the late ripening stages in *Mangifera indica* cv Keitt. *Postharvest Biol. Technol.* **2022**, *183*, 111746. [CrossRef]
7. Rastegar, S.; Hassanzadeh Khankahdani, H.; Rahimzadeh, M. Effectiveness of alginate coating on antioxidant enzymes and biochemical changes during storage of mango fruit. *J. Food Biochem.* **2019**, *43*, e12990. [CrossRef] [PubMed]

8. Ntsoane, M.L.; Zude-Sasse, M.; Mahajan, P.; Sivakumar, D. Quality assesment and postharvest technology of mango: A review of its current status and future perspectives. *Sci. Hortic.* **2019**, *249*, 77–85. [CrossRef]
9. Díaz-Montes, E.; Castro-Muñoz, R. Edible films and coatings as food-quality preservers: An overview. *Foods* **2021**, *10*, 249. [CrossRef] [PubMed]
10. Berihu, H.; Zegeye, A. Enhancement of quality and storability of Avocado (*Persea americana*) fruit using a blend of aloe vera gel and corn starch as surface coating. *Int. J. Food Eng. Technol.* **2022**, *6*, 21. [CrossRef]
11. Oyom, W.; Xu, H.; Liu, Z.; Long, H.; Li, Y.; Zhang, Z.; Bi, Y.; Tahergorabi, R.; Prusky, D. Effects of modified sweet potato starchedible coating incorporated with cumin essential oil on storage quality of 'early crisp'. *LWT* **2022**, *153*, 12475. [CrossRef]
12. Ghoshal, G.; Chopra, H. Impact of apricot oil incorporation in tamarind starch/gelatin based edible coating on shelf life of grape fruit. *J. Food Meas. Charact.* **2022**, *16*, 1274–1290. [CrossRef]
13. Rather, J.A.; Makroo, H.A.; Showkat, Q.A.; Majid, D.; Dar, B.N. Recovery of gelatin from poultry waste: Characteristics of the gelatin and lotus starch-based coating material and its application in shelf-life enhancement of fresh cherry tomato. *Food Packag. Shelf Life* **2022**, *31*, 100775. [CrossRef]
14. Cazón, P.; Velazquez, G.; Ramírez, J.A.; Vázquez, M. Polysaccharide-based films and coatings for food packaging: A review. *Food Hydrocoll.* **2017**, *68*, 136–148. [CrossRef]
15. Oladzadabbasabadi, N.; Nafchi, A.M.; Ariffin, F.; Wijekoon, M.J.O.; Al-Hassan, A.A.; Dheyab, M.A.; Ghasemlou, M. Recent advances in extraction, modification, and application of chitosan in packaging industry. *Carbohydr. Polym.* **2022**, *277*, 118876. [CrossRef]
16. Yan, D.; Li, Y.; Liu, Y.; Li, N.; Zhang, X.; Yan, C. Antimicrobial properties of chitosan and chitosan derivatives in the treatment of enteric infections. *Molecules* **2021**, *26*, 7136. [CrossRef]
17. Román-Doval, R.; Torres-Arellanes, S.P.; Tenorio-Barajas, A.Y.; Gómez-Sánchez, A.; Valencia-Lazcano, A.A. Chitosan: Properties and its application in agriculture in context of molecular weight. *Polymers* **2023**, *15*, 2867. [CrossRef]
18. Hesami, A.; Kavoosi, S.; Khademi, R.; Sarikhani, S. Effect of chitosan coating and storage temperature on shelf-life and fruit quality of Ziziphus mauritiana. *Int. J. Fruit Sci.* **2021**, *21*, 509–518. [CrossRef]
19. Saki, M.; ValizadehKaji, B.; Abbasifar, A.; Shahrjerdi, I. Effect of chitosan coating combined with thymol essential oil on physicochemical and qualitative properties of fresh fig (*Ficus carica* L.) fruit during cold storage. *J. Food Meas. Charact.* **2019**, *13*, 1147–1158. [CrossRef]
20. Parvin, N.; Rahman, A.; Roy, J.; Rashid, M.H.; Paul, N.C.; Mahamud, M.A.; Kader, M.A. Chitosan coating improves postharvest shelf-life of mango (*Mangifera indica* L.). *Horticulturae* **2023**, *9*, 64. [CrossRef]
21. Ojeda, G.A.; Arias Gorman, A.M.; Sgroppo, S.C.; Zaritzky, N.E. Application of composite cassava starch/chitosan edible coating to extend the shelf life of black mulberries. *J. Food Process. Preserv.* **2021**, *45*, e15073. [CrossRef]
22. Shapi'i, R.A.; Othman, S.H.; Nordin, N.; Basha, R.K.; Naim, M.N. Antimicrobial properties of starch films incorporated with chitosan nanoparticles: In vitro and in vivo evaluation. *Carbohydr. Polym.* **2020**, *230*, 115602. [CrossRef] [PubMed]
23. da Costa, J.C.M.; Miki, K.S.L.; da Silva Ramos, A.; Teixeira-Costa, B.E. Development of biodegradable films based on purple yam starch/chitosan for food application. *Heliyon* **2020**, *6*, 4.
24. Lee, H.M.; Kim, M.H.; Yoon, Y.I.; Park, W.H. Fluorescent property of chitosan oligomer and its application as a metal ion sensor. *Mar. Drugs* **2017**, *15*, 105. [CrossRef] [PubMed]
25. *ASTM D882-12*; Standard Test Method for Tensile Properties of Thin Plastic Sheeting. ASTM International: West Conshohocken, PA, USA, 2012.
26. Wongkaew, M.; Sangta, J.; Chansakaow, S.; Jantanasakulwong, K.; Rachtanapun, P.; Sommano, S.R. Volatile profiles from over-ripe purée of Thai mango varieties and their physiochemical properties during heat processing. *PLoS ONE* **2021**, *16*, e0248657. [CrossRef] [PubMed]
27. Kim, K.M.; Son, J.H.; Kim, S.K.; Weller, C.L.; Hanna, M.A. Properties of Chitosan Films as a Function of pH and Solvent Type. *J. Food Sci. E Food Eng. Phys. Prop.* **2006**, *71*, 119–124. [CrossRef]
28. Xu, J.; Liu, K.; Chang, W.; Chiou, B.S.; Chen, M.; Liu, F. Regulating the Physicochemical Properties of Chitosan Films through Concentration and Neutralization. *Foods* **2022**, *11*, 1657. [CrossRef]
29. Abidin, M.Z.A.Z.; Julkapli, N.M.; Juahir, H.; Azaman, F.; Abidin, I.Z.; Sulaiman, N.H. Fabrication and properties of chitosan with starch for packaging application. *Malays. J. Anal. Sci.* **2015**, *19*, 1032–1042.
30. Juliano, B.O. The chemical basis of rice grain quality. In *Proceedings of the Workshop on Chemical Aspects of Rice Grain Quality*; International Rice Research Institute: Los Baños, Laguna, Philippines, 1979; pp. 69–90.
31. Pfister, B.; Zeeman, S.C.; Rugen, M.D.; Field, R.A.; Ebenhöh, O.; Raquin, A. Theoretical and experimental approaches to understand the biosynthesis of starch granules in a physiological context. *Photosynth. Res.* **2020**, *145*, 55–70. [CrossRef] [PubMed]
32. Mathew, S.; Brahmakumar, M.; Abraham, T.E. Microstructural imaging and characterization of the mechanical, chemical, thermal, and swelling properties of starch–chitosan blend films. *Biopolymers* **2006**, *82*, 176–187. [CrossRef]
33. Seo, S. Depolymerization and Decolorization of Chitosan by Ozone Treatment. Master's Thesis, Louisiana State University and Agricultural and Mechanical College, Baton Rouge, LA, USA, 2006.
34. Bilbao-Sainz, C.; Sen Chiou, B.; Williams, T.; Wood, D.; Du, W.X.; Sedej, I.; Ban, Z.; Rodov, V.; Poverenov, E.; Vinokur, Y.; et al. Vitamin D-fortified chitosan films from mushroom waste. *Carbohydr. Polym.* **2017**, *167*, 97–104. [CrossRef]

35. Ibrahim, M.I.J.; Sapuan, S.M.; Zainudin, E.S.; Zuhri, M.Y.M. Physical, thermal, morphological, and tensile properties of cornstarch-based films as affected by different plasticizers. *Int. J. Food Prop.* **2019**, *22*, 925–941. [CrossRef]
36. Sharma, M.; Kulshrestha, S. Colletotrichum gloeosporioides: An anthracnose causing pathogen of fruits and vegetables. *Biosci. Biotechnol. Res. Asia* **2015**, *12*, 1233–1246. [CrossRef]
37. Le, T.D.; Viet Nguyen, T.; Muoi, N.V.; Toan, H.T.; Lan, N.M.; Pham, T.N. Supply chain management of Mango (*Mangifera indica* L.) fruit: A review with a focus on product quality during postharvest. *Front. Sustain. Food Syst.* **2022**, *5*, 799431. [CrossRef]
38. Marques, K.M.; Galati, V.C.; Fernandes, J.D.R.; Guimarães, J.E.R.; Silva, J.P.; Mattiuz, B.H.; Mattiuz, C.F.M. Use of chitosan for the control of postharvest anthracnose and quality in avocados. *Acta Hortic.* **2016**, *1120*, 225–231. [CrossRef]
39. Edirisinghe, M.; Ali, A.; Maqbool, M.; Alderson, P.G. Chitosan controls postharvest anthracnose in bell pepper by activating defense-related enzymes. *J. Food Sci. Technol.* **2014**, *51*, 4078–4083. [CrossRef]
40. Ayón Reyna, L.E.; Uriarte Gastelum, Y.G.; Camacho Díaz, B.H.; Tapia Maruri, D.; López López, M.E.; López Velázquez, J.G.; Vega Garcia, M.O. Antifungal activity of a chitosan and mint essential oil coating on the development of Colletotrichum gloeosporioides in papaya using macroscopic and microscopic analysis. *Food Bioprocess Technol.* **2022**, *15*, 368–378. [CrossRef]
41. Limon, T.; Birke, A.; Monribot-Villanueva, J.L.; Guerrero-Analco, J.A.; Altúzar-Molina, A.; Carrión, G.; Aluja, M. Chitosan coatings reduce fruit fly (*Anastrepha obliqua*) infestation and development of the fungus Colletotrichum gloeosporioides in Manila mangoes. *J. Sci. Food Agric.* **2021**, *101*, 2756–2766. [CrossRef]
42. Qin, Y.; Li, P.; Guo, Z. Cationic chitosan derivatives as potential antifungals: A review of structural optimization and applications. *Carbohydr. Polym.* **2020**, *236*, 116002. [CrossRef]
43. Lopez-Moya, F.; Suarez-Fernandez, M.; Lopez-Llorca, L.V. Molecular mechanisms of chitosan interactions with fungi and plants. *Int. J. Mol. Sci.* **2019**, *20*, 332. [CrossRef]
44. Goehring, L. Drying and cracking mechanisms in a starch slurry. *Phys. Rev. E* **2009**, *80*, 036116–036125. [CrossRef]
45. Izawa, H.; Ishisaka, S.; Saimoto, H.; Ifuku, S. Drying-Induced Surface Wrinkles Generated on Chitosan Films Having Polyion Complex Skin Layers: Effects of Physical Properties of Skin Layers and Substrates on Surface Wrinkling upon Drying. *Bull. Chem. Soc. Jpn.* **2022**, *95*, 1289–1295. [CrossRef]
46. Morawetz, H. On the versatility of fluorescence techniques in polymer research. *J. Polym. Sci. Part A Polym. Chem.* **1999**, *37*, 1725–1735. [CrossRef]
47. Roman, M.; Nechita, P.; Vasile, M.-A.; Cantaragiu Ceoromila, A.-M. Barrier and Antimicrobial Properties of Coatings Based on Xylan Derivatives and Chitosan for Food Packaging Papers. *Coatings* **2023**, *13*, 1761. [CrossRef]
48. Wongs-Aree, C.; Nguyen, H.T.; Noichinda, S. Improved Postharvest Techniques for Fruit Coatings. In *New Advances in Postharvest Technology*; İbrahim Kahramanoğlu; IntechOpen: Rijeka, Croatia, 2023; ISBN 9781837685424.
49. Bertuzzi, M.A.; Castro Vidaurre, E.F.; Armada, M.; Gottifredi, J.C. Water vapor permeability of edible starch based films. *J. Food Eng.* **2007**, *80*, 972–978. [CrossRef]
50. Mangaraj, S.; Goswami, T.K.; Panda, D.K. Modeling of gas transmission properties of polymeric films used for MA packaging of fruits. *J. Food Sci. Technol.* **2015**, *52*, 5456–5469. [CrossRef]
51. Luchese, C.L.; Pavoni, J.M.F.; Dos Santos, N.Z.; Quines, L.K.; Pollo, L.D.; Spada, J.C.; Tessaro, I.C. Effect of chitosan addition on the properties of films prepared with corn and cassava starches. *J. Food Sci. Technol.* **2018**, *55*, 2963–2973. [CrossRef]
52. Fadiji, T.; Rashvand, M.; Daramola, M.O.; Iwarere, S.A. A Review on Antimicrobial Packaging for Extending the Shelf Life of Food. *Processes* **2023**, *11*, 590. [CrossRef]
53. Mitchell, J.R.; MacNaughtan, W.; Foster, T.J.; Harabagiu, V.; Song, Y.; Zheng, Q. Comparison of the Mechanical Properties of Cellulose and Starch Films. *Biomacromolecules* **2010**, *11*, 126–132.
54. Bertoft, E. Understanding Starch Structure: Recent Progress. *Agronomy* **2017**, *7*, 56. [CrossRef]
55. D'Angelo, G.; Elhussieny, A.; Faisal, M.; Fahim, I.S.; Everitt, N.M. Mechanical Behavior Optimization of Chitosan Extracted from Shrimp Shells as a Sustainable Material for Shopping Bags. *J. Funct. Biomater.* **2018**, *9*, 37. [CrossRef]
56. Prasad, K.; Sharma, R.R.; Srivastav, M. Postharvest treatment of antioxidant reduces lenticel browning and improves cosmetic appeal of mango (*Mangifera indica* L.) fruits without impairing quality. *J. Food Sci. Technol.* **2016**, *53*, 2995–3001.
57. Bansal, H.; Singh, S.; Sharma, A.; Sundaramurthy, S.; Mehta, S.K. Polymer nanocomposite films and coatings for antimicrobial and antifungal applications. In *Polymer Nanocomposite Films and Coatings*, 1st ed.; Pandey, M., Deshmukh, K., Eds.; Woodhead Publishing: Cambridge, UK, 2024; pp. 785–815.
58. Heredia-Guerrero, J.A.; Benítez, J.J.; Domínguez, E.; Bayer, I.S.; Cingolani, R.; Athanassiou, A.; Heredia, A. Infrared and Raman spectroscopic features of plant cuticles: A review. *Front. Plant Sci.* **2014**, *5*, 305–310. [CrossRef]
59. Prinsloo, L.C.; du Plooy, W.; van der Merwe, C. Raman spectroscopic study of the epicuticular wax layer of mature mango (*Mangifera indica*) fruit. *J. Raman Spectrosc.* **2004**, *35*, 561–567. [CrossRef]
60. Dubis, E.N.; Dubis, A.T.; Morzycki, J.W. Comparative analysis of plant cuticular waxes using HATR FT-IR reflection technique. *J. Mol. Struct.* **1999**, *511*, 173–179. [CrossRef]
61. Kizil, R.; Irudayaraj, J.; Seetharaman, K. Characterization of irradiated starches by using FT-Raman and FTIR spectroscopy. *J. Agric. Food Chem.* **2002**, *50*, 3912–3918. [CrossRef]
62. Soe, M.T.; Pongjanyakul, T.; Limpongsa, E.; Jaipakdee, N. Films Fabricated with Native and Ball-Milled Modified Glutinous Rice Starch: Physicochemical and Mucoadhesive Properties. *Starch* **2021**, *73*, 2000012. [CrossRef]

63. Wang, S.; Wang, J.; Zhang, W.; Li, C.; Yu, J.; Wang, S. Molecular order and functional properties of starches from three waxy wheat varieties grown in China. *Food Chem.* **2015**, *181*, 43–50. [CrossRef]
64. Xu, Y.X.; Kim, K.M.; Hanna, M.A.; Nag, D. Chitosan–starch composite film: Preparation and characterization. *Ind. Crops Prod.* **2005**, *21*, 185–192. [CrossRef]
65. Rakkapao, N.; Vao-Soongnern, V.; Masubuchi, Y.; Watanabe, H. Miscibility of chitosan/poly (ethylene oxide) blends and effect of doping alkali and alkali earth metal ions on chitosan/PEO interaction. *Polymers* **2011**, *52*, 2618–2627. [CrossRef]
66. Gieroba, B.; Sroka-Bartnicka, A.; Kazimierczak, P.; Kalisz, G.; Lewalska-Graczyk, A.; Vivcharenko, V.; Przekora, A. Surface chemical and morphological analysis of chitosan/1, 3-β-d-glucan polysaccharide films cross-linked at 90 °C. *Int. J. Mol. Sci.* **2022**, *23*, 5953. [CrossRef]
67. Queiroz, M.F.; Melo, K.R.T.; Sabry, D.A.; Sassaki, G.L.; Rocha, H.A.O. Does the use of chitosan contribute to oxalate kidney stone formation? *Mar. Drugs* **2014**, *13*, 141–158. [CrossRef]
68. Hadthamard, N.; Chaumpluk, P.; Buanong, M.; Boonyaritthongchai, P.; Wongs-Aree, C. Effects of Multilayer Coating of Chitosan and Polystyrene Sulfonate on Quality of 'Nam Dok Mai No. 4' Mango. *Int. J. Agric. Biosyst. Eng.* **2019**, *13*, 42–48.
69. Felicia, W.X.L.; Rovina, K.; Nur'Aqilah, M.N.; Vonnie, J.M.; Erna, K.H.; Misson, M.; Halid, N.F.A. Recent advancements of polysaccharides to enhance quality and delay ripening of fresh produce: A review. *Polymers* **2022**, *14*, 1341. [CrossRef] [PubMed]
70. Kumari, M.; Mahajan, H.; Joshi, R.; Gupta, M. Development and structural characterization of edible films for improving fruit quality. *Food Packag. Shelf-Life* **2017**, *12*, 42–50. [CrossRef]
71. Jongsri, P.; Wangsomboondee, T.; Rojsitthisak, P.; Seraypheap, K. Effect of molecular weights of chitosan coating on postharvest quality and physicochemical characteristics of mango fruit. *LWT-Food Sci. Technol.* **2016**, *73*, 28–36. [CrossRef]
72. Wongkhot, A.; Rattanapanone, N.; Chanasut, U. BrimA, total acidity and total soluble solids correlate to total carotenoid content as indicators of the ripening process of six thai mango fruit cultivars. Chiang Mai Univ. *J. Nat. Sci.* **2012**, *11*, 97–103.
73. Silva, G.M.C.; Silva, W.B.; Medeiros, D.B.; Salvador, A.R.; Cordeiro, M.H.M.; da Silva, N.M.; Santana, D.B.; Mizobutsi, G.P. The chitosan affects severely the carbon metabolism in mango (*Mangifera indica* L. cv. Palmer) fruit during storage. *Food Chem.* **2017**, *237*, 372–378. [CrossRef]

polymers

MDPI

Article

Lyophilized Emulsions of Thymol and Eugenol Essential Oils Encapsulated in Cellulose

Koranit Shlosman [1,2], Dmitry M. Rein [3], Rotem Shemesh [2] and Yachin Cohen [3,*]

[1] The Interdepartmental Program in Polymer Engineering, Technion-Israel Institute of Technology, Haifa 32000, Israel; skoranit@bazan.co.il
[2] R&D and Customer Service Department Carmel Olefins Ltd., Haifa 31014, Israel; srotem@bazan.co.il
[3] Department of Chemical Engineering, Technion-Israel Institute of Technology, Haifa 32000, Israel; cerycdr@technion.ac.il
* Correspondence: yachinc@technion.ac.il; Tel.: +972-77-8872010

Abstract: Efforts to tap into the broad antimicrobial, insecticidal, and antioxidant activities of essential oils (EOs) are limited due to their strong odor and susceptibility to light and oxidation. Encapsulation of EOs and subsequent drying overcome these limitations and extend their applications. This study characterized freeze-dried (lyophilized) emulsions of eugenol (EU) and thymol (TY) EOs, encapsulated by chemically unmodified cellulose, a sustainable and low-cost resource. High-resolution scanning electron microscopy showed successful lyophilization. While the observed "flake-like" structure of the powders differed significantly from that of the emulsified microcapsules, useful properties were retained. Fourier transform infrared spectroscopy confirmed the presence of EOs in their corresponding powders and thermo-gravimetric analysis demonstrated high encapsulation efficiency (87–88%), improved thermal stability and resistance to evaporation, and slow EO release rates in comparison to their free forms. The lightweight and low-cost cellulose encapsulation, together with the results showing retained properties of the dried powder, enable the use of EOs in applications requiring high temperatures, such as EO incorporation into polymer films, that can be used to protect agricultural crops from microbial infections.

Keywords: cellulose capsules; essential oils; lyophilization; volatiles release rate; biopolymer; bio-based material

1. Introduction

Essential oils (EOs) are natural compounds present in aromatic and medicinal plants known for their antimicrobial and pesticidal activities and have been the subject of many studies [1–10]. EOs are safe for humans, animals, and the environment, and are used in many applications, including food, cosmetics, personal protective preparations, pharmaceutics, and agriculture [4,11–13]. For example, polymer nanofibers containing an inclusion complex of prochloraz in hydroxypropyl-γ-cyclodextrin was recently reported [14]. EOs are also very volatile, an advantage in applications requiring no direct contact between the target surface/microorganism and the EO-releasing device, for example, insect repellants for crop protection [4]. Yet, EOs are limited by vulnerability to light, high temperature, moisture, and oxygen [12,15,16], and their distinctive strong odor, and poor solubility in water [17,18]. These characteristics limit their use in industrial applications [19]. Encapsulation of EOs can overcome these barriers by decreasing the interactions between the "core" (i.e., encapsulated ingredient) and the environment [2,18,20–22]. It is an effective technique to decrease their volatility and sensitivity to environmental conditions, to enable their controlled release [2,23], decrease their degradation [1], mask their strong organoleptic characteristics [18], and enhance their antimicrobial activity [3,15,24].

Drying is a common and significant process to convert encapsulated particle emulsions into powders. It widens the application range of encapsulated EOs, increases their shelf life, and reduces the weight of the final product, resulting in a more economical transportation [12,25]. Spray-drying and freeze-drying (also referred to as lyophilization) are common drying methods used in the pharmaceutical and food industries. Spray-drying is the oldest and most robust method, and is industrially favorable as it enables production of large quantities at minimal costs [26]. However, the high temperatures used in the spray-drying process can damage the sensitive ingredients in the sample and may require a more complex geometry of the high-pressure spraying nozzle. The spray-drying process also involves high operation and installation costs [27]. Consequently, the lyophilization process, which is slower and entails higher energy consumption, is more suitable for delicate applications [28,29]. During lyophilization, low temperatures (below the freezing point of water) and a very low pressure (high vacuum) are maintained to enable the sublimation of water (ice) [22]. The lyophilization process is composed of mainly four stages: (1) freezing, (2) ice sublimation (primary drying), (3) desorption of unfrozen water (secondary drying), and (4) storage [26,30–32]. Lyophilization preserves natural ingredients by minimizing their deterioration due to oxidation [20] and maintains existing emulsion-like cell structures and characteristics of the original natural ingredients. In addition, due to the ice crystals formed during the freezing stage, lyophilization results in a porous structure, which may provide better control of the encapsulated materials' release [33]. Lyophilization was shown to be a useful technique for the formation of cellulose aerogels from hydrogels [32]. It was shown to induce a microstructure with useful properties for advanced applications, such as water removal from waste emulsions, where the aerogel is actually re-introduced into a liquid environment [34,35].

Our previous work described a novel method to encapsulate thymol (TY) and eugenol (EU) EOs using chemically unmodified cellulose [36]. Regenerated cellulose hydrogel obtained from micro-crystalline cellulose (MCC) was used to form a continuous encapsulating shell around the EOs to form micro-particles with diameters of 1–5 μm. For EO encapsulation, a high-pressure homogenization process (HPH) was applied, while testing three pressures (5000, 10,000, and 20,000 PSI) and two cellulose:EO weight ratios (1:1 and 1:8). It was found that the pressure applied during the HPH process did not affect the capsule size, while the cellulose:EO ratio substantially affected both the capsules size and anti-mold activity of the emulsions. The 1:8 cellulose:EO ratio (optimally prepared by HPH at 10,000 PSI) yielded capsules of a diameter of about 5 μm that exhibited excellent anti-mold activity for both EU and TY in the alfalfa plant, used as a model system for hay.

To further study aqueous emulsions of cellulose-encapsulated EOs, it is of interest to investigate the structure, properties, and the possible applications of dried emulsions. While encapsulation can alleviate some limitations of EOs, the drying of the water-based emulsions may further contribute to their applicability in industrial uses. For example, their significantly lighter weight lowers transport costs and extends shelf life. Cellulose-encapsulated EO powders may be more stable in higher-temperature processes, such as compounding with polymers and film formation, thus presenting a significant bio-based alternative to the use of synthetic chemical pesticides. The current work aimed to characterize the structure and anti-mold capacities of freeze-dried TY and EU encapsulated by unmodified cellulose.

2. Materials and Methods

2.1. Materials

Eugenol (98%), thymol (98.5%), NaOH, and microcrystalline cellulose (MCC) powder (batch No. MKCJ3230, particle size 70–250 μm) were purchased from Sigma–Aldrich Co. (Rehovot, Israel). A similar MCC was previously characterized as having molecular weight of 50 kDa [37]. Deionized water was used to prepare the emulsion samples.

2.2. Methods

2.2.1. Encapsulation of Essential Oil Emulsions and Their Lyophilization

EO–cellulose emulsions were prepared by homogenizing cellulose hydrogel suspension with EOs, as previously reported by Shlosman et al. [36]. Briefly, the hydrogel suspension was prepared by regeneration from a solution of MCC aqueous 7 wt.% NaOH at −17 °C. Emulsions were prepared at 1:8 cellulose: EO ratio was made by homogenizing a mixture consisting of 36.2% regenerated cellulose hydrogel (containing 2 gr of cellulose), 10.7% (16 gr) EO (EU or TY), and 53.1% water using a mechanical homogenizer (T18 digital Ultra-Turrax, IKA Works, Staufen, Germany), followed by high-pressure homogenization (HPH, Model LM-20 microfluidizer, Microfluidics, Newton, MA, USA) at 10,000 PSI. The content of emulsified medium EO in the emulsion corresponds to 88.9%. The obtained emulsions were centrifuged at a speed of 6000 rpm (relative centrifugal force about 2400× *g*) for 7 min in an MRC benchtop centrifuge (SCEN-206, MRC laboratory Instruments Ltd., Holon, Israel). Excess water and EOs were removed manually (by pipette) and the dense emulsion was then lyophilized.

Approximately 1.5 g centrifuged emulsions were placed in a 10 mL glass vial and positioned on the shelves of a Labconco FreeZone® stoppering tray dryer (model 7948030, Labconco, Kansas, MI, USA). The lyophilization coil temperature was set to −50 °C and pressure was set to 0.2 mbar. The samples were cooled to −40 °C and maintained at this temperature for 48 h, after which, they were heated to room temperature and removed from the lyophilizer. For reference, a sample of cellulose hydrogel suspension without EOs underwent homogenizations, centrifugation, and lyophilization. The identification of the studied samples is given in Table 1.

Table 1. Tested capsules.

Sample ID	EO [1] Type
Hydrogel_lyo	None
EU-1:8-10k_lyo	EU
TY-1:8-10k_lyo	TY

[1] EO—essential oil.

2.2.2. Characterization

Morphological Analysis via Scanning Electron Microscopy and Light Microscopy

The morphology of lyophilized emulsions was studied by scanning electron microscopy (SEM) and light microscopy. A Zeiss Ultra Plus high-resolution scanning electron microscope (Carl Zeiss, Jena, Germany) equipped with a Schottky field-emission gun was used. Specimens were placed on specimen holders and imaged at a low acceleration voltage of 1.0–1.1 kV without metal coating using Everhart–Thornley ("SE2") and the In-the-column (InLens) secondary electron imaging detectors. Images were acquired using SmartSEM softwareV7 (Carl Zeiss, Jena, Germany) and analyzed by ImageJ 1.53 (U.S. National Institutes of Health, Bethesda, MD, USA). The overall structure of the lyophilized specimens and after slight milling (placed on a glass slide) was observed using an Olympus BX60 light microscope equipped with an MPLFLN lens unit, using a transmission light source (Evident Corp., Waltham, MA, USA). Stream Essentials 2.4 software (Olympus Scientific Solutions, Evident Corp., Waltham, MA, USA) was used to record and analyze images.

Chemical Composition of the Lyophilized Powders

Attenuated total reflectance Fourier transform infrared spectroscopy (ATR-FTIR) spectroscopy was used to identify the functional groups of EOs and cellulose in respective powders. A Perkin Elmer Fourier transform infrared spectrophotometer (Perkin Elmer FTIR Spectrum BX II, Waltham, MA, USA) was used in attenuated total reflectance (ATR) mode and acquired at a spectral range of 4000–500 cm^{-1} at a resolution of 4 cm^{-1}, for 32 cycles. Perkin Elmer spectrum IR version 10.6.1 software was used for the analysis.

Thermal Analysis

The thermal stability and composition of the micro-capsules, as well as the EO releasing rates from the capsules, were studied using thermo-gravimetric analysis (TGA) (TGA Q5000 system, TA instruments, New Castle, DE, USA) equipped with a ceramic pan. Dynamic TGA was performed to assess thermal stability and % encapsulation efficiency (%EE), and for compositional analysis. Samples were heated from room temperature up to 800 °C at a heating rate of 10 °C/min, under nitrogen atmosphere. To study the EO release rates from the capsules, a static TGA program was used: the samples were heated to 40 °C and 50 °C (for EU and TY, respectively) at a heating rate of 20 °C/min and maintained at this temperature for 900 min. The weight loss in both dynamic and static TGA modes was monitored throughout the experiment and EO content in each sample was calculated. Each measurement point represents an average of two measurements and a bar representing the upper and lower values is specified in the corresponding graphs.

Encapsulation efficiency: Different procedures have been reported in the literature for calculation of the % encapsulation efficiency (EE), most of which are based on different extraction and filtration methods [26,38–41]. All the methods require knowledge of the initial amount of EO incorporated into the formula and the actual amount of EO present in the sample, as given by Equation (1) [39]:

$$EE(\%) = \frac{\text{EO content in powder}}{\text{EO content in feed liquid}} \times 100\% \quad (1)$$

3. Results and Discussion

3.1. Morphology of Lyophilized Emulsions

Figure 1a–c depicts the morphology of the lyophilized emulsions, as viewed by SEM. A morphology of irregular porous flakes with sharp edges was observed and is considered to be made of the aggregated particles. In contrast to the emulsions [36], the cryo-electron microscopy found no individual micron-sized particles. A similar morphology of freeze dried capsules (not ground) was previously reported by several researchers, who described the structure as "cake-like" [26], "slab-like" [42], "flake-like" [43], and "broken glass-like" [28,39], implying that lyophilized material is characterized by non-spherical particles. This structure likely resulted from the concentration of emulsion droplets and remaining free hydrogel particles, first by centrifugation, followed by the front of crystallizing water during freezing. During coalescence and vitrification at low temperatures, the forces between the impinging crystal fronts are not strong enough to break the vitrified mass into droplets. The pores may have formed by the sublimation of ice crystals [26,28,30,39,42,43]. Agglomeration of cellulose fibrils into sheet-like structures, induced by the ice crystal formation, has been observed in freeze-dried bacterial cellulose (BC) cryogels [32]. Freeze-dried aqueous BC suspensions were shown to exhibit "interlinked sheets" with "irregularly shaped macro-pores" [35].

SEM analyses performed at higher magnification are presented in Figures S1–S3 in the Supplementary Materials. The higher resolution enabled imaging the hydrogel structure as a porous network, similar to that observed by cryogenic SEM analysis of the original emulsions [36]. Flake thickness was found to be between 80 and 300 nm.

Light microscopy images of the lyophilized powder structure before (Figure 1d–f) and after light manual milling (Figure 1g–i) showed irregular-shaped agglomerates with a flaky texture, regardless of the application of milling. Light manual milling had no effect on the observed particles' morphology. Thus, it was concluded that SEM images captured directly after sample lyophilization well represent the actual sample morphology.

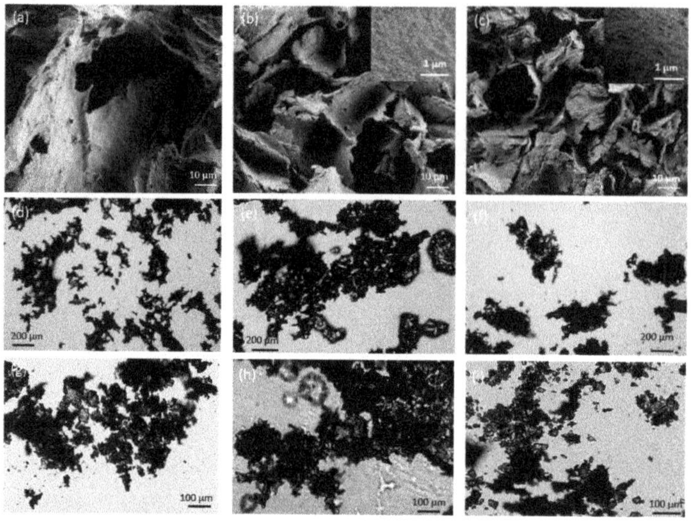

Figure 1. SEM (**a–c**) and light microscope (**d–i**) images of lyophilized samples of (**a,d,g**) hydrogel_lyo, (**b,e,h**) EU-1:8-10k_lyo and (**c,f,i**) TY-1:8-10k_lyo. Lyophilized specimens (**d–f**) before and (**g–i**) after light milling.

3.2. Chemical Composition of the Lyophilized Powders Attenuated Total Reflectance Fourier Transform Infrared Spectroscopy (ATR-FTIR)

ATR-FTIR spectra of the lyophilized samples of cellulose hydrogel, cellulose-encapsulated EO and TY, and free EU and TY, are presented in Figure 2. Absorption peaks which are characteristic of cellulose, EU, and TY [36] were observed, and indicated successful encapsulation of EU and TY. The characteristic bands of cellulose attributed to the presence of OH groups (3000–3700 cm^{-1}) and C–H stretching vibration (893 cm^{-1}, 2877 cm^{-1} and 1030 cm^{-1}) were clearly seen in the spectra of lyophilized cellulose. The O–H bending vibration (1636 cm^{-1}), due to adsorbed water molecules, was also observed in the lyophilized cellulose FTIR-ATR spectra [44,45]. Characteristic EU FTIR bands included stretching vibration of the aromatic C–C bond (1510, 1607, 1638, and 3525 cm^{-1}), asymmetric stretch of C–O–C bonds (1121, 1147, 1181, and 1202 cm^{-1}), alcoholic C–O vibrations (1033 cm^{-1}), and C–H bonding vibrations originated in the CH_2 and CH_3 groups (792, 994, 3000, and 3062 cm^{-1}) [24,36,46–49]. TY FTIR was characterized by phenolic O–H stretching vibration (3000–3500 cm^{-1}), C–O stretching vibration (1157, 1195 cm^{-1}), C–H bonds (2865, 2926, 2958 cm^{-1}), C–H stretching vibrations (1418, 1459 cm^{-1}), and aromatic C=C stretching of benzene ring (1622 cm^{-1}) bands [36,50–52]. Similar bands were also observed in the corresponding lyophilized samples EU-1:8-10k_lyo and TY-1:8-10k_lyo indicating the presence of the respective EO in the lyophilized samples. Slight shifts of absorption bands were also observed, indicating different interactions between constituents [24]. The presence of cellulose and absorbed water in both lyophilized samples was also apparent from the wide band at 2980–3700 cm^{-1}. Overall, FTIR analysis indicated successful lyophilization, whereby the EOs content was preserved in the capsules, in agreement with FTIR results previously reported for EO emulsions [36], which showed that no chemical changes occurred during encapsulation of EU and TY EOs.

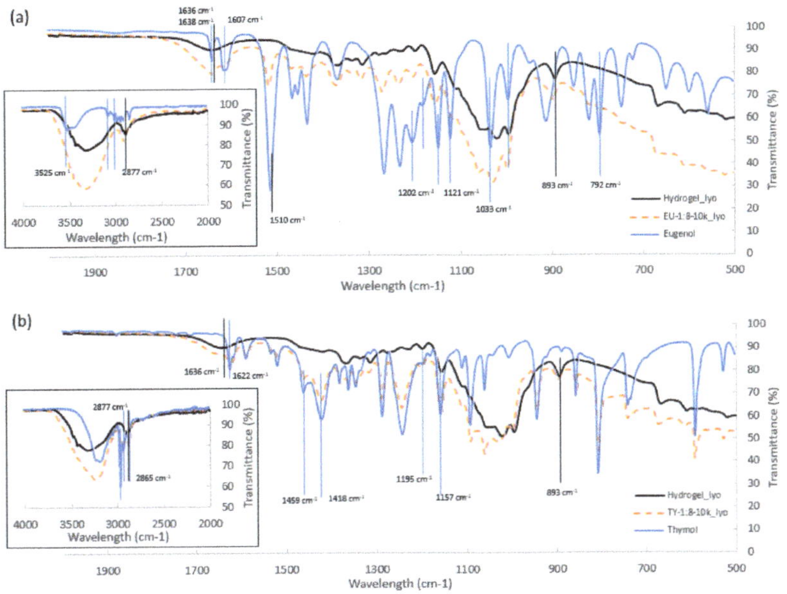

Figure 2. FTIR-ATR spectra of Hydrogel_lyo together with (**a**) free EU and EU-1:8-10k_lyo and (**b**) free TY and TY-1:8-10k_lyo.

3.3. Thermal Stability and EO Content of the Lyophilized Powders

Thermo-gravimetric analysis (TGA) is a powerful tool that can be used to obtain information on the thermal stability, composition (including moisture content), % encapsulation efficiency (%EE) and release rates of encapsulated EOs after lyophilization [50,53,54]. TGA graphs taken in dynamic mode reflect the sample's weight loss during its heating at a constant rate. Figure 3 presents TGA and derivative TG (DTG) of Hydrogel_lyo, free EU and TY, EU-1:8-10k_lyo and TY-1:8-10k_lyo. The thermograms of free EU [24] and TY [52] are characterized by a single peak in the temperature range of 25–100 °C, attributed to their evaporation, due to their high volatility. The differential analysis indicated that the peak temperatures of maximum rate for this phase transition were 104.6 and 84.2 °C for free EU and TY, respectively. The thermogram of Hydrogel_lyo was characterized by two weight loss stages. The first, in the temperature range of 25–100 °C, can be attributed to the loss of surface-absorbed water [55,56] and corresponded to 7.6% of the sample weight. The second weight loss stage of about 70%, seen in the temperature range of 100–700 °C, can be attributed to cellulose decomposition, by the breakdown of glycosyl units. A 21.6% residue (quantified at 800 °C) was present in the sample. The thermograph was slightly different than that of MCC thermograms published in the literature, due to the more amorphous structure and lower molecular weight of the cellulose hydrogel compared to crystalline MCC [56,57], as was shown in several studies characterizing regenerated cellulose by viscosity and by GPC [58], and its structure by X-ray diffraction [59,60]. The thermograms of EU-1:8-10k_lyo and TY-1:8-10k_lyo were characterized by three weight loss stages. The first, attributed to moisture contents of 5.2% and 1%, was recorded in the temperature range of 25–65 °C and 25–50 °C for EU-1:8-10k_lyo and TY-1:8-10k_lyo, respectively. The second weight loss, attributed to the evaporation of the encapsulated EO, measured 77.5% and 78.5% for EU-1:8-10k_lyo and TY-1:8-10k_lyo, respectively. This quantification was performed up to the temperature of 200 °C, starting at the temperature at which water ceased to evaporate. The third decomposition stage, observed in the temperature range of 200–500 °C, can be attributed to the decomposition of cellulose, and measured 13.9% and 16.7% for EU-1:8-10k_lyo and TY-1:8-10k_lyo, respectively. The residual content, quantified at 800 °C, measured 2.5% and 3.5% for EU-1:8-10k_lyo and TY-1:8-10k_lyo, respectively. These results complement the FTIR analysis and indicate that EU and TY were well encapsulated by the regenerated cellulose hydrogel. When considering the temperature at which the EOs begin to evaporate and the peak temperature of maximum evaporation, it can be seen that the peak temperature at the maximum evaporation rate increased from 104.6 °C to 110 °C and from 82.4 °C to 91.7 °C for EU-1:8-10k_lyo and TY-1:8-10k_lyo, respectively. These results support the suggested use of the dried capsules in applications demanding high temperature, such as compounding and extrusion with polymers, used in packaging [61,62] and post-harvest disease protection [63]. An interesting effect was observed regarding the maximum evaporation rate temperature for the decomposition of cellulose: while the peak temperature for Hydrogel_lyo was 303 °C, when used as a shell material, the peak temperature increased to about 340 °C for both encapsulated EO samples. This can be explained by interactions between cellulose and EOs, leading to a more thermally stable structure. The weight percentages of each ingredient in the different samples and peak temperature of EO evaporation are summarized in Table 2.

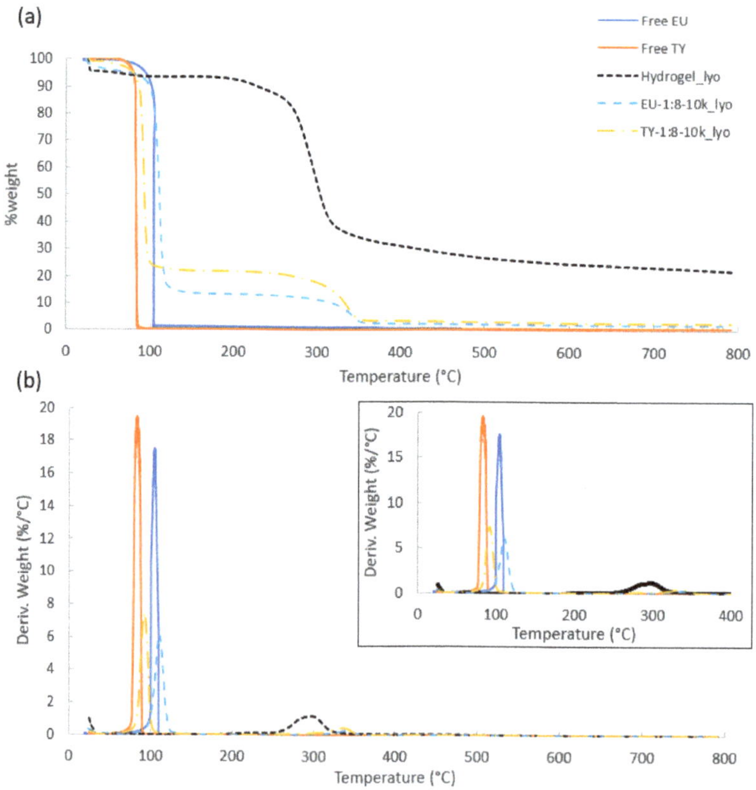

Figure 3. Thermal analyses of Hydrogel_lyo, free EU, free TY, EU-1:8-10k_lyo and TY-1:8-10k_lyo: (**a**) TGA and (**b**) DTG thermographs Inset: zoom-in of DTG in the range of 25–400 °C.

By using Equation (1), a very high %EE of 87% and 88% was achieved for EU-1:8-10k_lyo and TY-1:8-10k_lyo, respectively. While high %EE can also be achieved using spray-drying, the high %EE of lyophilization with the use of crop-waste encapsulating material can provide effective alternative for industrial-scale processes [64–66].

Table 2. Summary of compositions and thermal stability of tested samples [1].

Sample	Moisture Content (%)	EO Content (%)	Decomposed Cellulose Content (%)	Peak Temp. of EO Maximum Evaporation Rate (°C)
Hydrogel_lyo	7.6 (2.3)	0	70 (0.1)	-
Free EU	0	100	0	104.6
Free TY	0	100	0	82.4
EU-1:8-10k_lyo	5.2 (1.7)	77.5 (9.4)	13.9 (3.8)	110
TY-1:8-10k_lyo	1 (0.8)	78.5 (3.2)	16.7 (5.9)	91.7

[1] Average of two measurements (difference between the two measurements).

The thermal analyses complement the FTIR tests, demonstrating the successful inclusions of EOs in an unmodified cellulose shell, and their retention in the lyophilized structure. Increased thermal stability observed for encapsulated EU and TY compared to their corresponding free EOs, indicates that their encapsulation increases their resistance to evaporation [41,50,52,67,68].

Release rate: Static TGA can also provide valuable information on EO release rates from the capsules. In such assessments, the samples were heated to a desired temperature (40 °C for EU and EU-1:8-10k_lyo and 50 °C for TY and TY-1:8-10k_lyo) and retained at that temperature for 900 min, while the weight loss was monitored. In order to optimize the test temperature for the different samples, a preliminary test was performed on free TY at 40, 50 and 60 °C, as shown in Figure 4. It can be seen that while maintained at 60 °C, all the TY evaporated within 300 min, while at 40 °C, there was almost no loss of TY (6.1% weight loss after 480 min), attributed to the fact that TY is solid at this temperature (its melting temperature is 52 °C [51]), with low vapor pressure. Thus, the temperature for the experiment testing TY release was set to 50 °C. Unlike TY, EU is liquid at room temperature, enabling the above test to be performed at 40 °C in an acceptable time frame; hydrogel_lyo was also analyzed at 40 °C.

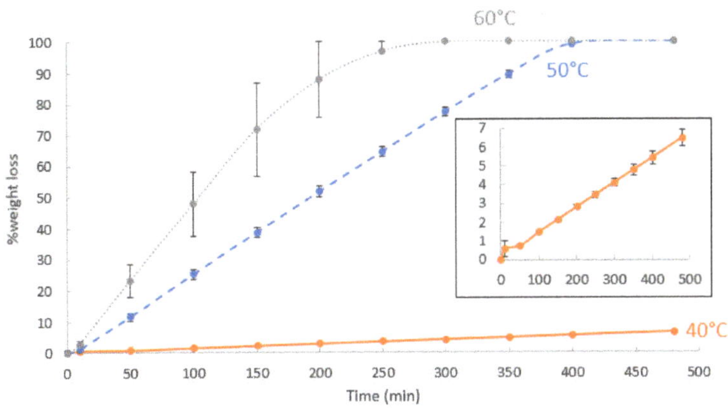

Figure 4. Static TGA of free TY performed at 40, 50 and 60 °C. Each point represents an average of two measurements; the error bars represent the upper and lower values. Inset—zoom-in of the measurements at 40 °C.

Figure 5a–c shows the static TGA measurements of the studied samples. Figure 5a depicts the weight loss % of Hydrogel_lyo at 40 °C. Neither thermal degradation nor evaporation occurred at this temperature, therefore the weight loss observed can be fully attributed to moisture absorbed on the cellulose surface (corresponding to the amount observed in Figure 3 and listed in Table 2). After 10 min, 50 min, and 100 min, 76, 90, and 94% of the moisture content had evaporated, respectively, and a plateau was reached within 150 min. For the lyophilized samples, the initial weight loss (during the first

150 min) includes both water and EO evaporation. Therefore, in this time frame, the weight loss of water was considered to be equal to the water weight loss % observed in the Hydrogel_Lyo, Figure 5a (raw data and water-subtracted data are given in Tables S1 and S2, respectively). The water-subtracted weight loss of lyophilized samples was further normalized by dividing with the EO content of the respective lyophilized powders and presented in Figure 5b,c (normalized data are given in Table S3 in SI).

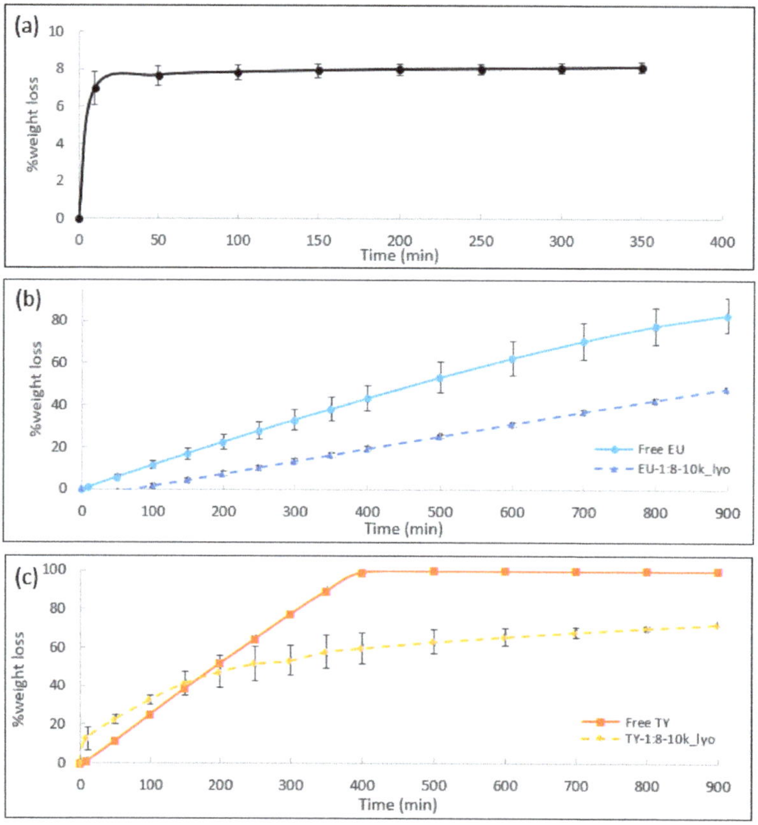

Figure 5. Static TGA; % weight loss over time for (**a**) Hydrogel_lyo performed at 40 °C, (**b**) free EU and EU-1:8-10k_lyo performed at 40 °C, and (**c**) free TY and TY-1:8-10k_lyo performed at 50 °C. An average of two measurements is presented. The error bars connect the measured values, when shown, otherwise the measurements were identical. In (**b**,**c**), for comparison purpose with free EO, the weight loss of lyophilized samples is presented after water subtraction and normalization to EO content, as explained in the text and SI.

The main quantitative analysis was made within a time frame of 400 min, but to further understand the release rates at much longer times, weight loss was monitored for 900 min at the predefined temperature. Free EU and EU-1:8-10k_lyo exhibited a linear rate of weight loss (R^2 = 0.99, slope = 0.1%/min and R^2 = 0.99, slope = 0.05%/min, respectively). Weight loss of free EU measured 43% after 400 min and 83% after 900 min, while EU-1:8-10k_lyo exhibited only 19% weight loss after 400 min and 48% weight loss after 900 min. Taken together, the encapsulated EU exhibited a significantly lower weight loss rate than free EU (by about 45% after 400 and by 58% after 900 min). Free TY also exhibited a linear rate of weight loss (R^2 = 0.99, slope = 0.3%/min), with 99% of the total weight lost within 400 min. Unlike neat TY, TY-1:8-10k_lyo presented a non-linear release rate, with two successive nearly linear release processes: first, a quick release stage within a time interval of 0–150 min (R^2 = 0.93, slope = 0.2%/min) with a total weight loss of 33%, and a second stage, characterized by a much slower release rate after 200 min (R^2 = 0.95, slope = 0.03%/min), ending with a weight loss of 57% after 900 min.

Such forms of weight loss and the difference between the release profiles of the encapsulated EU and TY can be explained by consideration of the components' thermodynamic properties and the capsule structure. The higher boiling point of EU compared to TY (254 °C and 233 °C, respectively) [69,70] implies that TY possess a higher vapor pressure in the liquid form in comparison to EU. Thus, the initial release rate TY is higher than that of EU at the tested temperatures (50 °C and 40 °C, respectively). Additionally, the encapsulating shells of the initial emulsions present another barrier for EO diffusion. Owing to breakage of a certain portion of the capsules during lyophilization, free EO can be considered to be present in the lyophilized powder in addition to encapsulated EO, which affects the release profile. The static TGA results complement thermal analysis performed by dynamic TGA and further demonstrate the successful encapsulation of EOs in the powder form. The release rate of EOs is a function of their volatility (i.e., vapor pressure); by increasing EO resistance to evaporation by their encapsulation (as shown in Figure 3), their release rate is slowed as EO diffusion through the capsule shell is slower than its diffusion in the free form. Such decreased release rates improve EO retention, which is a particular benefit for certain applications, e.g., incorporation of EO-loaded capsules into polymeric materials for enhanced antimicrobial activity of plastic films.

4. Conclusions

In this work, emulsions of EU and TY encapsulated in cellulose were successfully dried by lyophilization. The combination of the previously reported EO encapsulation method, with a subsequent lyophilization process resulted in high encapsulation efficiency and fabrication of well-dried emulsions, with extremely low moisture content, as well as maintenance of the porous structure. The lyophilized powder exhibits a "flake-like" structure, significantly different from that of the emulsified microcapsules. The chemically unmodified cellulose shell material increased thermal stability and decreased the release of the encapsulated EOs. When compared to the liquid emulsions, the dry form of encapsulated EOs extends potential industrial applications, such as EO incorporation into products that are subjected to intense thermal treatment, including compounding with polymer melts and fabrication of plastic articles. This can provide prolonged antimicrobial activity in household and agricultural applications with a reduced harmful impact on health and the environment.

Supplementary Materials: The following supporting information can be downloaded at: https://www.mdpi.com/article/10.3390/polym16101422/s1, Figure S1: HR-SEM micrographs of hydrogel_lyo at different magnifications. Dashed yellow circles indicate the area focused in the subsequent image; Figure S2: HR-SEM micrographs of EU-1:8-10k_lyo at different resolutions magnifications. Dashed yellow circles indicate the area focused in the subsequent image; Figure S3: HR-SEM micrographs of TY-1:8-10k_lyo at different resolutions magnifications. Dashed yellow circles indicate the area focused in the subsequent image; Table S1: TGA Raw data for weight loss; Table S2: TGA water-reduced data for weight loss: Eugenol @ 40 C, Thymol @ 50 C; Table S3: TGA normalized data for weight loss.

Author Contributions: Conceptualization, K.S., D.M.R. and Y.C.; methodology, K.S., D.M.R. and Y.C.; formal analysis, K.S.; investigation, K.S.; resources, D.M.R.; writing—original draft preparation, K.S.; writing—review and editing, D.M.R., R.S. and Y.C; supervision, Y.C.; funding acquisition, R.S. and Y.C. All authors have read and agreed to the published version of the manuscript.

Funding: This work was supported by the Israel Innovation Authority, SMART MAGNET consortium. D. M. Rein's work was supported by a joint grant from the Center for Adsorption in Science of the Ministry of Immigrant Absorption and the Committee for Planning and Budgeting of the Council for Higher Education under the framework of the KAMEA program.

Institutional Review Board Statement: Not applicable.

Data Availability Statement: The original contributions presented in the study are included in the article and Supplementary Materials, further inquiries can be directed to the corresponding author.

Acknowledgments: SEM specimen preparation and imaging were performed by Inbal Weisboard at the Technion Center for Electron Microscopy of Soft Matter and by Maria Koifman Khristosov at the Electron Microscopy Center at the Technion Department of Materials Science and Engineering, and supported by the Technion Russell Berrie Nanotechnology Institute (RBNI). Lyophilization was performed at the Laboratory for Biomaterials and Regenerative Medicine at the Technion Faculty of Biomedical Engineering. The authors thank Dror Seliktar for the use of the lyophilizer.

Conflicts of Interest: Author Koranit Shlosman and Rotem Shemesh were employed by the company R&D and Customer Service department Carmel Olefins Ltd. Koranit Shlosman and Rotem Shemesh and the remaining authors declare that the research was conducted in the absence of any commercial or financial relationships that could be construed as a potential conflict of interest.

References

1. Froiio, F.; Ginot, L.; Paolino, D.; Lebaz, N.; Bentaher, A.; Fessi, H.; Elaissari, A. Essential oils-loaded polymer particles: Preparation, characterization and antimicrobial property. *Polymers* **2019**, *11*, 1017. [CrossRef] [PubMed]
2. Gandra, E.A.; Radünz, M.; Helbig, E.; Borges, C.D.; Gandra, T.K.V. A Mini-Review on Encapsulation of Essential Oils. *J. Anal. Pharm. Res.* **2018**, *7*, 00205. [CrossRef]
3. Donsì, F.; Annunziata, M.; Sessa, M.; Ferrari, G. Nanoencapsulation of essential oils to enhance their antimicrobial activity in foods. *LWT—Food Sci. Technol.* **2011**, *44*, 1908–1914. [CrossRef]
4. Isman, M.B. Bioinsecticides based on plant essential oils: A short overview. *Zeitschrift fur Naturforsch. Sect. C J. Biosci.* **2020**, *75*, 179–182. [CrossRef] [PubMed]
5. Sharma, H.; Mendiratta, S.K.; Agarwal, R.K.; Gurunathan, K. Bio-preservative effect of blends of essential oils: Natural anti-oxidant and anti-microbial agents for the shelf life enhancement of emulsion based chicken sausages. *J. Food Sci. Technol.* **2020**, *57*, 3040–3050. [CrossRef] [PubMed]
6. Chouhan, S.; Sharma, K.; Guleria, S. Antimicrobial Activity of Some Essential Oils—Present Status and Future Perspectives. *Medicines* **2017**, *4*, 58. [CrossRef] [PubMed]
7. Maes, C.; Bouquillon, S.; Fauconnier, M.L. Encapsulation of essential oils for the development of biosourced pesticides with controlled release: A review. *Molecules* **2019**, *24*, 2539. [CrossRef] [PubMed]
8. Hossain, F.; Follett, P.; Vu, K.D.; Salmieri, S.; Fraschini, C.; Jamshidian, M. Lacroix, MAntifungal activity of combined treatments of active methylcellulose-based films containing encapsulated nanoemulsion of essential oils and γ–irradiation: In vitro and in situ evaluations. *Cellulose* **2019**, *26*, 1335–1354. [CrossRef]
9. Akturk, A. Enrichment of Cellulose Acetate Nanofibrous Scaffolds with Retinyl Palmitate and Clove Essential Oil for Wound Healing Applications. *ACS Omega* **2023**, *8*, 5553–5560. [CrossRef]
10. Zikeli, F.; Vinciguerra, V.; Sennato, S.; Mugnozza, G.S.; Romagnoli, M. Preparation of Lignin Nanoparticles with Entrapped Essential Oil as a Bio-Based Biocide Delivery System. *ACS Omega* **2020**, *5*, 358–368. [CrossRef]
11. FSousa, L.; Santos, M.; Rocha, S.M.; Trindade, T. Encapsulation of essential oils in SiO2 microcapsules and release behaviour of volatile compounds. *J. Microencapsul.* **2014**, *31*, 627–635. [CrossRef] [PubMed]
12. Hazarika, U.; Gosztola, B. Lyophilization and its Effects on the Essential Oil Content and Composition of Herbs and Spices—A Review. *Acta Sci. Pol. Technol. Aliment.* **2020**, *19*, 467–473. [CrossRef]
13. Gomes, D.S.; da Costa, A.; Pereira, A.M.; Casal, M.; Machado, R. Biocomposites of Silk-Elastin and Essential Oil from Mentha piperita Display Antibacterial Activity. *ACS Omega* **2022**, *7*, 6568–6578. [CrossRef] [PubMed]
14. Yang, G.; Li, F.; Zhang, H.; Yan, H.; Gao, S.; Fu, Y.; Ye, F. Electrospinning for producing antifungal nanofibers consisting of prochloraz/hydroxypropyl-γ-cyclodextrin inclusion complex. *Ind. Crops Prod.* **2024**, *211*, 118282. [CrossRef]
15. Majeed, H.; Bian, Y.-Y.; Ali, B.; Jamil, A.; Majeed, U.; Khan, Q.F.; Iqbal, K.J.; Shoemaker, C.F.; Fang, Z. Essential oil encapsulations: Uses, procedures, and trends. *RSC Adv.* **2015**, *5*, 58449–58463. [CrossRef]
16. Cadena, M.B.; Preston, G.M.; Van der Hoorn, R.A.L.; Townley, H.E.; Thompson, I.P. Species-specific antimicrobial activity of essential oils and enhancement by encapsulation in mesoporous silica nanoparticles. *Ind. Crops Prod.* **2018**, *122*, 582–590. [CrossRef]
17. Guzmán, E.; Lucia, A. Essential oils and their individual components in cosmetic products. *Cosmetics* **2021**, *8*, 114. [CrossRef]
18. Plati, F.; Paraskevopoulou, A. Micro- and Nano-encapsulation as Tools for Essential Oils Advantages' Exploitation in Food Applications: The Case of Oregano Essential Oil. *Food Bioprocess Technol.* **2022**, *15*, 949–977. [CrossRef]
19. Enciso-Sáenz, S.; Borrás-Enriquez, A.J.; Ventura-Canseco, L.M.; Gutiérrez-Miceli, F.; Dendooven, L.; Grajales-Lagunes, A.; Ruiz-Cabrera, M.A.; Ruíz-Valdiviezo, V.; Abud-Archila, M. Lemongrass (Cymbopogon citratus (DC) Stapf) essential oil encapsulation by freeze-drying. *Rev. Mex. Ing. Quim.* **2018**, *17*, 407–420. [CrossRef]
20. Souza, J.M.; Caldas, A.L.; Tohidi, S.D.; Molina, J.; Souto, A.P.; Fangueiro, R.; Zille, A. Properties and controlled release of chitosan microencapsulated limonene oil. *Rev. Bras. Farmacogn.* **2014**, *24*, 691–698. [CrossRef]

21. Mehran, M.; Masoum, S.; Memarzadeh, M. Microencapsulation of Mentha spicata essential oil by spray drying: Optimization, characterization, release kinetics of essential oil from microcapsules in food models. *Ind. Crops Prod.* **2020**, *154*, 112694. [CrossRef]
22. Al-Hamayda, A.; Abu-Jdayil, B.; Ayash, M.; Tannous, J. Advances in microencapsulation techniques using Arabic gum: A comprehensive review. *Ind. Crops Prod.* **2023**, *205*, 117556. [CrossRef]
23. Kasiri, N.; Fathi, M. Entrapment of peppermint oil using cellulose nanocrystals. *Cellulose* **2018**, *25*, 319–329. [CrossRef]
24. Piletti, R.; Bugiereck, A.; Pereira, A.; Gussati, E.; Magro, J.D.; Mello, J.; Dalcanton, F.; Ternus, R.; Soares, C.; Riella, H.; et al. Microencapsulation of eugenol molecules by β-cyclodextrine as a thermal protection method of antibacterial action. *Mater. Sci. Eng. C* **2017**, *75*, 259–271. [CrossRef] [PubMed]
25. Yousefi, S.; Weisany, W.; Hosseini, S.E.; Ghasemlou, M. Mechanisms of nanoencapsulation to boost the antimicrobial efficacy of essential oils: A review. *Food Hydrocoll.* **2024**, *150*, 109655. [CrossRef]
26. Karthik, P.; Anandharamakrishnan, C. Microencapsulation of Docosahexaenoic Acid by Spray-Freeze-Drying Method and Comparison of its Stability with Spray-Drying and Freeze-Drying Methods. *Food Bioprocess Technol.* **2013**, *6*, 2780–2790. [CrossRef]
27. Bhandari, B.; Patel, K.C.; Chen, X.D. Spray drying of liquid foods. In *Drying Technologies in Food Processing*; Chen, X.D., Mujumdar, A.S., Eds.; Blackwell Pub.: Oxford, UK, 2008; Chapter 4; pp. 113–157.
28. Cano-Higuita, D.M.; Malacrida, C.R.; Telis, V.R.N. Stability of Curcumin Microencapsulated by Spray and Freeze Drying in Binary and Ternary Matrices of Maltodextrin, Gum Arabic and Modified Starch. *J. Food Process. Preserv.* **2015**, *39*, 2049–2060. [CrossRef]
29. Luo, X.; Sedman, J.; Ismail, A.A. Microencapsulation of oregano (*Origanum vulgare* L.), rosemary (*Rosmarinus officinalis* L.) and sage (*Salvia officinalis* L.) essential oils in β-lactoglobulin. *SDRP J. Food Sci. Technol.* **2019**, *4*, 970–985. [CrossRef]
30. Degobert, G.; Aydin, D. Lyophilization of nanocapsules: Instability sources, formulation and process parameters. *Pharmaceutics* **2021**, *13*, 1112. [CrossRef]
31. Stratta, L.; Capozzi, L.C.; Franzino, S.; Pisano, R. Economic analysis of a freeze-drying cycle. *Processes* **2020**, *8*, 1399. [CrossRef]
32. Beaumont, M.; Potthast, A.; Rosenau, T. Cellulose Nanofibrils: From Hydrogels to Aerogels. In *Cellulose Science and Technology: Chemistry, Analysis, and Applications*; Rosenau, T., Potthast, A., Hell, J., Eds.; John Wiley & Sons: Hoboken, NJ, USA, 2019; Chapter 13; pp. 277–339. [CrossRef]
33. Bakry, A.M.; Abbas, S.; Ali, B.; Majeed, H.; Abouelwafa, M.Y.; Mousa, A.; Liang, L. Microencapsulation of Oils: A Comprehensive Review of Benefits, Techniques, and Applications. *Compr. Rev. Food Sci. Food Saf.* **2016**, *15*, 143–182. [CrossRef] [PubMed]
34. Wang, M.; Li, Z.; Chen, A.; Tian, Q.; Liu, X.; Luo, Q.; Pei, C. Influence of the freezing and lyophilization of bacterial cellulose hydrogel on water removal from both water-in-oil and oil-in-water emulsion. *Cellulose* **2022**, *29*, 5979–5990. [CrossRef]
35. Yan, Z.; Zhu, K.; Li, X.; Wu, X. Recyclable Bacterial Cellulose Aerogel for Oil and Water Separation. *J. Polym. Environ.* **2022**, *30*, 2774–2784. [CrossRef]
36. Shlosman, K.; Rein, D.M.; Shemesh, R.; Koifman, N.; Caspi, A.; Cohen, Y. Encapsulation of Thymol and Eugenol Essential Oils Using Unmodified Cellulose: Preparation and Characterization. *Polymer* **2023**, *15*, 95. [CrossRef] [PubMed]
37. DRein, M.; Khalfin, R.; Szekely, N.; Cohen, Y. True molecular solutions of natural cellulose in the binary ionic liquid-containing solvent mixtures. *Carbohydr. Polym.* **2014**, *112*, 125–133. [CrossRef] [PubMed]
38. Manaf, M.A.; Subuki, I.; Jai, J.; Raslan, R.; Mustapa, A.N. Encapsulation of Volatile Citronella Essential Oil by Coacervation: Efficiency and Release Study. *IOP Conf. Ser. Mater. Sci. Eng.* **2018**, *358*, 012072. [CrossRef]
39. Ezhilarasi, P.N.; Indrani, D.; Jena, B.S.; Anandharamakrishnan, C. Freeze drying technique for microencapsulation of Garcinia fruit extract and its effect on bread quality. *J. Food Eng.* **2013**, *117*, 513–520. [CrossRef]
40. Lopes, S.; Afonso, C.; Fernandes, I.; Barreiro, M.F.; Costa, P.; Rodrigues, A.E. Chitosan-cellulose particles as delivery vehicles for limonene fragrance. *Ind. Crops Prod.* **2019**, *139*, 111407. [CrossRef]
41. Kotronia, M.; Kavetsou, E.; Loupassaki, S.; Kikionis, S.; Vouyiouka, S.; Detsi, A. Encapsulation of oregano (Origanum onites l.) essential oil in β-cyclodextrin (β-CD): Synthesis and characterization of the inclusion complexes. *Bioengineering* **2017**, *4*, 74. [CrossRef] [PubMed]
42. Hasani, S.; Ojagh, S.M.; Ghorbani, M. Nanoencapsulation of lemon essential oil in Chitosan-Hicap system. Part 1: Study on its physical and structural characteristics. *Int. J. Biol. Macromol.* **2018**, *115*, 143–151. [CrossRef]
43. Buljeta, I.; Pichler, A.; Šimunović, J.; Kopjar, M. Polysaccharides as Carriers of Polyphenols: Comparison of Freeze-Drying and Spray-Drying as Encapsulation Techniques. *Molecules* **2022**, *27*, 5069. [CrossRef] [PubMed]
44. Kane, S.N.; Mishra, A.; Dutta, A.K. Preface: International Conference on Recent Trends in Physics (ICRTP 2016). *J. Phys. Conf. Ser.* **2016**, *755*, 011001. [CrossRef]
45. Doncea, S.M.; Ion, R.M.; Fierascui, R.C.; Bacalum, E.; Bunaciu, A.A.; Aboul-Enein, H.Y. Spectral methods for historical paper analysis: Composition and age approximation. *Instrum. Sci. Technol.* **2010**, *38*, 96–106. [CrossRef]
46. Rodríguez, J.W.; Peyron, S.; Rigou, P.; Chalier, P. Rapid quantification of clove (*Syzygium aromaticum*) and spearmint (Mentha spicata) essential oils encapsulated in a complex organic matrix using an ATR-FTIR spectroscopic method. *PLoS ONE* **2018**, *13*, e0207401. [CrossRef]
47. Taraj, K.; Andoni, A.; Yllı, F.; Yllı, A.; Hoxha, R.; Llupa, J.; Malollari, İ. Spectroscopic Investigation of *Syzygium aromaticum* L. Oil by Water Distillation Extraction. *J. Int. Environ. Appl. Sci.* **2020**, *15*, 122–126.
48. Mohammed, K.A.K.; Abdulkadhim, H.M.; Noori, S.I. Chemical Composition and Anti-bacterial Effects of Clove (*Syzygium aromaticum*) Flowers. *Int. J. Curr. Microbiol. Appl. Sci.* **2016**, *5*, 483–489. [CrossRef]

49. Yadava, R.N.; Saini, V.K. UV and IR spectral studies of essentila oil of A. indica, M. hortensis and E. triplinerve leaves. *Asian J. Chem.* **1994**, *6*, 77–80.
50. Chaib, S.; Benali, N.; Arhab, R.; Ajmi, I.S.; Bendaoued, H.; Romdhane, M. Preparation of Thymus vulgaris Essential Oil Microcapsules by Complex Coacervation and Direct Emulsion: Synthesis, Characterization and Controlled Release Properties. *Arab. J. Sci. Eng.* **2021**, *46*, 5429–5446. [CrossRef]
51. Zamani, Z.; Alipour, D.; Moghimi, H.R.; Mortazavi, S.A.R.; Saffaryc, M. Development and evaluation of thymol microparticles using cellulose derivatives as controlled release dosage form. *Iran. J. Pharm. Res.* **2015**, *14*, 1031–1040. [CrossRef]
52. Zhu, Z.; Min, T.; Zhang, X.; Wen, Y. Microencapsulation of thymol in poly(lactide-co-glycolide) (PLGA): Physical and antibacterial properties. *Materials* **2019**, *12*, 1133. [CrossRef]
53. Aguiar, M.C.; Denadae, B.; Silva, M.F.; Fernandes, J.; Bittencourt, P.; Scremin, F.; Forim, M. Quality Control for Lignin and Gelatin Microcapsules Loaded with Orange Essential Oil. *J. Braz. Chem. Soc.* **2022**, *33*, 85–95. [CrossRef]
54. Vishwakarma, G.S.; Gautam, N.; Babu, J.N.; Mittal, S.; Jaitak, V. Polymeric Encapsulates of Essential Oils and Their Constituents: A Review of Preparation Techniques, Characterization, and Sustainable Release Mechanisms. *Polym. Rev.* **2016**, *56*, 668–701. [CrossRef]
55. Himed, L.; Merniz, S.; Monteagudo-Olivan, R.; Barkat, M.; Coronas, J. Antioxidant activity of the essential oil of citrus limon before and after its encapsulation in amorphous SiO_2. *Sci. Afr.* **2019**, *6*, e00181. [CrossRef]
56. Katakojwala, R.; Mohan, S.V. Microcrystalline cellulose production from sugarcane bagasse: Sustainable process development and life cycle assessment. *J. Clean. Prod.* **2020**, *249*, 119342. [CrossRef]
57. Yeng, L.C.; Wahit, M.U.; Othman, N. Thermal and flexural properties of regenerated cellulose(RC)/poly(3- hydroxybutyrate)(PHB)biocomposites. *J. Teknol.* **2015**, *75*, 107–112. [CrossRef]
58. Zhou, Y.; Zhang, X.; Zhang, J.; Cheng, Y.; Wu, J.; Yu, J.; Zhang, J. Molecular weight characterization of cellulose using ionic liquids. *Polym. Test.* **2021**, *93*, 106985. [CrossRef]
59. Rein, D.M.; Khalfin, R.; Cohen, Y. Cellulose as a novel amphiphilic coating for oil-in-water and water-in-oil dispersions. *J. Colloid Interface Sci.* **2012**, *386*, 456–463. [CrossRef] [PubMed]
60. Alfassi, G.; Rein, D.M.; Cohen, Y. Enhanced hydrolysis of cellulose hydrogels by morphological modification. *Bioprocess Biosyst. Eng.* **2017**, *40*, 1635–1641. [CrossRef] [PubMed]
61. Sharma, S.; Barkauskaite, S.; Jaiswal, A.K.; Jaiswal, S. Essential oils as additives in active food packaging. *Food Chem.* **2021**, *343*, 128403. [CrossRef]
62. Shemesh, R.; Krepker, M.; Goldman, D.; Danin-Poleg, Y.; Kashi, Y.; Nitzan, N.; Vaxman, A.; Segal, E. Antibacterial and antifungal LDPE films for active packaging. *Polym. Adv. Technol.* **2015**, *26*, 110–116. [CrossRef]
63. Phala, K.; Mapossa, A.B.; Augustyn, W.; Combrinck, S.; Both, B.a. Development of EVA and LLDPE polymer-based carvone and spearmint essential oil release systems for citrus postharvest diseases applications. *Arab. J. Chem.* **2023**, *16*, 104458. [CrossRef]
64. Yamashita, C.; Chung, M.M.S.; Santos, C.D.; Mayer, C.R.M.; Moraes, I.C.F.; Branco, I.G. Microencapsulation of an anthocyanin-rich blackberry (Rubus spp.) by-product extract by freeze-drying. *LWT* **2017**, *84*, 256–262. [CrossRef]
65. Pasrija, D.; Ezhilarasi, P.N.; Indrani, D.; Anandharamakrishnan, C. Microencapsulation of green tea polyphenols and its effect on incorporated bread quality. *LWT* **2015**, *64*, 289–296. [CrossRef]
66. Cheng, A.-W.; Xie, H.-X.; Qi, Y.; Liu, C.; Guo, X.; Sun, J.-Y.; Liu, L.-N. Effects of storage time and temperature on polyphenolic content and qualitative characteristics of freeze-dried and spray-dried bayberry powder. *LWT* **2017**, *78*, 235–240. [CrossRef]
67. Shemesh, R.; Krepker, M.; Natan, M.; Danin-Poleg, Y.; Banin, E.; Kashi, Y.; Nitzan, N.; Vaxman, A.; Segal, E. Novel LDPE/halloysite nanotube films with sustained carvacrol release for broad-spectrum antimicrobial activity. *RSC Adv.* **2015**, *5*, 87108–87117. [CrossRef]
68. Bezerra, F.M.; Carmona, O.G.; Carmona, C.G.; Lis, M.J.; de Moraes, F.F. Controlled release of microencapsulated citronella essential oil on cotton and polyester matrices. *Cellulose* **2016**, *23*, 1459–1470. [CrossRef]
69. Thermo Fisher Scientific. Eugenol, 4-Allyl-2-methoxyphenol CAS No. 97-53-0. 2021. Available online: https://www.fishersci.com/store/msds?partNumber=AC119111000&countryCode=US&language=en (accessed on 10 September 2023).
70. National library of medicine. Thymol, 2-Isopropyl-5-methylphenol CAS No. 89-83-8. 2021. Available online: https://pubchem.ncbi.nlm.nih.gov/compound/Thymol (accessed on 10 September 2023).

polymers

MDPI

Article

New Materials from the Integral Milk Kefir Grain Biomass and the Purified Kefiran: The Role of Glycerol Content on the Film's Properties

Yuly A. Ramírez Tapias [1,2], Guillermo D. Rezzani [1,2], Juan F. Delgado [1,2], Mercedes A. Peltzer [1,2] and Andrés G. Salvay [1,*]

1 Laboratorio de Obtención, Modificación, Caracterización y Evaluación de Materiales, Departamento de Ciencia y Tecnología, Universidad Nacional de Quilmes, Roque Sáenz Peña 352, Bernal B1876BXD, Argentina; yuly.tapias@unq.edu.ar (Y.A.R.T.); guillermo.rezzani@unq.edu.ar (G.D.R.); mercedes.peltzer@unq.edu.ar (M.A.P.)

2 Consejo Nacional de Investigaciones Científicas y Técnicas (CONICET), Godoy Cruz 2290, Buenos Aires C1425FQB, Argentina

* Correspondence: asalvay@unq.edu.ar

Abstract: Microbial exopolymers are gaining attention as sources for the development of biodegradable materials. Milk kefir, a fermented dairy product produced by a symbiotic community of microorganisms, generates milk kefir grains as a by-product, consisting of the polysaccharide kefiran and proteins. This study develops two materials, one from whole milk kefir grains and another from purified kefiran. Film-forming dispersions were subjected to ultrasonic homogenisation and thermal treatment, yielding homogeneous dispersions. Kefiran dispersion exhibited lower pseudoplastic behaviour and higher viscous consistency, with minimal effects from glycerol. Both films exhibited continuous and homogeneous microstructures, with kefiran films being transparent and milk kefir films displaying a yellowish tint. Analysis revealed that milk kefir films comprised approximately 30% proteins and 70% kefiran. Kefiran films demonstrated stronger interpolymeric interactions, as evidenced using thermogravimetric and mechanical tests. Glycerol increased hydration while decreasing thermal stability, glass transition temperature, elastic modulus, and tensile strength in both films. However, in kefiran films, elongation at the break and water vapour permeability decreased at low glycerol content, followed by an increase at higher plasticiser contents. This suggests an unusual interaction between glycerol and kefiran in the absence of proteins. These findings underscore differences between materials derived from the whole by-product and purified kefiran, offering insights into their potential applications.

Keywords: polysaccharide-based films; biomass-based films; milk kefir; kefiran; plasticisation; antiplasticisation; glycerol

1. Introduction

Natural polymers have been continuously explored for applications in materials science due to their biodegradable nature and renewability, which makes them attractive in the context of the development of more environmentally friendly materials and processes [1–3]. Despite their hydrophilicity, which presents challenges in certain applications, biopolymeric materials derived from proteins and polysaccharides have been extensively studied. These investigations have shown them to be promising alternatives, particularly in the field of food packaging, to helping develop eco-friendly solutions [4–6]. Materials derived from microbial exopolysaccharides, either purified or combined with the biomass components from which they originate, have garnered significant attention due to their versatility and potential to replace conventional non-biodegradable materials [7–10]. This growing interest reflects a continuing shift towards packaging solutions that are not only functional, but are also sustainable, helping to reduce the use of traditional plastic packaging.

Plasticisers are essential for enhancing the integrity and mechanical properties of biopolymeric films [1,4,6]. Typically, plasticisers are small molecules, with glycerol—a single triol compound—being the most commonly used in biopolymer-based films due to its good miscibility and low cost [1,11]. It has been suggested that plasticisers destabilise interpolymeric hydrogen bonds, reducing intermolecular forces, and thereby increasing the mobility and the space between polymer chains [6,7,11]. Consequently, plasticisers can modify mechanical properties, such as reducing tensile strength and hardness, while increasing the elongation at break and the flexibility and fracture resistance of the polymer matrix [6,9]. In contrast, plasticisers usually increase the hydration of the film and decrease the water vapour barrier properties [7,12]. Despite these general trends, further investigation is needed to fully understand the complex interactions between biopolymers and plasticisers, which could clarify unresolved questions regarding film behaviour [11], including the mechanisms behind antiplasticisation [13–15].

Exopolysaccharides are extracellular polysaccharides produced by many bacteria and secreted by specific membrane proteins [16,17]. These exopolysaccharides typically have a high average molecular weight and high polydispersity, accumulating extracellularly and imparting a gelatinous appearance to the culture [17]. Milk kefir, a fermented dairy beverage produced by a symbiotic community of bacteria and yeasts, generates a by-product known as milk kefir grains, primarily consisting of kefiran exopolysaccharide with a secondary fraction of proteins [10,18]. Kefiran is a branched heteropolysaccharide, slightly yellow, water-soluble, and composed of approximately equal amounts of D-glucose and D-galactose [8,10,19,20]. It is typically produced by several *Lactobacillus* species including *L. kefiranofaciens*, *L. kefirgranum*, and *L. parakefir*, as well as by other unidentified *Lactobacillus* species [19]. These bacteria excrete kefiran polysaccharides with elevated molecular weights and polydispersity [8,18], ranging from 50 to 15,000 kDa [8]. Kefiran has attracted significant attention due to its unique properties, including rheological and mechanical behaviours, antioxidant and biocide activities, and health-promoting effects [20].

Kefiran is a neutral polysaccharide with numerous hydroxyl groups, which gives it a polar and hydrophilic character [21]. Previous research has demonstrated the capacity of kefiran to form films with promising functional properties, making it suitable for applications such as food packaging [21–26]. However, several aspects regarding the development and properties of kefiran-based films require further improvements and clarification. For instance, these films were produced from dilute kefiran dispersions of 1 wt% [21,22] and 2 wt% [23–26], which may lead to excessively thin films or high drying energy demands [27]. Furthermore, kefiran-based films have been reported to exhibit unexpected behaviours, such as a decrease in water vapour permeability with the addition of glycerol [22] and an increase in elastic modulus at low glycerol concentrations [26]. Moreover, the question of utilising the whole biomass of milk kefir grains for film production remains unexplored. This is relevant, as using the integral by-product of fermentations for film production could prove to be cost-effective, eliminating the need for separation and purification, and may result in films with remarkable physical-chemical properties and the retention capacity of natural bioactive substances [7,9,10].

To address these challenges, this study aimed to develop films using both the integral biomass of milk kefir grains and purified kefiran. The films were obtained from film-forming dispersions containing 5 wt% of milk kefir grains and 5 wt% kefiran, which were subjected to sequential physical treatments including ultrasonic homogenisation and thermal processing. To study the role of the plasticiser, both types of films were produced with varying glycerol concentrations. The rheological properties of the film-forming dispersions were studied, and the resulting films were characterised in terms of colour, microstructure, spectroscopy, thermal properties, mechanical strength, hydration, and water vapour permeability.

2. Materials and Methods

2.1. Materials

Milk kefir grains LOMCEM SMK1 were acquired from a household in La Plata, Argentina, and stored frozen at −20 °C. Homogenised commercial milk with a standardised fat content of 3% (Ilolay, Santa Fe, Argentina) and pharmaceutical-grade ethanol 96% *v*/*v* (Bialcohol, Córdoba, Argentina) was purchased at a local market. Silica gel, analytical grade glycerol, and analytical-grade salts used for the preparation of saturated solutions were acquired from Biopack (Zárate, Argentina). Kjeldahl and thin-layer chromatography (TLC) reagents were obtained from Sigma (St. Louis, MO, USA).

2.2. Milk Kefir Grains and Culture Conditions

Milk kefir grains were reactivated through successive subcultures at 22 °C in 1 L of milk cultivation medium containing around 100 g of kefir grains. The medium was exchanged daily with fresh culture medium to sustain grain viability. This process was repeated multiple times to increase the biomass of milk kefir grain, resulting in a fivefold increase after ten subcultures. For film preparation and kefiran purification, the grains were separated from the fermented liquid by filtration using a plastic sieve, and were washed five times by immersion in 2 L of distilled water. The grains were then lightly pressed to remove excess water. The dry matter (d.m.) content of the washed and pressed milk kefir grains was 0.15 g per g, determined by drying at 105 °C. The protein content in the dried milk kefir grains was 30 ± 1%, measured with the Kjeldahl method [28] using a Kjeltec® 8100 distillation module (Foss, Hillerød, Denmark) coupled with a DT2508 digestor module and a SR210 scrubber (Foss, Hillerød, Denmark).

2.3. Kefiran Isolation and Purification

Kefiran was extracted from the washed and pressed milk kefir grains, and was purified using a procedure based on the method developed by Rimada and Abraham [29], with slight modifications. Briefly, the washed and pressed milk kefir grains were treated in boiling water at a ratio of 1:10 *w*/*w* for 45 min with continuous stirring. The mixture was then cooled and centrifuged (Avanti J-26 XP centrifuge, Beckman, Brea, CA, USA) at 10,000× *g* for 20 min at 20 °C to precipitate microbial cells and proteins. The supernatant was collected, and two volumes of ethanol cooled to −20 °C were added to precipitate the polysaccharide. The entire mixture was stored overnight at −20 °C. Then, the precipitated polysaccharide was separated by centrifugation at 10,000× *g* and 4 °C for 20 min. The resulting pellets were dissolved in hot distilled water, and the precipitation procedure was repeated twice. Finally, the precipitate was dissolved in hot distilled water, and the resulting kefiran solution was freeze-dried. The absence of other simple sugars (mono or di saccharides) in the samples was confirmed using qualitative thin-layer chromatography (TLC) following the methodology published by Piermaria et al. [18]. The protein concentration in the freeze-dried kefiran, determined using the Kjeldahl method [28], was found to be less than 0.1%.

2.4. Preparation of Film-Forming Dispersions of Milk Kefir and Kefiran

The washed and pressed milk kefir grains were used to prepare a dispersion containing 5 wt% d.m. in distilled water. The procedure for obtaining the milk kefir film-forming dispersion was similar to that previously developed in our laboratory for the production of water kefir film-forming dispersions [7,30]. Briefly, the dispersion was subjected to high-speed homogenisation at 18,000 rpm for 5 min using an Ultraturrax T-25 device (IKA Works, Inc., Staufen, Germany) to disrupt the grain structure. This was followed by ultrasonic homogenisation at 80 W with an Ultrasonic processor VCX-750 (Sonics and Materials, Inc., Newtown, CT, USA) for 15 min, using cycles of 30 s of pulsing and 30 s of rest to ensure complete disintegration of the grains. Thermal treatment at 90 °C in a water bath was then applied for 20 min to unfold the biopolymers and deactivate any residual enzymes and microorganisms. Following this, the dispersion underwent a second high-speed homogenisation at 15,000 rpm for 1 min to break down any aggregates formed during the thermal treatment. Finally, a second ultrasonic homogenisation was performed under the same conditions as the first to ensure the production of a fine dispersion.

The freeze-dried kefiran was used to prepare a dispersion containing 5 wt% d.m. in distilled water. This dispersion was subjected to the same ultrasound–temperature–ultrasound treatments as the milk kefir dispersion to obtain the kefiran film-forming dispersion.

Both film-forming dispersions were exposed to a complete degassing process for 30 min using a vacuum pump. Subsequently, pure glycerol was added to dispersions at levels of 0, 10, 20, and 30 wt% with respect to d.m., followed by stirring for 15 min.

2.5. *Preparation of Milk Kefir Films and Kefiran Films*

To achieve films with a thickness of approximately 0.15 mm, 20 g of the film-forming dispersion were placed in each plastic Petri dish of 86 mm in diameter. Water evaporation was performed at 40 °C and 40% relative humidity (r.h.) by casting in a ventilated oven (Sanyo MOV 212F, Moriguchi, Japan) until the water content in the films reached between 10 and 15%, a process that took around 12 h. Subsequently, the films were stored at 22 °C and 43% r.h. For experimental purposes, the films were then equilibrated at 22 °C in desiccators at various r.h. levels, achieved using saturated solutions of NaOH, $MgCl_2$, K_2CO_3, $Mg(NO_3)_2$, NaBr, NaCl, and $BaCl_2$, to generate atmospheres of 10, 33, 43, 52, 57, 75, and 90% r.h., respectively. Dried atmospheres were created using silica gel.

2.6. *Characterisation*

2.6.1. Rotational Rheology Measurements of the Film-Forming Dispersions

Flow curves were acquired through rotational experiments utilising an AR-G2 rheometer (TA Instruments, New Castle, DE, USA) fitted with a 2° steel cone geometry (40 mm in diameter and 55 µm of truncation gap). Measurements were conducted in triplicate at 22 °C using 0.6 mL of film-forming dispersions, with shear rates ranging from 0.005 to 1000 s^{-1}. Shear stress τ (Pa) as a function of shear rate γ (s^{-1}) was recorded, and the resulting curves were fitted with the Herschel–Bulkley model as described in Equation (1):

$$\tau = \tau_0 + K\,\gamma^n \qquad (1)$$

where τ_0 (Pa) is the yield stress, representing the maximum value of τ for a strain rate equal to zero, *K* is the fluid consistency index, linked to the apparent viscosity of the dispersion, and *n* is the flow behaviour index indicating the deviation to Newtonian flow type ($n > 1$ for dilatant and $n < 1$ for pseudoplastic fluids). The experimental data were fitted using OriginPro 8 (OriginLab Corporation, Northampton, MA, USA).

2.6.2. Visual Appearance, Quality Evaluation, Thickness Measurements, and Density of the Films

The visual examination of the films was conducted using a Samsung SM-A145M camera, capturing photographs of the sample surfaces from a height of 20 cm under natural light. The quality assessment of the films, focused on handleability, homogeneity, and continuity, was evaluated in three independent replicates [30,31]. The thickness of the films was measured with a digital calliper ($\pm 10^{-6}$ m; 3109-25-E, Insize Co., Suzhou, China). Measurements were taken at 20 distinct positions on each film, resulting in an average thickness of around 0.15 mm per specimen. For density determination of the dried films, circular samples with an area of 58 cm^2 were dehydrated in containers with silica gel for 10 days. The films were then weighed using an analytical balance ($\pm 10^{-4}$ g). The density of the dried film $\rho_{d.f.}$ (g m^{-3}) was calculated using Equation (2):

$$\rho_{d.f.} = m/(AL) \qquad (2)$$

where *m* is the dry mass (g), *A* the area (m^2), and *L* is the thickness (m).

2.6.3. Colour Determination (CIELab Coordinates) of the Films

The colour of the samples was assessed with a Konica Minolta CR-400 colourimeter (Tuscaloosa, AL, USA) using illuminant C and a 2° observer following the CIE 1931 standard [32]. The films were placed on a white background, and the CIELab colour space was utilised to determine the parameters L^* (lightness), a^* (green to red), and b^* (blue to yellow). The total colour change ΔE was calculated using Equation (3):

$$\Delta E = \sqrt{(L^* - L_0)^2 + (a^* - a_0)^2 + (b^* - b_0)^2} \tag{3}$$

where L_0, a_0, and b_0 represent the coordinates corresponding to the unplasticised kefiran film, considered the reference to determine the colour change. Measurements were taken at five different points for each formulation.

2.6.4. Microstructural Characterisation Using Scanning Electron Microscopy (SEM)

The microstructure of the films was investigated by observing the surfaces and cross-sections using a scanning electron microscope FEI-Quanta 200 (Fei Co., Hillsboro, OR, USA) operated at 15 kV. Cross-sections were prepared by cutting the samples at −20 °C with a sharp blade. Subsequently, all of the samples were placed in the specimen holder and stored at 22 °C and 43% r.h. To enhance visibility under the microscope, the samples were coated with a layer of gold. Images of the surfaces (magnification 3000×) and cross-sections (magnification 1000× and 20,000×) of the films were acquired under high-vacuum conditions.

2.6.5. Attenuated Total Reflectance-Fourier Transform Infrared (ATR-FTIR) Spectroscopy Analyses of the Films

Infrared spectra of films were recorded in the range of 4000–500 cm^{-1} using a Fourier-Transform Infrared Analyser (FTIR) Shimadzu IR-Affinity (Shimadzu Co., Kyoto, Japan) equipped with an attenuated total reflectance monolithic diamond crystal module (GladiATR, Pike Technologies, Madison, WI, USA). The spectra were obtained by averaging 48 scans with a resolution of 4.0 cm^{-1} and Happ-Genzel apodisation. A blank spectrum was recorded before each test to account for humidity and carbon dioxide in the air by subtracting their effects from the spectra. Each measurement was performed in triplicate.

2.6.6. Thermogravimetric Analyses (TGA) of the Films

To investigate the thermal degradation of the films, the weight loss of the samples as a function of temperature was recorded using a TA Instruments Q-500 thermogravimetric analyser (New Castle, DE, USA). Before the experiment, the samples were equilibrated at 22 °C and 52% r.h. Approximately 3 mg of each sample was placed in a platinum sample pan and heated from 40 to 800 °C at a rate of 10 °C per minute. The experiments were carried out in triplicate under a nitrogen atmosphere with a flow rate of 60 mL per minute. The initial degradation temperature (T_0) was defined as the temperature at which 15% of the weight had been lost. The temperature corresponding to the maximum degradation rate (T_{max}) was obtained from the peak of the derivative of the weight loss with respect to temperature using the TA Universal Analysis software (v4.5, TA Instruments, New Castle, DE, USA).

2.6.7. Differential Scanning Calorimetry (DSC) Studies of the Films

The thermal behaviour of films under controlled heating was analysed using a differential scanning calorimeter (TA Instruments Q200, New Castle, DE, USA). Film samples were placed into Tzero® aluminium pans (TA Instruments, New Castle, DE, USA) and dehydrated at 22 °C in containers with silica gel for 10 days. Subsequently, the pans were rapidly sealed with hermetic lids, each containing approximately 3 mg of dried samples. Thermograms were obtained within a range from −80 °C to 180 °C, with an initial equilibration step at −80 °C for 1 min, followed by a temperature increase at a rate of 10 °C per minute. Glass transition temperatures (T_g) were determined at the midpoint using the TA Universal Analysis software (v4.5, TA Instruments, New Castle, DE, USA). The experiments were conducted in triplicate.

2.6.8. Mechanical Uniaxial Tensile Experiments of the Films

Uniaxial tensile tests of the films were performed using a Universal Testing Machine (TC-500 II-Series, Micrometric, Buenos Aires, Argentina) equipped with a 300 N cell. Rectangular samples of 50 mm in length and 10 mm in width were prepared. The samples were equilibrated in an atmosphere at 52% of r.h. and then placed between the machine's jaws with an effective gauge length of 25 mm. The testing speed was set at 5 mm per minute, and ten specimens of each composition were tested at 22 °C. Elongation at break ($e_{\%}$, %), maximum tensile strength ($TS_{\max}$, MPa), and elastic modulus (Y, MPa) were calculated from the resulting stress–strain curves, with averages taken from ten replicates, following the guidelines of ASTM D882 1997 [33].

2.6.9. Water Sorption Isotherms of the Films

Water sorption isotherms were determined gravimetrically at 22 °C, following the standard procedure previously described [20]. Dried film samples, each with a superficial area of 58 cm^2, were placed in 3 L desiccators at different water activities (a_w). The samples were periodically weighed using an analytical balance ($\pm 10^{-4}$ g), and the moisture content was monitored at each condition until a constant weight was achieved. The water content, or hydration (h), was expressed in units of g of water per g of d.m. and was evaluated as a function of a_w (a_w = % r.h./100). Experiments were conducted in triplicate, and the resulting isotherms were fitted using the Guggenheim–Anderson–De Boer (GAB) model [34], as shown in Equation (4):

$$h(a_w) = \frac{Ncka_w}{[(1 + (c-1)ka_w)(1 - ka_w)]} \tag{4}$$

where N represents the monolayer water content (g of water per g of d.m.), which is associated with the primary binding sites of hydration water molecules; c is a parameter related to the sorption heat of the monolayer, reflecting the binding force of water to the monolayer; while k is related to sorption heat multilayer, which is linked to the ability of water to bind to the multilayer. For each sample formulation, one water sorption isotherm was obtained, and the water content h corresponding to each a_w was determined as the mean of three experimental measurements, including their respective errors. The values of the parameters N, c, and k, along with their associated errors, were derived by fitting the experimental data points to the GAB model using OriginPro 8 (OriginLab Corporation, Northampton, MA, USA).

2.6.10. Experimental Water Vapour Permeability Measurements

The experimental water vapour permeability (${P_w}^{exp}$) of the films was determined using the cup method described in ASTM-E96 2016 [35], with some modifications [12]. The films were sealed on the top of cups containing a saturated salt solution of $BaCl_2$, providing a high r.h. of 90%. The test cups were placed in 7 L desiccators maintained at a constant temperature of 22 °C and 10% r.h. To ensure uniform conditions inside the desiccators and over the films, a fan was installed, as previously recommended [36]. The weight loss of the test cups, indicative of water vapour transport through the film, was monitored using an analytical balance ($\pm 10^{-3}$ g). Weight loss m (g) versus time t (min) was plotted, and once the steady state (indicated by a linear trend) was established, data were recorded for 40 h further. ${P_w}^{exp}$ was determined as displayed in Equation (5):

$$P_w^{exp} = \left(\frac{1}{A}\frac{\Delta m}{\Delta t}\right)\frac{L}{\Delta p_w} \tag{5}$$

where A = 2.2 $\times 10^{-3}$ m^2 is the effective area of the exposed film, $\Delta m/\Delta t$ denotes the slope of a linear regression of the weight loss versus time, L (m) is the film thickness, $\Delta p_w = (p_{w2} - p_{w1})$ is the differential water vapour pressure across the film, and p_{w2} and p_{w1} are the partial pressures (Pa) of water vapour at the film surfaces inside and outside the cup, respectively [37]. $P_w{}^{exp}$ is given in units of g s^{-1} m^{-1} Pa^{-1}. Experiments were performed in triplicate.

2.6.11. Determination of Effective Water Solubility and Effective Water Diffusion in the Films

When there are no pores, faults, or punctures in the film, $P_w{}^{exp}$ can be expressed as Equation (6) [38,39]:

$$P_w^{exp} = D_w^{eff} S_w^{eff} \tag{6}$$

where $S_w{}^{eff}$ (g m^{-3} Pa^{-1}) is the effective water solubility coefficient over the concentration range c_{w2} to c_{w1}, corresponding to water vapour pressures p_{w2} and p_{w1}, respectively, and $D_w{}^{eff}$ (m^2 s^{-1}) is the effective diffusion coefficient over the water concentration range c_{w2} to c_{w1}. Water sorption isotherms were used to evaluate the water concentration c_w ($c_w = h(a_w) \times \rho_{d.f.}$) of each film specimen surface in the permeability experiment. $S_w{}^{eff}$, corresponding to the pressure gradient or a_w interval [a_{w2} = 0.9; a_{w1} = 0.1], was obtained using Equation (7) [37]:

$$S_w^{eff} = [(h(a_{w2}) - h(a_{w1}))/(p_{w2} - p_{w1})] \times \rho_{d.f.} \tag{7}$$

where $h(a_{w2})$ and $h(a_{w1})$ are the water content of the film (g H_2O per g d.m.) at its underside surface at p_{w2} and its surface outside the cup at p_{w1}, respectively. According to Equation (6), $D_w{}^{eff}$ was then calculated using Equation (8):

$$D_w^{eff} = P_w^{exp}/S_w^{eff} \tag{8}$$

2.7. Statistical Analyses

Statistical analyses were performed using OriginPro 8 (OriginLab Corporation) and R software (version 3.4.4, R Foundation for Statistical Computing, Vienna, Austria). All results are expressed as means with standard deviation. An analysis of variance was performed on the data, and a post hoc test (Tukey HSD) was used to compare the means. Differences were deemed significant at $p < 0.05$. The errors in the parameters from the Herschel–Bulkley and GAB models, derived from the flow curves and water sorption isotherms, respectively, were estimated using a fit analysis.

3. Results and Discussion

3.1. Rotational Rheology of Film-Forming Dispersions

Understanding the rheological properties of film-forming dispersions is crucial, as they dictate the processing conditions and machinability necessary for large-scale film production [22]. The flow behaviour of milk kefir and kefiran film-forming dispersions is shown in Figure 1. All of the studied dispersions were homogeneous, exhibiting no phase separation. The Herschel–Bulkley model (Equation (1)) was used to analyse the shear stress τ versus the shear rate γ data, and the resulting model parameters are summarised in Table 1. High values of the statistical parameter R^2 indicate a good fit of the model to the experimental data.

All of the studied dispersions exhibited pseudoplastic behaviour ($n < 1$) beyond their yield stress τ_0. As illustrated in Table 1, this shear-thinning behaviour was more pronounced in dispersions prepared from milk kefir grains. Milk kefir dispersions displayed significantly lower shear stress values compared to kefiran dispersions, as shown in Figure 1a,b. The consistency index K, directly related to the apparent viscosity, was at least three times higher for kefiran dispersions (Table 1). Additionally, kefiran dispersions exhibited higher τ_0 compared to milk kefir dispersions (Figure 1 and Table 1). Fluids with

$\tau_0 > 0$ display solid-like behaviour until the yield stress is surpassed, after which they transition to liquid-like behaviour when $\tau > \tau_0$ [40]. The higher τ_0 values suggest stronger interpolymeric interactions within the dispersion, consistent with observations reported in polysaccharide dispersions [30,32,41]. These results imply that the presence of proteins in milk kefir samples weakens the interactions between kefiran chains, leading to dispersions with a lower consistency index and yield stress.

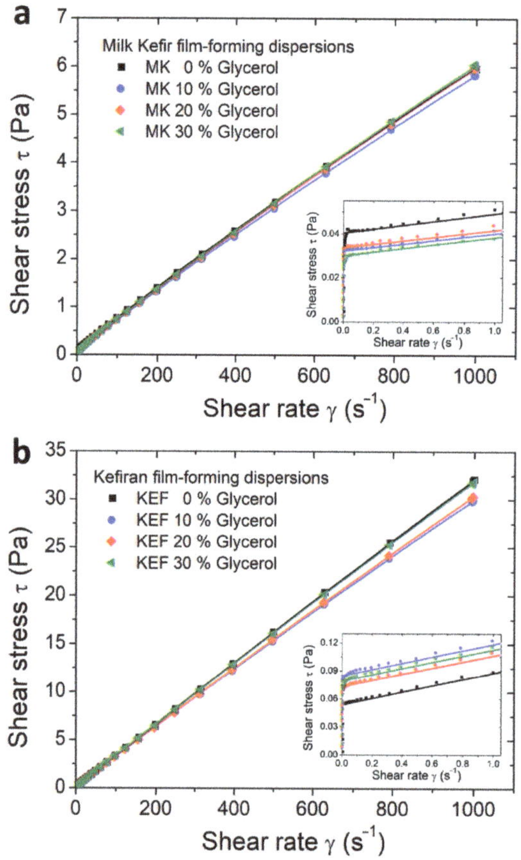

Figure 1. Rotational rheology of film-forming dispersions with different contents of glycerol. (**a**) Flow curves of milk kefir dispersions. (**b**) Flow curves of kefiran dispersions. Flow curves were fitted using the Herschel–Bulkley model (Equation (1)). Fitted parameters are shown in Table 1. The low-shear-rate region is shown in the insert of the figures.

As observed in Table 1, the incorporation of glycerol into the dispersions had minimal effects on the K and n parameters of both dispersions. Previous studies on 1 wt% kefiran dispersions reported that the parameters K and n remained unchanged upon the addition of glycerol at concentrations of 0, 25, and 50 wt% relative to d.m. [22]. Similarly, other researchers reported no change in the apparent viscosity for a 2 wt% kefiran dispersion after the addition of glycerol or sorbitol at 25 wt% relative to d.m. [42]. Interestingly, the effect of glycerol content on τ_0 differed between milk kefir and kefiran dispersions (Table 1): τ_0 increased with increasing glycerol content for kefiran dispersions, whereas it decreased for milk kefir dispersions.

Table 1. Parameters obtained from fitting the Herschel–Bulkley model (Equation (1)) to the flow curves of Figure 1. MK denotes milk kefir and KEF. Errors in the parameters were estimated from the fit analysis.

Sample	τ_0 (10^{-3} Pa)	K (10^{-3} Pa·s)	n	R^2
MK 0% Glycerol	42 ± 2	10.9 ± 0.2	0.916 ± 0.002	0.9999
MK 10% Glycerol	32 ± 2	9.6 ± 0.2	0.928 ± 0.003	0.9999
MK 20% Glycerol	33 ± 2	9.4 ± 0.2	0.932 ± 0.003	0.9999
MK 30% Glycerol	30 ± 2	9.3 ± 0.2	0.933 ± 0.003	0.9999
KEF 0% Glycerol	54 ± 3	34.8 ± 0.2	0.988 ± 0.001	0.9999
KEF 10% Glycerol	84 ± 5	36.3 ± 0.3	0.972 ± 0.001	0.9999
KEF 20% Glycerol	73 ± 4	34.8 ± 0.2	0.980 ± 0.001	0.9999
KEF 30% Glycerol	79 ± 4	34.7 ± 0.2	0.987 ± 0.001	0.9999

A more pronounced shear-thinning behaviour (lower n-values) was observed in 1 wt% kefiran dispersions studied by other authors [22] compared to the kefiran dispersions investigated in the present work. However, these authors reported consistency index K values similar to those shown in Table 1 for 5 wt% kefiran dispersions. While K typically increases with polymer concentration [43], it is important to note that the 5 wt% kefiran dispersions studied in the present work were subjected to sonication. Sonication is known to break polysaccharide chains and reduce their molecular weight [44,45], thereby weakening their gelling ability and enhancing dispersibility [46]. As a result, the ultrasound treatments likely reduced K in the 5 wt% kefiran dispersions, bringing it in line with that of non-sonicated 1 wt% kefiran dispersions.

High viscosity in film-forming dispersions can hinder air bubble removal, resulting in film defects such as pores and holes [4,47]. Conversely, low viscosity in dilute dispersions can lead to excessively thin films or require higher drying energy [27]. This study aims to strike a balance by formulating a dispersion with a high biopolymer concentration, yet a consistency index K comparable to that of a low-concentration dispersion, thus minimising these issues.

3.2. Assessment of Quality, Visual Appearance, and Microstructure of the Films

Visual inspection revealed homogeneous, continuous, and smooth films with uniform surfaces, free of air bubbles, cracks, and pores (Figure 2a,b). Plasticised films containing 10, 20, and 30 wt% of glycerol (relative to d.m.) displayed similar visual appearances compared to unplasticised films. However, film flexibility increased with rising glycerol content. Unplasticised milk kefir films were brittle, difficult to handle, and required careful removal from the Petri dish. Conversely, milk kefir films with added glycerol, as well as both unplasticised and plasticised kefiran films, were flexible and easy to peel and manipulate. This enhanced flexibility is likely due to the effective dispersion of the small glycerol molecules. It has been reported that glycerol intercalates between polymer chains, disrupting interpolymer interactions and increasing chain separation, leading to a more flexible film [7,48].

Milk kefir films displayed a yellow tint while remaining translucent (Figure 2a), whereas kefiran films exhibited high transparency (Figure 2b). Transparency and colour are crucial attributes of packaging materials, as they influence consumer perception of the product. Transparent films allow for product visibility, while coloured films could offer protection from radiation [30]. The CIELab colour parameters and colour representations for unplasticised milk kefir films and kefiran films are shown in Figure 2c,d, respectively. The addition of glycerol did not significantly alter these parameters in either film type. High L^* values, particularly in kefiran films, indicated excellent clarity and transparency. Milk kefir films showed a slight reduction in L^*, exhibiting a trend towards yellow (increased b^* values), likely due to Maillard reactions between milk kefir proteins and carbohydrates. Milk kefir films had ΔE values greater than 6, indicating that the colour difference with kefiran films is perceptible to the human eye [30], which matched visual observations.

Notably, the remarkable transparency of the kefiran films in this study exceeded that reported for films made from non-sonicated 2 wt% kefiran dispersions [23,24]. This suggests that sonication of the kefiran film-forming dispersion before casting resulted in a highly translucent matrix.

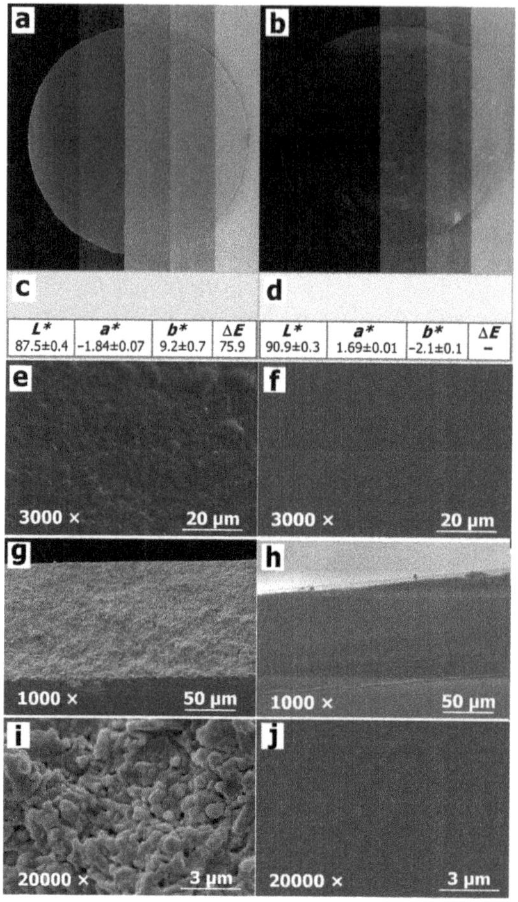

Figure 2. Visual appearance, colour, and microstructure of the unplasticised films. Photographs of (**a**) milk kefir film and (**b**) kefiran film; CIELab colour parameters and colour representation of (**c**) milk kefir film and (**d**) kefiran film; SEM micrographs of the surface of films at 3000× of (**e**) milk kefir film and (**f**) kefiran film; SEM micrographs of the cross-sections of films at 1000× of (**g**) milk kefir film and (**h**) kefiran film; SEM micrographs of the cross-sections of films at 20,000× of (**i**) milk kefir film and (**j**) kefiran film.

SEM was utilised to analyse the microstructure of the films, as shown in Figure 2e–j. The micrographs revealed continuous and homogeneous surfaces and cross-sections in both film types, devoid of pores or punctures. However, notable morphological differences were observed. Milk kefir films exhibited a rougher texture, possibly due to the presence of proteins, indicating a more rubber-like character. In contrast, kefiran films displayed a smoother surface and a more uniform structure. The concentration of glycerol did not appear to influence the microstructure as observed in SEM, indicating good compatibility between the plasticiser and both film matrices. This observation is consistent with findings from previous studies which reported no significant microstructural differences in kefiran films containing different amounts of glycerol or sorbitol [23,24]. Furthermore, X-ray diffraction analyses reported similar amorphous structures with crystallinity below 3% for both plasticised and unplasticised kefiran films [21,22]. It has been suggested that the amorphous nature of kefiran likely contributes to its excellent miscibility with glycerol within the film matrix [42].

3.3. Infrared Spectroscopy of the Films

ATR-FTIR spectroscopy was employed to identify characteristic functional groups within the films. Figure 3 shows the ATR-FTIR spectra of unplasticised milk kefir film and kefiran films. For clarity, the spectra have been normalised to the major peak at 1028 cm^{-1}. Both films displayed common bands, but significant differences were observed in the 1800–1200 cm^{-1} range, likely due to proteins present in the milk kefir films. Plasticisation with 10, 20, or 30 wt% glycerol resulted in similar features for both unplasticised milk kefir and kefiran films, with only increased band intensities observed. This aligns with findings for glycerol-plasticised kefiran films [21,42] and water kefir grain-based films [7], attributed to both increased water content and the presence of glycerol itself [7,8].

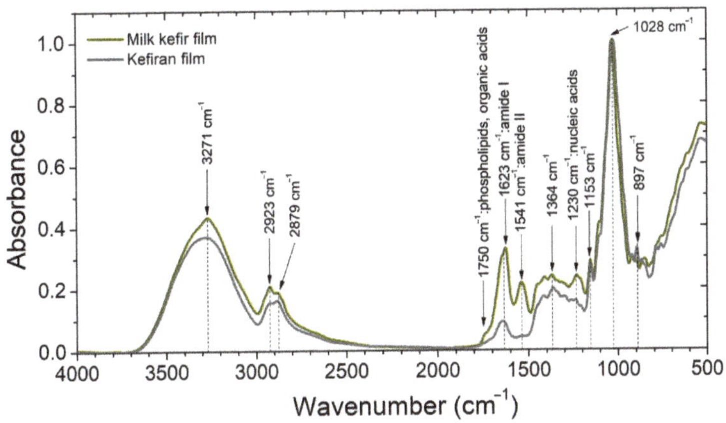

Figure 3. ATR-FTIR spectra of unplasticised films.

The kefiran spectrum shown in Figure 3 is similar to those reported for purified kefiran [26,49], kefiran films [21,42], and kefiran electrospun nanofibre materials [49]. This observation confirms the integrity of the kefiran units' chemical structure throughout the ultrasound treatment, casting, and electrospinning processes.

Analysis of the ATR-FTIR spectra (Figure 3) revealed several key features. The broad peak at 3271 cm^{-1}, assigned to the O-H stretching of water and polysaccharides [7], spanned the 3700–3000 cm^{-1} region. Milk kefir films exhibited a slightly sharper peak in this region, due to the presence of N-H bonds from proteins [30]. The bands observed at 2923 and 2879 cm^{-1} correspond to the symmetric and anti-symmetric stretching of C-H bonds in the methyl (CH_3) and methylene (CH_2) groups [21]. The differences observed between kefiran films and milk kefir films in the 3700–2700 cm^{-1} region can be attributed to the presence of proteins in milk kefir films. Similar observations have been reported in water kefir/yeast biomass blend films, suggesting that protein–polysaccharide interactions alter the chemical environment in this region, thereby affecting the stretching vibrations [30].

The most significant spectral differences were observed between 1800 and 1200 cm^{-1}, primarily due to proteins in the milk kefir films. These films displayed bands at 1623 and 1541 cm^{-1}, characteristic of amide I and II in proteins [30]. Additionally, peaks at 1750 and 1230 cm^{-1} were attributed to organic acids remnants of the fermentation, phospholipids, and nucleic acids from residual cell material in milk kefir films. Conversely, the kefiran film spectrum lacked these protein and cell-related peaks, but displayed a peak near 1640 cm^{-1}, likely due to O-H bending vibrations in water molecules [50]. Both films exhibited small peaks and shoulders between 1400 and 1300 cm^{-1}, assigned to variations in C–H bond angles [42,49].

The region between 1200 and 900 cm^{-1} displayed a prominent peak at 1028 cm^{-1} in both samples, a characteristic feature of kefiran [21]. Additional peaks corresponding to carbohydrate ring vibrations and functional groups (C–O–C, C–OH, C–H) were also detected. These peaks likely originate from the vibrational modes of glucose and galactose units within the kefiran structure, which is the primary component of milk kefir grains. Specifically, peaks at 1153 and 897 cm^{-1} were identified and assigned to vibrational modes of glucose, galactose, and β-linkages, indicative of the pure kefiran structure [21,26,51].

3.4. Thermogravimetric Analysis of the Films

To investigate thermal degradation, a thermogravimetric analysis was conducted on films that had been previously hydrated to equilibrium at 52% r.h. Figure 4 presents the mass loss curves as a function of the temperature, revealing distinct thermal degradation zones. The initial stage, occurring below 150 °C, was associated with dehydration [30] and resulted in approximately 10% weight loss.

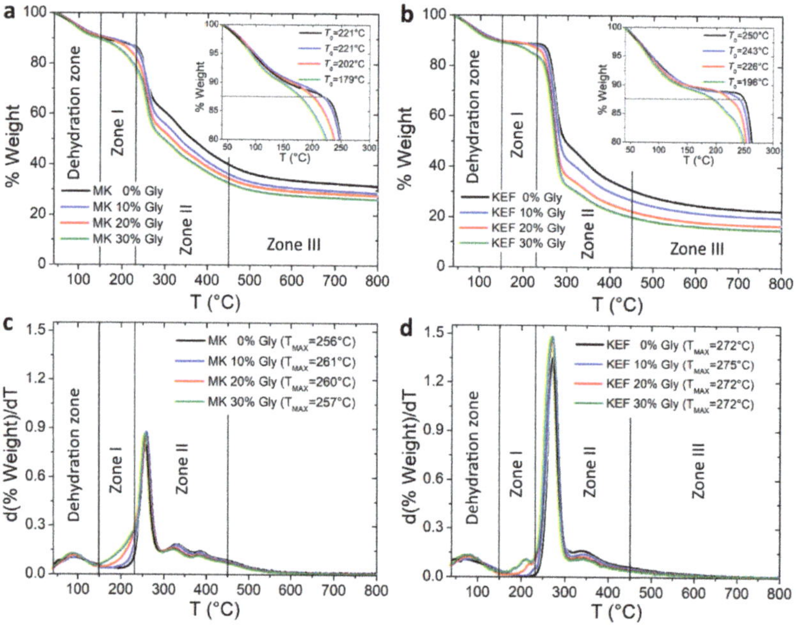

Figure 4. Thermogravimetric analysis of previously hydrated films at 43% r.h. (**a**) Mass loss of milk kefir (MK) films with different contents of glycerol. (**b**) Mass loss of kefiran (KEF) films with different contents of glycerol. The onset degradation temperatures (T_0) are displayed in the insert of the figures. (**c**) Derivative of mass loss of milk kefir films. (**d**) Derivative of mass loss of kefiran films.

A primary degradation zone (Zone I) was observed between 150 °C and 230 °C. Kefiran films exhibited higher thermal stability than milk kefir films, with this trend reflected in the initial degradation temperature (T_0). Glycerol incorporation reduced thermal stability in both film types. Thermal degradation in this zone was intensified with increasing glycerol content and was more pronounced for milk kefir films (Figure 4c,d). These results can be attributed to the degradation in glycerol that occurs within this temperature range [7,32] and to the onset of protein decomposition [30,52,53] present in milk kefir films.

The second degradation zone (Zone II), spanning from 230 °C to 450 °C, marked the maximum degradation rate of the samples. T_{max} values of 256 °C and 272 °C were determined for the unplasticised milk kefir and kefiran films, respectively (Figure 4c,d). The lower T_{max} of milk kefir films suggests the presence of weakened interpolymeric interactions due to the presence of proteins, which favour thermal degradation. This type of interaction between proteins and kefiran was also observed in the rheological tests of the dispersions, as discussed in Section 3.1. Additionally, the T_{max} value of unplasticised kefiran film in the present study was lower than the reported T_{max} values of 300 °C for purified kefiran [54] and 306 °C for unplasticised kefiran film obtained from a film-forming dispersion that was not sonicated [26]. It is noteworthy that the films in our study were prepared from sonicated dispersions. Ultrasound treatment is known to induce chain scission in polysaccharides [44], potentially leading to a decrease in polymer network entanglement. This reduction in entanglement density could lead to a less-rigid film matrix, thereby promoting thermal degradation and decreasing T_{max}. On the other hand, the incorporation of glycerol did not substantially affect T_{max}. Following T_{max}, both samples exhibit minor and scattered peaks of thermal degradation, especially in the milk kefir films. These secondary peaks were also reported for purified kefiran [54] and kefiran films [26].

A final weight loss occurred between 450 °C and 800 °C (Zone III), attributed to the pyrolysis of carbohydrates [32,55] and massive protein degradation [30]. As illustrated in Figure 4a,b, the final residue at 800 °C was influenced by the initial glycerol content. This gradual decrease in final carbon with the glycerol content was more pronounced in kefiran films and reflects the effect of glycerol on the film's structure. As described previously, the glycerol makes the film's structure more susceptible to thermal degradation under dynamic conditions [7].

3.5. Differential Scanning Calorimetry of the Films

To investigate thermal transitions, dehydrated films were subjected to DSC analysis. Figure 5a,b presents the DSC thermograms for milk kefir and kefiran films, respectively. The thermograms of milk kefir films show two thermal transitions for the unplasticised sample and three for the plasticised ones. All of the samples display two T_g values: one ranging from 28 to −23 °C, corresponding to the protein-rich zone of the blend, and the other from 54 to 48 °C, corresponding to the polysaccharide-rich zone. The decrease in the T_g value with glycerol addition aligns with previous studies that show a reduction in T_g when glycerol is incorporated into biopolymeric matrices [7,9,56,57]. The protein-rich zone, which is absent in kefiran films (Figure 4b), seems to be more influenced by glycerol addition. At higher temperatures, plasticised samples displayed an endotherm peak with a low heat of fusion, corresponding to the melting point of the blend. This finding is consistent with reports that kefiran has a small percentage of crystallinity [21,22]. The presence of glycerol affects this transition. It has been proposed that plasticisers reduce interpolymeric forces, increase segmental chain mobility, and improve polymer flexibility, thereby decreasing the melting point [57].

The thermograms of kefiran films (Figure 5b) confirm that the glass transition in the range from 58 °C to 46 °C corresponds to the polysaccharide fraction. In kefiran films, the addition of glycerol decreased both the T_g and the endothermic peak observed at higher temperatures. As previously described, this effect is attributed to the capacity of the glycerol to reduce hydrogen bonding within the polymer network, thereby increasing macromolecular mobility [7,57].

It is important to note that DSC experiments in the present study were performed in hermetically sealed capsules, with samples fully dehydrated prior to the analysis. Therefore, the plasticising effect of water was not expected to influence these results, unlike other studies on kefiran films [23,26]. Many previous studies have reported an endothermic peak at 84 °C for neat kefiran film and from 82 °C to 89 °C for plasticised samples [26]. Similarly, endothermic peaks have been observed at 84.7 °C (neat kefiran films) and up to 86.6 °C for γ-irradiated kefiran films [25]. The higher temperatures of endothermic peaks observed in the present study are likely due to the sample's preparation and the conditions of the capsules used (hermetically sealed in the present work versus non-specified conditions). Therefore, it is plausible that the endothermic transitions reported in other works were influenced by the plasticising effect of water hydration and, possibly, by water evaporation during the test.

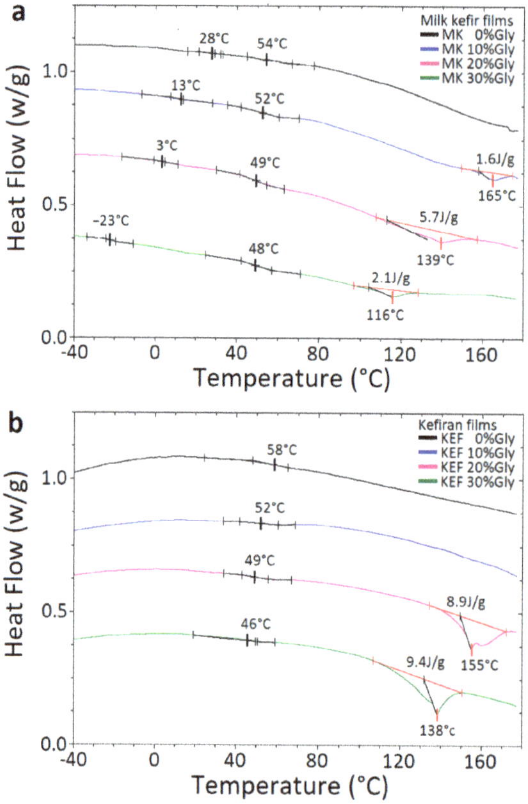

Figure 5. DSC thermograms and glass transition temperature of dried films with different contents of glycerol. (**a**) Milk kefir films. (**b**) Kefiran films.

Furthermore, the T_g values of kefiran films in the present study were higher than those reported in the literature [23,26]. Ghasemlou et al. reported a T_g value around 5 °C for kefiran films, which decreased to −20 °C when glycerol was added at 35 wt% relative to d.m. [23]. Montoille et al. found a T_g around −20 °C for neat kefiran films, which increased to −17 °C with the addition of 5 wt% of glycerol [26]. These differences may be due to the presence of hydration water in those studies, which acts as a plasticiser, thereby lowering the T_g [7,9].

3.6. Mechanical Properties of the Films

The mechanical properties of a material, specifically the elastic modulus (Y), maximum tensile strength (TS_{max}), and elongation at break ($e_{\%}$), are determined by its composition, structure, and interpolymeric interactions. These parameters were determined from the stress–strain curves (Figure 6) and are summarised in Table 2. Unplasticised milk kefir films were too brittle to undergo testing under the specified conditions (22 °C and 52% r.h.).

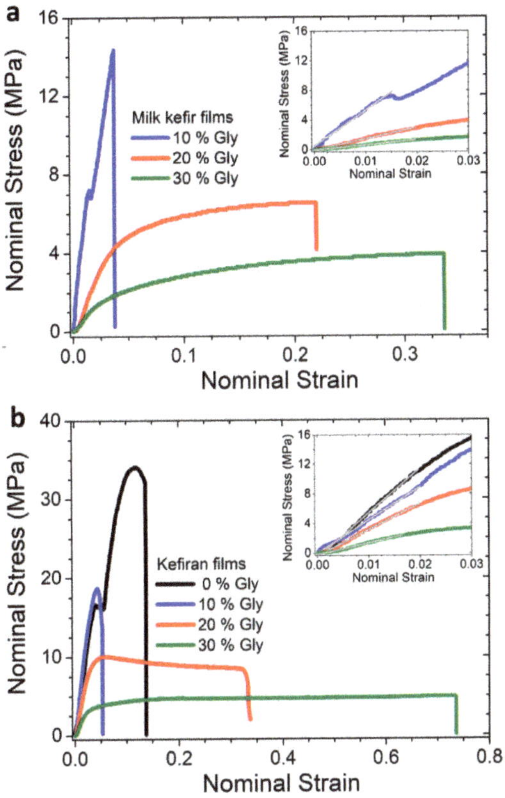

Figure 6. Representative stress–strain curves for one of ten replications of the mechanical test performed for films with different contents of glycerol. (**a**) Milk kefir films. (**b**) Kefiran films. Elastic modulus Y was calculated from the slope in the linear region (insert of the figure), maximum tensile strength $TS_{\max}$ from the maximum value of the nominal stress, and $e_{\%}$ from the maximum value of the nominal strain. Values of the mechanical parameters are shown in Table 2.

Table 2. Mechanical parameters of milk kefir films (MK) and kefiran films (KEF). The mean and standard deviation are reported. The different letters assigned in each column refer to significant differences ($p \leq 0.05$). n.d. denotes not determined.

Sample	*Y (MPa)*	*TS_{max} (MPa)*	*$e_{\%}$ (%)*
MK 0% Glycerol	n.d.	n.d.	n.d.
MK 10% Glycerol	443 ± 33 b	14 ± 2 c	4 ± 1 f
MK 20% Glycerol	155 ± 19 d	7 ± 1 d	20 ± 4 c
MK 30% Glycerol	63 ± 9 e	4 ± 1 e	34 ± 3 b
KEF 0% Glycerol	567 ± 61 a	34 ± 3 a	13 ± 2 d
KEF 10% Glycerol	544 ± 48 a	20 ± 4 b	8 ± 2 e
KEF 20% Glycerol	313 ± 25 c	9 ± 2 d	38 ± 8 b
KEF 30% Glycerol	146 ± 21 d	5 ± 1 e	72 ± 10 a

Table 2 shows that increasing glycerol content reduced both Y and $TS_{\max}$ in both film types. The parameter Y reflects the strength of intermolecular bonds, while $TS_{\max}$ is associated with the quantity of these intermolecular bonds [9]. The observed reduction in Y and TS_{max} with glycerol addition in both films demonstrates the plasticiser's role in decreasing interpolymer interactions. Conversely, glycerol addition led to a monotonic increase in $e_{\%}$ in milk kefir films, whereas in kefiran films, $e_{\%}$ initially decreased slightly at 10 wt% glycerol before increasing significantly at 20 and 30 wt% glycerol. Although biopolymeric materials typically show a correlation between the reduction in Y and T_{max} and the increase in $e_{\%}$ with the addition of glycerol [9,58], kefiran films displayed a non-monotonic increase in $e_{\%}$ response, suggesting a more complex mechanism.

Previous studies reported an unusual increase in Y for kefiran films at glycerol content below 5 wt% [26]. The term "antiplasticisation" has been used to describe the mechanical behaviour observed when low-molecular-weight compounds are added to certain polymer materials, leading to an increase in Y or a decrease in $e_{\%}$ [13]. Other studies demonstrated the antiplasticisation effect in the mechanical properties of kappa-carrageenan films plasticized with either glycerol or sorbitol [59]. In that work, the authors suggested that this behaviour could be due to the strong interaction of the polymer chains with the plasticiser. Accordingly, the mechanical performance observed in kefiran films suggests the presence of an antiplasticisation effect at low glycerol concentrations. These findings indicate the existence of anomalous interactions between glycerol and kefiran in the absence of proteins.

Milk kefir films displayed lower Y and TS_{max} compared to kefiran films (Figure 6, Table 2) due to the weakening of interpolymer interactions induced by the presence of proteins. This is consistent with the findings from the rheological and thermogravimetric experiments discussed in Sections 3.1 and 3.4, respectively. Additionally, kefiran films exhibited higher $e_{\%}$ values.

Piermaria et al. reported values of Y, TS_{max}, and $e_{\%}$ of ~1300 MPa, 41 $\pm$ 8 MPa, and 3 $\pm$ 1%, respectively, for unplasticised kefiran films prepared from a 1 wt% kefiran dispersion without sonication [23]. Comparable values have been obtained by other researchers for unplasticised kefiran films prepared under similar conditions [26]. A comparison of these reported values with those of the unplasticised kefiran films displayed in Table 2 suggests that the ultrasound treatment of the film-forming dispersions results in a more ductile material. This could be due to the sonication-induced breakdown of polysaccharide chains, resulting in a film matrix with reduced entanglement density. This observation agrees with the findings from thermogravimetric studies (Section 3.4) and translates to a more flexible material, potentially offering advantages for specific packaging applications.

3.7. Hydration Properties of the Films

Biopolymer-based films are hydrophilic matrices that are susceptible to moisture uptake, which significantly affects their storage conditions, shelf life, and overall performance. Hydration water plays a crucial role in influencing the structural and functional properties of hydrophilic films. It acts as a plasticiser by inserting itself between the polymer chains, increasing space between them, lowering the glass transition temperature T_g, and enhancing flexibility [7,9]. Water sorption isotherms provide essential insights into the interactions between water and the film, as well as the distribution of water within the material [32,34,60]. Figure 7a,b presents the water sorption isotherms for milk kefir and kefiran films with varying glycerol content. The experimental data were fitted to the GAB model (Equation (4)), with the corresponding parameters summarised in Table 3.

Figure 7a shows that the water sorption isotherms of milk kefir films exhibit a slight increase in hydration water content at low a_w values, followed by a pronounced rise for a_w > 0.6. This pattern is typical for most hydrophilic films made from biopolymers [37,60–62]. Similar behaviour was observed in the glycerol-plasticised kefiran films, although it was less prominent in the unplasticised kefiran film (Figure 7b). Unfortunately, no prior studies on sorption isotherms for kefiran films have been found in the literature, with existing research focusing only on the water content of the films after casting [21–24], which provides limited information on hydration.

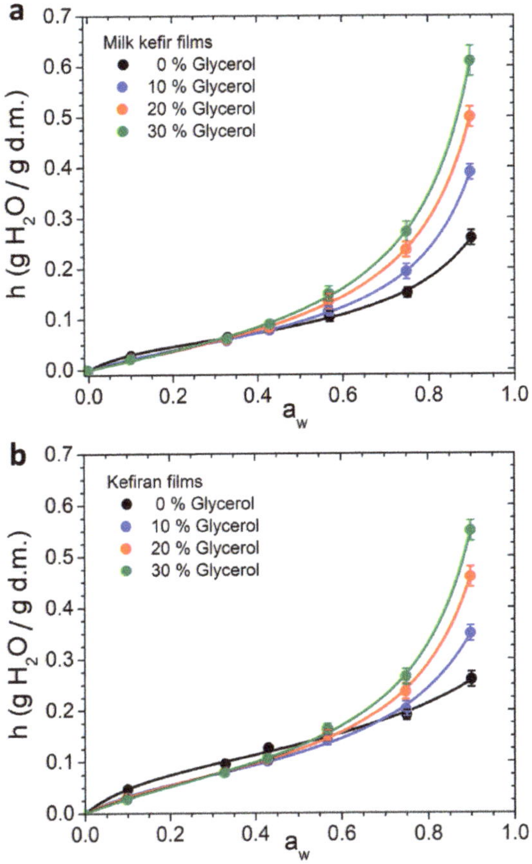

Figure 7. Water vapour sorption isotherms of films with different content of glycerol. (**a**) Milk kefir films. (**b**) Kefiran films. Experimental data were fitted with the GAB model using Equation (2). The fitting parameters are shown in Table 3.

Table 3. Parameters obtained from fitting the GAB model (Equation (2)) to the water sorption isotherms of Figure 7. Errors in the GAB parameters were estimated from the fit analysis. $h_{90\%rh}$ refers to hydration equilibrium values measured at 90% r.h. MK denotes milk kefir and KEF. The different letters assigned to each column refer to significant differences ($p \leq 0.05$).

Sample	$h_{90\%rh}$ (g·g^{-1})	*GAB Parameters*			
		N (g·g^{-1})	*c*	*k*	R^2
MK 0% Glycerol	0.26 ± 0.01 [a]	0.056 ± 0.002	9.6 ± 0.9	0.876 ± 0.009	0.999
MK 10% Glycerol	0.39 ± 0.01 [c]	0.064 ± 0.001	4.2 ± 0.4	0.937 ± 0.003	0.999
MK 20% Glycerol	0.50 ± 0.02 [e]	0.082 ± 0.004	2.5 ± 0.4	0.941 ± 0.007	0.999
MK 30% Glycerol	0.61 ± 0.03 [g]	0.094 ± 0.005	2.0 ± 0.3	0.954 ± 0.006	0.999
KEF 0% Glycerol	0.26 ± 0.01 [a]	0.104 ± 0.009	10 ± 1	0.690 ± 0.008	0.996
KEF 10% Glycerol	0.35 ± 0.01 [b]	0.079 ± 0.005	6.5 ± 0.7	0.871 ± 0.007	0.998
KEF 20% Glycerol	0.46 ± 0.01 [d]	0.080 ± 0.003	5.4 ± 0.5	0.925 ± 0.006	0.999
KEF 30% Glycerol	0.55 ± 0.02 [f]	0.088 ± 0.004	3.9 ±0.5	0.941 ± 0.007	0.999

As demonstrated in Figure 7a, the addition of glycerol increased the water content h for $a_w > 0.6$, while maintaining the overall shape of the isotherms. This behaviour has also been observed in films based on polysaccharides [62], proteins [61], and natural multicomponent films such as integral water kefir grains biomass [7] and integral cellulosic Kombucha tea by-product biomass [9]. On the other hand, the isotherms for the unplasticised kefiran film were found to be more concave for $a_w < 0.6$ compared to the glycerol-plasticised films (Figure 7b). A similar behaviour was observed in potato-starch-based films plasticised with varying glycerol concentrations [14].

The observed convexity at $a_w > 0.6$ in all isotherms in Figure 7 suggests the predominance of multilayer hydration within these films [37]. The increase in hydration as a function of the glycerol content was particularly notable at $a_w = 0.9$, as evidenced by the measured $h_{90\%rh}$ values (Table 3). Comparing the $h_{90\%rh}$ values for unplasticised milk kefir and kefiran films reveals no differences. However, when plasticised with glycerol, milk kefir films show significantly higher $h_{90\%rh}$ values. This suggests that the interaction of glycerol with proteins in milk kefir films enhances hydrophilicity compared to pure kefiran films plasticised with glycerol.

The hydration equilibrium value at 90% r.h. ($h_{90\%rh}$) for unplasticised milk kefir and kefiran films was found to be lower than of other unplasticised biopolymeric films, such as cellulose films [63], integral Kombucha tea by-product biomass films [9], starch films [60], myofibrillar protein films [64], and sodium caseinate film [65]. This indicates that films obtained from milk kefir grains and the purified kefiran exhibit less hydrophilicity compared to other biopolymer-based materials. Consistent with these findings, other studies have reported higher water contact angles for kefiran films compared to other polysaccharide-based films, indicating greater surface hydrophobicity and reduced wettability [23].

Table 3 summarises the parameters obtained from fitting sorption isotherms to the GAB model. The parameter N, representing the number of primary binding sites of hydration, increased with glycerol content in milk kefir films but decreased in kefiran films. On the other hand, as the glycerol content increased in both milk kefir and kefiran films, the parameter c, which reflects the strength of the water binding at primary sites, decreased, while k, associated with the capacity of water binding to the multilayers, increased. Consequently, since most of the hydration water forms multilayers, the $h_{90\%rh}$ values increased with rising glycerol content in both types of films. Glycerol interacts with kefiran and protein chains by forming hydrogen bonds with the reactive groups of these polymers. This disrupts interpolymer interactions, reduces the attractive forces between polymer chains, and increases the free volume available for water molecules, leading to an overall increase in hydration water content.

3.8. Water Vapour Permeability of the Films

Hydration water is closely related to the water vapour transport properties of hydrophilic polymeric films [12,37]. The water vapour permeability of biopolymer-based films is a critical property that determines their ability to regulate water vapour transport between a system, such as food, and its environments. SEM studies on milk kefir and kefiran films revealed a continuous and homogenous matrix, without pores, faults, or punctures (Section 3.2) These findings indicate that water transport in milk kefir and kefiran films does not occur through pores, but via a sorption–diffusion–desorption mechanism [39]. Consequently, the water vapour permeability is influenced by the hydration or water solubility within the film, as well as the mobility of water molecules in the matrix [37].

Water vapour permeability in biopolymer films is significantly influenced by experimental conditions, particularly film thickness [12,37] and water vapour pressure gradient [37,39]. To ensure accurate comparisons, all of the films in this study were maintained at a consistent thickness (see Table 4) and identical water vapour pressure gradient conditions. Table 4 presents experimental water vapour permeability values (P_w^{exp}) for milk kefir and kefiran films with varying glycerol content. It can be seen that for milk kefir films, P_w^{exp} increased with glycerol content. Glycerol addition typically increases water vapour permeability, as observed in various biopolymer-based films, including those made from corn starch [60], myofibrillar protein [64], cellulose [66], β-lactoglobulin [67], water kefir grains biomass [7], and yeast biomass [12]. It has been suggested that the addition of glycerol increases the permeability of the films mainly due to the increase in hydration and solubility of water within the material [12]. Interestingly, as shown in Table 4, unplasticised kefiran films exhibited the highest P_w^{exp} values among all kefiran samples. Kefiran films containing 10 wt% glycerol presented the lowest P_w^{exp} values. Subsequent glycerol additions led to increased P_w^{exp}, although values remained below those of the unplasticised film.

Table 4. Experimental water vapour permeability (P_w^{exp}), water solubility (S_w^{eff}), and water diffusion (D_w^{eff}). L and $\rho_{d.f.}$ refer to the thickness and density measurements of the dried film, respectively. MK denotes milk kefir and KEF. The same letters in the data reported in a column mean non-significant differences ($p < 0.05$).

Sample	L $(10^{-5} g\ m)$	$\rho_{d.f}$ $(10^4 g\ m^3)$	P_w^{exp} $(10^{-10}\ g\ s^{-1} m^{-1} Pa^{-1})$	S_w $(g\ m^{-3} Pa^{-1})$	D_w $(10^{-13} m^2 s^{-1})$
MK 0% Glycerol	15.6 ± 0.7 [ab]	113 ± 9 [ab]	3.2 ± 0.1 [a]	124 ± 6 [b]	26 ± 2 [d]
MK 10% Glycerol	15.1 ± 0.8 [ab]	121 ± 10 [ab]	4.6 ± 0.1 [c]	211 ± 10 [d]	22 ± 1 [c]
MK 20% Glycerol	14.5 ± 0.8 [ab]	131 ± 9 [ab]	5.6 ± 0.1 [e]	295 ± 15 [f]	19 ± 1 [b]
MK 30% Glycerol	14.1 ± 0.7 [b]	139 ± 9 [b]	6.1 ± 0.1 [f]	388 ± 19 [g]	16 ± 1 [a]
KEF 0% Glycerol	16.0 ± 0.7 [a]	104 ± 9 [a]	5.0 ± 0.1 [d]	105 ± 8 [a]	48 ± 4 [f]
KEF 10% Glycerol	15.4 ± 0.8 [ab]	113 ± 10 [ab]	4.0 ± 0.1 [b]	172 ± 10 [c]	23 ± 2 [cd]
KEF 20% Glycerol	14.9 ± 0.8 [ab]	121 ± 10 [ab]	4.7 ± 0.1 [c]	247 ± 11 [e]	19 ± 1 [b]
KEF 30% Glycerol	14.3 ± 0.8 [b]	129 ± 9 [b]	4.8 ± 0.1 [cd]	321 ± 12 [f]	15 ± 1 [a]

Table 4 additionally presents S_w^{eff} and D_w^{eff} values, which independently contribute to P_w^{exp}, as described in Equation (6). Both sample types exhibited a significant increase in water solubility S_w^{eff} and a decrease in water diffusion D_w^{eff} with rising glycerol content. Milk kefir films demonstrated a significantly greater increase in S_w^{eff} compared to the decrease in D_w^{eff} with increasing glycerol content, resulting in an overall increase in P_w^{exp}. On the other hand, D_w^{eff} for the unplasticised kefiran film was found to be remarkably high compared to the kefiran film plasticised with 10 wt% glycerol. However, with the addition of more glycerol, the D_w^{eff} value decreased only slightly, similar to what was observed in milk kefir samples. As a result, the unplasticised kefiran film exhibited the highest P_w^{exp} value, while the film with 10 wt% glycerol had the lowest. With each subsequent addition of glycerol, P_w^{exp} increased; however, it never reached the level observed in the absence of the plasticiser.

This anomalous effect of glycerol on the water vapour permeability of kefiran films was also observed by Piermaria et al. [22], who attributed this behaviour to the development of a more compact structure in plasticised films. As shown in Table 4, the density values ($\rho_{d.f.}$) of milk kefir and kefiran films exhibit a slight tendency to increase with the glycerol content; however, the anomaly was observed only in the kefiran film. For kefiran films, this anomaly could be the result of a stronger interaction between glycerol, water molecules, and the available (–OH) groups of kefiran. Given the structure of the kefiran film, these components can engage in multiple interactions via hydrogen bonds, thereby retarding water movement. Once the (–OH) groups of kefiran are saturated at 10 wt% glycerol, the addition of more plasticiser leads to a smaller decrease in D_w^{eff}, while S_w^{eff} increases notably. Consequently, the P_w^{exp} of kefiran film increased but did not reach the level observed in the absence of plasticiser. In contrast, in milk kefir films, the protein–kefiran matrix may inhibits these specific interactions between glycerol, water molecules, and the available

(–OH) groups of the composite matrix. As a result, the anomalous behaviour observed in kefiran films is not present in this sample.

Furthermore, the anomalous effect of glycerol on water vapour permeability was also observed in potato-starch-based films, which exhibited water sorption isotherm behaviour at varying glycerol concentrations similar to that of kefiran films [14]. It has been suggested that plasticisers may either retard or facilitate moisture transmission depending on their concentration [63]. The decrease in P_w^{exp} to a minimum value at low glycerol concentration, followed by an increase at higher plasticiser levels, can be attributed to the antiplasticisation effect at low plasticiser concentration described in Section 3.6. This effect arises from a specific interaction between the polymer and the plasticiser molecules, which reduces the molecular mobility of the polymer chains, as was observed in the mechanical properties of kefiran films (Section 3.6). A similar antiplasticisation effect has been observed in the mechanical behaviour and water permeability of cellulose acetate films [13].

Although the precise mechanism behind antiplasticisation at low plasticiser concentrations remains unclear, one hypothesis suggests that incorporating small amounts of plasticiser into a polymer increases the free volume, allowing for the polymer chains to rearrange into a more thermodynamically stable and compact structure [13]. Unfortunately, the scanning electron microscopy images obtained in the present study do not offer sufficient resolution to observe these structural changes. Further characterisation of these polymeric structures is required, as it could enhance our understanding of plasticisation and antiplasticisation phenomena, which would be of significant interest to the field of materials science. In conclusion, the observed P_w^{exp} behaviour as a function of the glycerol content in kefiran films is likely the result of structural modifications within the kefiran matrix induced by the glycerol, which restrict water transport at low concentrations.

4. Conclusions

Films derived from both the integral biomass of the water kefir grains and from the purified kefiran exhibited excellent continuity and homogeneity. While kefiran films were transparent, milk kefir films had a yellowish tint. The application of ultrasonic treatments to both film-forming dispersions allowed for the use of 5 wt% d.m. concentration, reducing the energy required for drying during the casting process.

Kefiran films showed stronger interpolymeric interactions compared to milk kefir films, as evidenced by thermogravimetric analyses and mechanical testing. These findings suggest that the presence of proteins in the milk kefir matrix weakens interpolymeric interactions. Glycerol increased ductility and hydration while decreasing the thermal stability and glass transition temperature of the films.

Additionally, kefiran films exhibited an antiplasticisation effect at low glycerol content, affecting elongation at break and water vapour permeability. This unexpected behaviour suggests specific characteristics in the interactions between glycerol and kefiran. The absence of this effect in milk kefir films indicates that the presence of proteins inhibits it.

Traditionally, biopolymeric materials are developed by purifying biopolymers from their original biomass and subsequently modifying them physically or chemically to enhance their film-forming properties. However, the results of this study highlight the potential for using integral milk kefir grains directly in the development of novel biodegradable materials. These findings emphasise the differences between materials derived from the integral milk kefir grains and those from purified kefiran, providing insights into their application potential.

The hydrophilicity and sensitivity to high humidity of these materials make them most suitable for food packaging applications involving dry or fatty foods. However, with further experimentation, this hydrophilicity could be exploited to develop delivery packaging for instant products, where the packaging dissolves in hot water and potentially modifies the rheology of the final product. While these materials may not fully replace petroleum-based polymers, they provide a viable alternative for specific applications, supporting a shift toward a more sustainable and circular economy.

Author Contributions: Conceptualization, A.G.S.; methodology, Y.A.R.T., G.D.R., J.F.D., M.A.P. and A.G.S.; formal analysis, Y.A.R.T., G.D.R., J.F.D., M.A.P. and A.G.S.; investigation, Y.A.R.T., G.D.R., J.F.D., M.A.P. and A.G.S.; data curation, Y.A.R.T., M.A.P. and A.G.S.; writing—original draft preparation, A.G.S.; writing—review and editing, Y.A.R.T., G.D.R., J.F.D., M.A.P. and A.G.S.; visualisation, Y.A.R.T., G.D.R., J.F.D., M.A.P. and A.G.S.; supervision, A.G.S.; funding acquisition, M.A.P. and A.G.S. All authors have read and agreed to the published version of the manuscript.

Funding: This work has been possible thanks to the support of the Agencia Nacional de Promoción Científica y Tecnológica (Argentina) through PICT 2021-92 and the Universidad Nacional de Quilmes (UNQ, Argentina) through R&D programme PUNQ R 990/19 Expte: 827-1300/19.

Institutional Review Board Statement: Not applicable.

Data Availability Statement: The data presented in this study are available on request from the corresponding author.

Acknowledgments: G.D.R. is a doctoral fellowship of CONICET, and Y.A.R.T., J.F.D., and M.A.P. are members of CONICET; A.G.S. is a researcher at the UNQ; all institutions are in Argentina. We thank Lola Laurent-Iranmehr and Manuela Da Silva from Institut Agro Montpellier, France, for their assistance with the purification of kefiran during their stay as student interns in our laboratory.

Conflicts of Interest: The authors declare no conflicts of interest.

References

1. Vieira, M.G.A.; Altenhofen Silva, M.; Oliveira dos Santos, L.; Beppu, M.M. Natural based plasticizers and biopolymer films: A review. *Eur. Polym. J.* **2011**, *47*, 254–263. [CrossRef]
2. George, A.; Sanjay, M.R.; Srisuk, R.; Parameswaranpillai, J.; Suchart Siengchin, S. A comprehensive review on chemical properties and applications of biopolymers and their composites. *Int. J. Biol. Macromol.* **2020**, *154*, 3292338. [CrossRef] [PubMed]
3. Janowicz, M.; Galus, S.; Ciurzyńska, A.; Nowacka, M. The Potential of Edible Films, Sheets, and Coatings Based on Fruits and Vegetables in the Context of Sustainable Food Packaging Development. *Polymers* **2023**, *15*, 4231. [CrossRef]
4. Han, J.H.; Gennadios, A. Edible films and coatings: A review. In *Innovations in Food Packagings*; Han, J.H., Ed.; Elsevier: Amsterdam, The Netherlands, 2005; pp. 239–262.
5. Ferreira, A.R.V.; Alves, V.D.; Coelhoso, I.M. Polysaccharide-Based Membranes in Food Packaging Applications. *Membranes* **2016**, *6*, 22. [CrossRef]
6. Cazón, P.; Velazquez, G.; Ramírez, J.A.; Vázquez, M. Polysaccharide-based films and coatings for food packaging: A review. *Food Hydrocoll.* **2017**, *68*, 136–148. [CrossRef]
7. Coma, M.E.; Peltzer, M.A.; Delgado, J.F.; Salvay, A.G. Water kefir Grains as an innovative source of materials: Study of plasticiser content on film properties. *Eur. Polym. J.* **2019**, *120*, 109234. [CrossRef]
8. Marangoni Júnior, L.; Vieira, R.P.; Rodrigues Anjos, C.A. Kefiran-based films: Fundamental concepts, formulation strategies and properties. *Carbohydr. Polym.* **2020**, *246*, 116609. [CrossRef]
9. Ramirez Tapias, Y.A.; Peltzer, M.A.; Delgado, J.F.; Salvay, A.G. Kombucha Tea By-product as Source of Novel Materials: Formulation and Characterization of Films. *Food Bioprocess Technol.* **2020**, *13*, 1166–1180. [CrossRef]
10. Cottet, C.; Ramirez-Tapias, Y.A.; Delgado, J.F.; de la Osa, O.; Salvay, A.G.; Peltzer, M.A. Biobased materials from microbial biomass and its derivatives. *Materials* **2020**, *13*, 1263. [CrossRef]
11. Özeren, H.D.; Wei, X.-F.; Nilsson, F.; Olsson, R.T.; Hedenqvist, M.S. Role of hydrogen bonding in wheat gluten protein systems plasticized with glycerol and water. *Polymer* **2021**, *232*, 124149. [CrossRef]
12. Delgado, J.F.; Peltzer, M.A.; Wagner, J.R.; Salvay, A.G. Hydration and water vapour transport properties in yeast biomass based films: A study of plasticizer content and thickness effects. *Eur. Polym. J.* **2018**, *99*, 9–17. [CrossRef]
13. Guo, J.H. Effects of plasticizers on water permeation and mechanical properties of cellulose acetate: Antiplasticization in slightly plasticized polymer film. *Drug Dev. Ind. Pharm.* **1993**, *19*, 1541–1555. [CrossRef]
14. Talja, R.A.; Helén, H.; Roos, Y.H.; Jouppila, K. Effect of various polyols and polyol contents on physical and mechanical properties of potato starch-based films. *Carbohydr. Polym.* **2007**, *67*, 288–295. [CrossRef]
15. Mascia, L.; Kouparitsas, Y.; Nocita, D.; Bao, X. Antiplasticization of Polymer Materials: Structural Aspects and Effects on Mechanical and Diffusion-Controlled Properties. *Polymers* **2020**, *12*, 769. [CrossRef]
16. Bianco, M.I.; Jacobs, M.; Salinas, S.R.; Salvay, A.G.; Ielmini, M.V.; Ielpi, L. Biophysical characterization of the outer membrane polysaccharide export protein and the polysaccharide co-polymerase protein from Xanthomonas campestris. *Protein Expr. Purif.* **2014**, *101*, 42–53. [CrossRef]
17. Mouro, C.; Gomes, A.P.; Gouveia, I.C. Microbial Exopolysaccharides: Structure, Diversity, Applications, and Future Frontiers in Sustainable Functional Materials. *Polysaccharides* **2024**, *5*, 241–287. [CrossRef]
18. Piermaria, J.A.; de la Canal, M.L.; Abraham, A.G. Gelling properties of kefiran, a food-grade polysaccharide obtained from kefir grain. *Food Hydrocoll.* **2008**, *22*, 1520–1527. [CrossRef]

19. Moradi, Z.; Kalanpour, N. Kefiran, a branched polysaccharide: Preparation, properties and applications: A review. *Carbohydr. Polym.* **2019**, *223*, 115100. [CrossRef]
20. Simonelli, N.; Gagliarini, N.; Medrano, M.; Piermaria, J.A.; Abraham, A.; Kefiran, G. *Polysaccharides of Microbial Origin*; Oliveira, J.M., Radhouani, H., Reis, R.L., Eds.; Springer Nature: Cham, Switzerland, 2022; pp. 210–216.
21. Piermaria, J.; Bosch, A.; Pinotti, A.; Yantorno, O.; Garcia, M.A.; Abraham, A.G. Kefiran films plasticized with sugars and polyols: Water vapor barrier and mechanical properties in relation to their microstructure analyzed by ATR/FT-IR spectroscopy. *Food Hydrocoll.* **2011**, *25*, 1261–1269. [CrossRef]
22. Piermaria, J.A.; Pinotti, A.; Garcia, M.A.; Abraham, A.G. Films based on kefiran, an exopolysaccharide obtained from kefir grain: Development and characterization. *Food Hydrocoll.* **2009**, *23*, 684–690. [CrossRef]
23. Ghasemlou, M.; Khodaiyan, F.; Oromiehie, A.; Yarmand, M.S. Development and characterisation of a new biodegradable edible film made from kefiran, an exopolysaccharide obtained from kefir grains. *Food Chem.* **2011**, *127*, 1496–1502. [CrossRef]
24. Ghasemlou, M.; Khodaiyan, F.; Oromiehie, A. Physical, mechanical, barrier, and thermal properties of polyol-plasticized biodegradable edible film made from kefiran. *Carbohydr. Polym.* **2011**, *84*, 477–483. [CrossRef]
25. Shahabi-Ghahfarrokhi, I.; Khodaiyan, F.; Mousavi, M.; Yousefi, H. Effect of γ-irradiation on the physical and mechanical properties of kefiran biopolymer film. *Int. J. Biol. Macromol.* **2015**, *74*, 343–350. [CrossRef] [PubMed]
26. Montoille, L.; Vicencio, C.M.; Fontalba, D.; Ortiz, J.A.; Moreno-Serna, V.; Peponi, L.; Matiacevich, S.; Zapata, P.A. Study of the effect of the addition of plasticizers on the physical properties of biodegradable films based on kefiran for potential application as food packaging. *Food Chem.* **2021**, *360*, 129966. [CrossRef]
27. Mayrhofer, A.; Kopacic, S.; Bauer, W. Extensive Characterization of Alginate, Chitosan and Microfibrillated Cellulose Cast Films to Assess their Suitability as Barrier Coating for Paper and Board. *Polymers* **2023**, *15*, 3336. [CrossRef]
28. A.O.A.C. *Official Methods of Analysis*, 15th ed.; Association of Official Analytical Chemists: Washington, DC, USA, 1990.
29. Rimada, P.S.; Abraham, A.G. Kefiran improves rheological properties of glucono-δ-lactone induced skim milk gels. *Int. Dairy J.* **2006**, *16*, 33–39. [CrossRef]
30. Lago, A.; Delgado, J.F.; Rezzani, G.D.; Cottet, C.; Ramírez Tapias, Y.A.; Peltzer, M.A.; Salvay, A.G. Multi-Component Biodegradable Materials Based on Water Kefir Grains and Yeast Biomasses: Effect of the Mixing Ratio on the Properties of the Films. *Polymers* **2023**, *15*, 2594. [CrossRef]
31. Moraes, J.O.; Scheibe, A.S.; Sereno, A.; Laurindo, J.B. Scale-up of the production of cassava starch based films using tape-casting. *J. Food Eng.* **2013**, *119*, 800–808. [CrossRef]
32. Ramirez Tapias, Y.A.; Di Monte, M.V.; Peltzer, M.A.; Salvay, A.G. Kombucha fermentation in yerba mate: Cellulose production, films formulation and its characterisation. *Carbohydr. Polym. Technol. Appl.* **2023**, *5*, 100310.
33. *ASTM D882 1997*; Standard Test Method for Tensile Properties of Thin Plastic Sheeting. ASTM International: West Conshohocken, PA, USA, 1997.
34. Guggenheim, E.A. *Applications of Statistical Mechanics*; Oxford University Press: London, UK, 1966.
35. *ASTM-E96 2016*; Standard Test Methods for Water Vapor Transmission of Materials. ASTM International: West Conshohocken, PA, USA, 2016.
36. McHugh, T.H.; Avena-Bustillo, R.; Krochta, J.M. Hydrophilic edible films: Modified procedure for water vapor permeability and explanation of thickness effects. *J. Food Sci.* **1993**, *58*, 899–903. [CrossRef]
37. Delgado, J.F.; Peltzer, M.A.; Salvay, A.G. Water vapour transport in biopolymeric materials: Effects of thickness and water vapour pressure gradient on yeast biomass-based films. *J. Polym. Environ.* **2022**, *30*, 2976–2989. [CrossRef]
38. Rogers, C.E. Permeation of gases and vapours in polymers. In *Polymer Permeability*; Comyn, J., Ed.; Springer: Dordrecht, The Netherlands, 1985; pp. 11–73.
39. Roy, S.; Gennadios, A.; Weller, C.L.; Testin, R.F. Water vapor transport parameters of a cast wheat gluten film. *Ind. Crops Prod.* **2000**, *11*, 43–50. [CrossRef]
40. Irgens, F. *Rheology and Non-Newtonian Fluids*, 1st ed.; Springer: Cham, Switzerland, 2014.
41. Robinson Lang, E.R.; Rha, C. Determination of the yield stress of hydrocolloid dispersions. *J. Texture Stud.* **1981**, *12*, 47–62. [CrossRef]
42. Ghasemlou, G.; Khodaiyan, F.; Oromiehie, A. Rheological and structural characterisation of film-forming solutions and biodegradable edible film made from kefiran as affected by various plasticizer types. *Int. J. Biol. Macromol.* **2011**, *49*, 814–821. [CrossRef]
43. Hamed, S.B.; Belhadri, M. Rheological properties of biopolymers drilling fluids. *J. Pet. Sci. Eng.* **2009**, *67*, 84–90. [CrossRef]
44. Peres, G.L.; Leite, D.C.; da Silveira, N.P. Ultrasound effect on molecular weight reduction of amylopectin. *Starch* **2015**, *66*, 407–414. [CrossRef]
45. Mohammed, A.A.B.A.; Hasan, Z.; Omran, A.A.B.; Kumar, V.V.; Elfaghi, A.M.; Ilyas, R.A.; Sapuan, S.M. Corn: Its Structure, Polymer, Fiber, Composite, Properties, and Applications. *Polymers* **2022**, *14*, 4396. [CrossRef]
46. Basedow, A.M.; Ebert, K. Ultrasonic degradation of polymers in solution. In *Advances in Polymer Science—Physical Chemistry*; Springer: Berlin/Heidelberg, Germany, 1977; pp. 83–148.
47. Peressini, D.; Bravin, B.; Lapasin, R.; Rizzotti, C.; Sensidoni, A. Starch–methylcellulose based edible films: Rheological properties of film-forming dispersions. *J. Food Eng.* **2003**, *59*, 25–32. [CrossRef]
48. Gontard, N.; Guilbert, S.; Cuq, J.-L. Water and glycerol as plasticizers affect mechanical and water vapor barrier properties of an edible wheat gluten film. *J. Food Sci.* **1993**, *58*, 206–211. [CrossRef]

49. Esnaashari, S.; Rezaei, S.; Mirzaei, E.; Afshari, H.; Rezayat, S.M.; Faridi-Majidi, R. Preparation and characterization of kefiran electrospun nanofibers. *Int. J. Biol. Macromol.* **2014**, *70*, 50–56. [CrossRef]
50. Pop, C.; Apostu, S.; Rotar, A.M.; Semeniuc, C.A.; Sindic, M.; Mabon, N. FTIR spectroscopic characterization of a new biofilm obtained from kefiran. *J. Agroaliment. Process. Technol.* **2013**, *19*, 157–159.
51. Wang, Y.; Ahmed, Z.; Feng, W.; Li, C.; Song, S. Physicochemical properties of exopolysaccharide produced by Lactobacillus kefiranofaciens ZW3 isolated from Tibet kefir. *Int. J. Biol. Macromol.* **2008**, *43*, 283–288. [CrossRef]
52. Guerrero, P.; Retegi, A.; Gabilondo, A.; de la Caba, K. Mechanical and thermal properties of soy protein films processed by casting and compression. *J. Food Eng.* **2010**, *100*, 145–151. [CrossRef]
53. Ramos, O.L.; Reinas, I.; Silva, S.I.; Fernandes, J.C.; Cerqueira, M.A.; Pereira, R.N.; Vicente, A.A.; Poças, M.F.; Pintado, M.E.; Malcata, F.X. Effect of whey protein purity and glycerol content upon physical properties of edible films manufactured therefrom. *Food Hydrocoll.* **2013**, *30*, 110–122. [CrossRef]
54. Ahmed, Z.; Wang, Y.; Anjum, N.; Ahmad, A.; Khan, S.T. Characterization of exopolysaccharide produced by Lactobacillus kefiranofaciens ZW3 isolated from Tibet kefir—Part II. *Food Hydrocoll.* **2013**, *30*, 343–350. [CrossRef]
55. Stawski, D.; Rabiej, S.; Herczyńska, L.; Draczyński, Z. Thermogravimetric analysis of chitins of different origin. *J. Therm. Anal. Calorim.* **2008**, *93*, 489–494. [CrossRef]
56. Sanyang, M.L.; Sapuan, S.M.; Jawaid, M.; Ishak, M.R.; Sahari, J. Effect of plasticizer type and concentration on physical properties of biodegradable films based on sugar palm (Arenga pinnata) starch for food packaging. *J. Food Sci. Technol.* **2016**, *53*, 326–336. [CrossRef]
57. Srithep, Y.; Pholharn, D. Plasticizer effect on melt blending of polylactide stereocomplex. *e-Polymers* **2017**, *17*, 409–416. [CrossRef]
58. Park, H.J.; Weller, C.L.; Vergano, P.J.; Testin, R.F. Permeability and mechanical properties of cellulose-based edible films. *J. Food Sci.* **1993**, *58*, 1361–1364. [CrossRef]
59. Farhan, A.; Hani, N.M. Characterization of edible packaging films based on semi-refined kappa-carrageenan plasticized with glycerol and sorbitol. *Food Hydrocoll.* **2017**, *64*, 48–58. [CrossRef]
60. Bertuzzi, M.A.; Castro Vidaurre, E.F.; Armada, M.; Gottifredi, J.C. Water vapor permeability of edible starch based films. *J. Food Eng.* **2007**, *80*, 972–978. [CrossRef]
61. Coupland, J.N.; Shaw, N.B.; Monahan, F.J.; O'Riordan, E.D.; O'Sullivan, M. Modeling the effect of glycerol on the moisture sorption behavior of whey protein edible films. *J. Food Eng.* **2000**, *43*, 25–30. [CrossRef]
62. Mali, S.; Sakanaka, L.S.; Yamashita, F.; Grossmann, M.V.E. Water sorption and mechanical properties of cassava starch films and their relation to plasticizing effect. *Carbohydr. Polym.* **2005**, *60*, 283–289. [CrossRef]
63. Chinnan, M.S.; Park, H.J. Effect of plasticizer level and temperature on water vapor transmission of cellulose-based edible films. *J. Food Process Eng.* **1995**, *18*, 417–429. [CrossRef]
64. Cuq, B.; Gontard, N.; Cuq, J.-L.; Guilbert, S. Selected functional properties of fish myofibrillar protein based films as affected by hydrophilic plasticizers. *J. Agr. Food Chem.* **1997**, *45*, 622–626. [CrossRef]
65. Kristo, E.; Koutsoumanis, K.P.; Biliaderis, C.G. Thermal, mechanical and water vapor barrier properties of sodium caseinate films containing antimicrobials and their inhibitory action on Listeria monocytogenes. *Food Hydrocoll.* **2008**, *22*, 373–386. [CrossRef]
66. Natanya, M.; Hansen, L.; Plackett, D. Sustainable films and coatings from hemicelluloses: A review. *Biomacromolecules* **2008**, *9*, 1493–1505.
67. Sothornvit, R.; Krochta, J.M. Plasticizer effect on mechanical properties of β-lactoglobulin films. *J. Food Eng.* **2001**, *50*, 149–155. [CrossRef]

MDPI AG
Grosspeteranlage 5
4052 Basel
Switzerland
Tel.: +41 61 683 77 34

Polymers Editorial Office
E-mail: polymers@mdpi.com
www.mdpi.com/journal/polymers

www.ingramcontent.com/pod-product-compliance
Lightning Source LLC
Chambersburg PA
CBHW061515210926
51850CB00008B/1931
9783725868186